Dealing with Disasters from Early Modern to Modern Times

Dealing with Disasters from Early Modern to Modern Times

Cultural Responses to Catastrophes

Edited by
Hanneke van Asperen and
Lotte Jensen

Routledge
Taylor & Francis Group
LONDON AND NEW YORK

The publication of this book is made possible by a grant from the Dutch Research Council NWO.

First published in 2023 by Amsterdam University Press Ltd.

Published 2025 by Routledge
4 Park Square, Milton Park, Abingdon, Oxon OX14 4RN
605 Third Avenue, New York, NY 10158

Routledge is an imprint of the Taylor & Francis Group, an informa business

ISBN: 9789463725798 (hbk)
ISBN: 9781041177951 (pbk)
ISBN: 9781003693727 (ebk)
NUR 696

Cover illustration: Detail of Gerard Overman, *Fire in the New Lutheran Church on 18 September 1822*, 1822–24, etching and aquatint, hand-coloured, 36 x 46.7 cm. Rijksmuseum, Amsterdam, RP-P-1935-1465

Cover design: Coördesign, Leiden

DOI 10.5117/9789463725798

For Product Safety Concerns and Information please contact our EU representative:
GPSR@taylorandfrancis.com
Taylor & Francis Verlag GmbH, Kaufingerstraße 24, 80331 München, Germany

Table of Contents

List of Illustrations

Images are in the public domain, unless otherwise stated.

Figures

Table

Donderdag 20 Augustus 1925 — 14e Jaargang No. 17.

DE ROSKAM

VERSCHIJNT OM DE VEERTIEN DAGEN

Nadruk verboden.

POLITIEK-SATYRIEK- EN HUMORISTISCH TIJDSCHRIFT

Abonnementsprijs: f 2.30 per jaar; f 1.20 per half jaar; f 0.65 per kwartaal; losse nummers f 0.10.

Redactie- en Administratie-adres: EMMAPLEIN 2, 's-HERTOGENBOSCH, TELEFOON No. 70.

DE STORMRAMP VAN 10 AUGUSTUS.

Gij allen!! Steunt de ongelukkigen van deze ramp, zonder onderscheid van geloof en richting der getroffenen.

Figure 0.1 Plate signed Graafland (probably Rob Graafland), *The Storm of August 1925*, cover of the bi-weekly *De roskam* 14 (20 August 1925). Katholiek Documentatie Centrum / Catholic Documentation Centre, Radboud University, Nijmegen, AFBK-4B33448

Introduction

Cultural Responses to Catastrophes from Early Modern to Modern Times

Hanneke van Asperen and Lotte Jensen

Abstract

By examining disasters via the thematic fields of 'emotion', 'blame', and 'time', this edited volume opens up new viewpoints within historical disaster research. The introduction places the individual contributions in a theoretical context. Primarily, it underlines the importance of cultural studies contending that media are crucial in shaping human perception of destruction. After all, dissemination, interpretation, and signification take place primarily through cultural media. To fully grasp the meaning of a disaster to contemporaries and later generations, scholars should analyse representations of these events, using multidisciplinary methods of history, art history, and literary studies. Only then can we identify voice, medium, discourse, and communities involved in production, perception, and transmission of disaster experiences. Subsequently, it becomes possible to identify general patterns and, more importantly, to carefully scrutinise historical sources.

Keywords

disaster studies – cultural representation – medialisation – identity – memory – community building

In August 1925, the Dutch political-satirical magazine *De roskam* published an issue with a cover depicting a devastating storm that had ravished

---

Asperen, van, Hanneke and Lotte Jensen (eds): *Dealing with Disasters from Early Modern to Modern Times. Cultural Responses to Catastrophes.* Taylor & Francis Group, 2023
DOI: 10.5117/9789463725798_INTRO

parts of the Netherlands ten days prior to publication (Fig. 0.1).[1] The artist Graafland who signed the lithograph plate depicted a scene of destruction against a pitch-black night sky.[2] Several forces of nature are at play: the wind rips doors and shutters from houses, rain pours from the heavens, and two lightning bolts simultaneously cleave the firmament as they strike buildings and cause fires. A church spire collapses in the process. Some people look in disbelief at the falling steeple as they recoil in reaction to the anticipated force of the impact. Holding firmly onto her child, a woman flees the scene of horror, while the storm is raging and rain is gushing from the heavens.

The scene seems straightforward enough in its depiction of tragedy, but we get a better sense of its meaning when we read the caption below. It directly addresses the magazine's readership: 'All of you!! Support the unfortunate sufferers of this disaster, without distinction as to faith and denomination of the victims'.[3] This caption tells us different things: On the one hand, the editors of *De roskam* felt the necessity to urge their readers to donate for the victims. At first glance, it highlights solidarity. On the other hand, it underlines that this solidarity was not self-evident. The town of Borculo, which was hit the hardest by the storm, is situated in a part of the Netherlands where most people adhered to the Reformed Church. The collapsing structure is in fact the church building of the Reformed community of Borculo, which was severely damaged during the 1925 storm. In contrast, *De roskam* served a Catholic readership, mainly living in the south of the Netherlands.

1 The work on this introduction was funded by the Dutch Research Council NWO and carried out with a Vici grant for the research project *Dealing with Disasters. The Shaping of Local and National Identities in the Netherlands* (*1421–1890*). The volume as a whole is the result of a two-day conference, titled 'Dealing with Disasters. Cultural Representations of Catastrophes, c. 1500–1900', held on 17–18 June 2021 at Radboud University. We thank the members of the research group, Adriaan Duiveman, Marieke van Egeraat, Fons Meijer, and Lilian Nijhuis, for their invaluable support in organising this event. Furthermore, our gratitude goes to Radboud Institute for Culture & History (RICH) for funds that allowed the editors to see this edited volume through. We would also like to thank Kate Delany for the editorial assistance in preparing this volume as well as the reviewers of the unpublished manuscript for their comments and suggestions.

2 This is probably the same Rob Graafland who was born in 1875 and active as illustrator of magazines and children's books until 1940. Cf. RKD explore, https://rkd.nl/explore/artists/33171 (latest update 30 March 2023; accessed 3 April 2023). The magazine *De roskam* was published between 1912 and 1929 as a bi-weekly supplement of *Het katholieke volk*, later *De volkskrant*.

3 The caption in Dutch reads: 'Gij allen!! Steunt de ongelukkigen van deze ramp, zonder onderscheid van geloof en richting der getroffenen'. Translation from Dutch is ours.

The cover was designed to activate an emotional response in *De roskam*'s principally Catholic audience.[4] It does so in various ways: Firstly, the image itself primarily focuses on the impact of the storm. In order to reach the magazine's audience, the artist Graafland tried to capture the tragedy in all its horrifying magnitude. Those depicted seem helpless against the destructive forces of darkness, rain, wind, and lightning. Their homes, their lives, their loved ones are at risk. The storm's violence breaks down the barriers that some of *De roskam*'s readers may have felt when confronted with people from different denominations. The disaster seems to be depicted as an equaliser. In the eye of the storm, everybody suffers regardless of their religious conviction.

Just as important in understanding the cover's message is the image's context, the added elements complementing the image. These frames can change the reception or its interpretation. On the cover of *De roskam*, the caption below the image points out an additional function that is more directive and instructing in nature: to offer support to the victims of the disaster. The image's context thus clarifies the intention behind it. However, the caption also calls attention to religious differences, to the dividing lines marking society. The people from Borculo are identified as Protestants and they are represented here as 'the Other', deserving of solidarity and assistance but different from Catholics nonetheless.[5] Moreover, the caption implies that helping others of different denominations is not self-evident but must be stimulated explicitly.

The cover of *De roskam* is but one example of a representation of a disaster and its (inadvertent) effects. In times of disaster, people expressed themselves through a wide variety of media, such as newspapers, chronicles, songs, sermons, poems, paintings, prints, and novels. This book deals with the materialised and surviving versions of these in early modern and modern times: the contributors from various disciplines – art history, literary and environmental history, and culture and media studies – chart and analyse how authors and artists have represented disasters in a wide variety of genres and art forms across time in different regions and countries, both textually and visually.

4 Much has been written on disasters and emotions. See Jennifer Spinks and Charles Zika (eds), *Disaster, Death and the Emotions in the Shadow of the Apocalypse, 1400–1700* (London: Palgrave Macmillan, 2016).

5 For more on otherness, see Fred Dervin, 'Cultural Identity, Representation and Othering', in Jane Jackson (ed.), *The Routledge Handbook of Language and Intercultural Representation* (London: Routledge, 2012).

In different geographical contexts and times, people have crafted specific representations of disasters for certain purposes. Although most chapters deal with European disasters, case studies on the Japanese, US American, and Mesoamerican contexts demonstrate that the general patterns are applicable more broadly. Artists and authors tried to make sense of the events, grasp their impact, and communicate moral, religious, or political messages. Representations of disasters strengthened group identities, on different levels. Differences in terms of religious or social background could either be bridged, as becomes evident in the above-mentioned example of the storm in Borculo, or be exacerbated, by scapegoating or setting groups apart.[6] Either action makes dividing lines in society visible.

As such, representations are never straightforward nor unambiguous. People commenting on disasters in written reports or visual depictions transformed the disasters in question. They re-arrange or even adapt details to serve their goals and intentions. They use tropes and narratives that – consciously or unconsciously – colour their disaster experiences. Thus, representations of disasters are inspired by memory as much as by history. Because memory is a dynamic process, cultural representations of disasters are always the result of manipulation and appropriation.[7] Part of memory – and therefore of representations of disasters – is rooted in history, but another part is 'forever open to the full range of its possible significations', as historian and memory researcher Pierre Nora eloquently phrased it.[8] This leaves a task for scholars to disentangle the different guiding forces that shape depictions of disasters; only then can we understand the reasons how and why disasters are represented and remembered in the first place.

6 For religious divides caused by crises in early modern England, see Alexandra Walsham, *Providence in Early Modern England* (Oxford: Oxford University Press, 2001). A framework for cultural history research on catastrophes is presented in Lotte Jensen, Hanneke van Asperen, Adriaan Duiveman, Marieke van Egeraat, Fons Meijer, and Lilian Nijhuis, 'Appropriating Disasters. A Framework for Cultural Historical Research on Catastrophes in Europe, 1500–1900', *Journal of Historical Geography* 76 (2022), 34–41.

7 Pierre Nora, 'Between Memory and History. Les Lieux de Mémoire', *Representations* 26 (1989), 7–24, at 8. Some recent examples of research focusing on memory and nature-induced disasters include Adam D. Sundberg, *Natural Disaster at the Closing of the Dutch Golden Age* (Cambridge: Cambridge University Press, 2022); Felix Mauch, 'Betwixt and Between. A Reflection on the Scale and Scope of Disaster Memories', *International Journal for History, Culture and Modernity* 7 (2017), 110–41; Marco Folin and Monica Preti (eds), *Wounded Cities. The Representation of Urban Disasters in European Art (14th–20th centuries)* (Leiden: Brill, 2015); Christian Pfister, '"The Monster Swallows You". Disaster Memory and Risk Culture in Western Europe, 1500–2000', *RCC Perspectives* 1 (2011), 1–23.

8 Nora, 'Between Memory and History', 24.

In various chapters in *Dealing with Disasters*, we focus on nature-induced disasters, including floods (van Asperen, Meijer, van Egeraat, and Duiveman), earthquakes (Jacoby, Koopmans, and Rohr), explosions and fires (Hulsenboom, van Egeraat, and Rohr), avalanches (Rohr), famine (van Os), and various epidemics including cholera and rinderpest (Jensen, Wieser, Scholten, and Dekker).[9] We do so fully acknowledging the many areas of overlap between man-made and nature-induced disasters: scholars in all disciplines agree that human factors play a role in most – if not all – disasters, including in those severely aggravated by natural forces.[10] Human beings may be responsible for the occurrence of 'natural' disasters (for example by not maintaining the dykes properly or ignoring warning signals). They may heighten the negative effects of a disaster by not responding swiftly or adequately, or even deliberately cause disasters (as was the case with certain floods or famines and subsequent epidemics in wartime).[11] Nevertheless, we exclude calamities that were a direct result of warfare, genocide, or terrorism. To limit the scope of this volume, we focus instead on unplanned catastrophes aggravated by biological, meteorological, or geographical forces. Humans may have facilitated and/or exacerbated the destruction, devastation, and suffering, but these effects were not intentional. Although a catastrophe could be explained as divine punishment for unvirtuous human behaviour and people could therefore have seen the disaster coming and prevented it, contemporaries did not envisage the catastrophes as the result of human *intention*. Still, we fully realise that there is a large grey area in between.

Cultural Approach

During the past fifteen years, historical disasters studies has developed into a flourishing area of research.[12] Historians have convincingly argued

9 On the term nature-induced disasters, see Gerrit Jasper Schenk and Monica Juneja, 'Viewing Disasters. Myth, History, Iconography and Media Across Europe and Asia', in Monica Juneja and Gerrit Jasper Schenk (eds), *Disaster as Image. Iconographies and Media Strategies Across Europe and Asia* (Regensburg: Schnell & Steiner, 2014), 7–40.

10 Bas van Bavel, Daniel R. Curtis, Jessica Dijkman, Matthew Hannaford, Maïka de Keyzer, Eline van Onacker, and Tim Soens, *Disasters and History. The Vulnerability and Resilience of Past Societies* (Cambridge: Cambridge University Press, 2020), 29–31.

11 At the far end, Niall Fergusson argues that all disasters are man-made and political, in *Doom. The Politics of Catastrophe* (New York: Penguin Press, 2021).

12 Foundational were articles by, amongst others, Monica Juneja and Franz Mauelshagen, 'Disasters and Pre-Industrial Societies. Historiographic Trends and Comparative Perspectives',

that historical sources can be used as a 'laboratory' to gather data that help us to understand and predict which societies cope with severe crises better than others.[13] Key concepts are 'vulnerability' and 'resilience', often interpreted from economic and socio-geographic perspectives: quantitative data on death toll, fluctuations in prosperity, and demographic patterns offer indications of how urban and rural societies dealt with crises in the past.

More recently, researchers have pointed out the necessity to include cultural responses as a topic of disaster research. Scholars in the fields of media, memory, culture, and literary studies have demonstrated that media play an essential role in our perception of destruction and catastrophe, because the dissemination of information and the interpretation of events takes place primarily through stories using narratives, tropes, and visualisation.[14] Furthermore, vulnerability and resilience also have social, moral, and religious dimensions, which are determined by cultural values.[15] Such beliefs and practices are also mirrored in and shaped by media which repeated and distributed them. As a result, there is a fascinating interaction at play between values, disasters, and media. Media transmitted and shaped interpretations of events, defined by cultural values. Subsequently, media influenced responses, which were in their turn infused and formed by cultural values. Therefore, the role of discursive patterns, narrative frameworks, and visual imagery must be included in order to fully understand how societies, past and present, have dealt with disasters. Cultural responses in literature and chronicles, paintings and prints, newspapers,

The Medieval History Journal 10: 1–2 (2007), 1–31 and Gerrit Jasper Schenk, 'Historical Disaster Research. State of Research, Concepts, Methods and Case Studies', *Historical Social Research* 32: 3 (2007), 9–31.

13 Bas van Bavel and Daniel R. Curtis, 'Better Understanding Disasters by Better Using History. Systematically Using the Historical Record as One Way to Advance Research into Disasters', in *International Journal of Mass Emergencies and Disasters* (2015), 1–32; van Bavel et al., *Disasters and History*, 43; Gerrit Jasper Schenk, '"Learning from History". Chances, Problems and Limits of Learning from Natural Disasters', in Fred Krüger, Greg Bankoff, Terry Cannon, Benedikt Orlowski, and E. Lisa F. Schipper (eds), *Cultures and Disasters. Understanding Cultural Framings in Disaster Risk Reduction* (New York: Routledge, 2015), 72–86.

14 Judith Pollmann, 'Of Living Legends and Authentic Tales. How to Get Remembered in Early Modern Europe', *Transactions of the RHS* 23 (2013), 103–25; Juneja and Schenk, *Disaster as Image*; Carla Meyer, Katja Patzel-Mattern, and Gerrit Jasper Schenk (eds), *Krisengeschichte(n). 'Krise' als Leitbegriff und Erzählmuster in kulturwissenschaftlicher Perspektive* (Stuttgart: Franz Steiner Verlag, 2013), 117–44; Raingard Esser, '"Ofter gheen water op en hadde gheweest". Narratives of Resilience on the Dutch Coast in the Seventeenth Century', *Dutch Crossing* 40: 2 (2016), 97–107.

15 See, for example, Christoph Mauch and Christian Pfister, *Natural Disasters, Cultural Responses. Case Studies Towards a Global Environmental History* (Lanham, MD: Lexington Books, 2009).

commemoration books, songs, and so on reveal much about the way societies have dealt with disruptive and destructive events such as floods, fires, storms, earthquakes, volcanic eruptions, and epidemics. Yet interpreting these cultural responses is not a simple or straightforward task. Every case needs cultural and historical contextualisation to fully grasp its meaning: words and pictures can be used with very different meanings and aims; these may vary in different contexts and change over time.[16] Furthermore, recipients may add further meaning to these cultural representations and re-use them to create new narratives with additional or entirely different connotations.

Generally speaking, the process of interpretation is at work on three different levels: perception, transmission, and reception. The first step is always perception. Disaster only becomes a disaster if it is perceived as such.[17] Natural violence then does not necessarily have to be a disaster. It gains its meaning from its social context. In other words, 'natural' disasters are the direct result of an interaction between the natural environment and the social realm. Only if humans perceive events as disasters – often because a group is deeply affected – will they represent them as such in the media. Considering that experience and perception are inherently subjective, disaster is inextricably linked with subjective interpretation. For example, one person or group may not perceive an event as a disaster, but this does not mean that others did not. Therefore, it is important to realise who does the 'perceiving' in order to expose individual, social, religious, or political inclinations, and even bias and prejudice.

The next stage in the process of cultural representation is transmission, a form of communication which may take many different forms. Here the

16 Jensen et al., 'Appropriating Disasters', 38–40.

17 Gerrit Jasper Schenk, 'Lektüren im Buch der Natur. Wahrnehmung, Beschreibung und Deutung von Naturkatastrophen', in Susanne Rau and Birgit Studt (eds), *Geschichte Schreiben. Ein Quellen- und Studienhandbuch zur Historiographie (ca. 1350–1750)* (Berlin: Akademie Verlag), 507–21; Gerrit Jasper Schenk, 'Historical Disaster Experiences. First Steps Toward a Comparative and Transcultural History of Disasters Across Asia and Europe in the Preindustrial Era', in Gerrit Jasper Schenk (ed.), *Historical Disaster Experiences. Transcultural Research – Heidelberg Studies on Asia and Europe in a Global Context* (Cham: Springer, 2017), 3–44, at 15–23; Gerrit Jasper Schenk and Monica Juneja, 'Viewing Disasters. Myth, History, Iconography and Media Across Europe and Asia', in Juneja and Schenk, *Disaster as Image*. Jörg Trempler has argued that images played a decisive role in the modern conceptualisation of 'catastrophe' in *Katastrophen. Ihre Entstehung aus dem Bild* (Berlin: Verlag Klaus Wagenbach, 2013); Jörg Trempler, 'Catastrophes and Their Images. Event and Pictorial Act', *RES. Journal of Anthropology and Aesthetics* 63–64 (2013), 201–14. Vera Fionie Koppenleitner argues that this conceptualisation through images started earlier, see her *Katastrophenbilder. Der Vesuvausbruch 1631 in den Bildkünsten der Frühen Neuzeit*, I Mandorli 22 (Berlin and Munich: Deutscher Kunstverlag, 2018).

process of transforming an experience into a representation takes shape: visibly or audibly, tangibly or intangibly. In the communication of the event, others learn how people lived through the events and reacted to them.[18] In the (paraphrased) words of music scholar Dorothea Redepenning: with the help of natural sciences and the study of historical documents, we can reconstruct what has happened, but via works of art we learn how people experienced these historical events.[19] In this way recipients learn even more than just about the disaster and its consequences. The person who communicates the disaster (orally, ritually, textually, or visually) to an audience adds a new layer of interpretation to the event. In the exchange the disaster is again transformed, because – as with perception – transmission is never neutral but always informed by context and loaded with new associations, sentiments, and convictions.

Finally, another layer can be added in the third phase: reception. In fact, reception is a subsequent form of perception. When an audience learns about a disaster via cultural representations, they add their layer of interpretation, mentally but also physically. With the physical expression of thoughts and emotions, the audience adds another representation, e.g. a person might express personal thoughts and feelings, add notes to a manuscript or pamphlet, or sometimes create an entirely new representation of their own informed by others.[20]

To understand the complex dynamics of this simplified description of the process, it is crucial to point out the many areas of overlap between perception, transmission (or communication), and reception (which is again a form of perception with the addition of a new layer of interpretation). The people involved in these three stages might be one and the same. Artists

18 For recurrent literary structures of disaster narrations, see Ansgar Nünning, 'Krise als Erzählung und Metapher. Literaturwissenschaftliche Bausteine für eine Metaphorologie und Narratologie von Krisen', in Meyer, Patzel-Mattern, and Schenk (eds), *Krisengeschichte(n)*, 117–44; Ansgar Nünning, 'Towards a Metaphorology of Crises, or: The Uses of Cognitive Metaphor Theory for the Study of Culture', in Ana Margarida Abrantes and Peter Hanenberg (eds), *Cognition and Culture* (Frankfurt am Main: Peter Lang, 2011), 71–98; Ansgar Nünning, 'Making Crises and Catastrophes. Metaphors and Narratives Shaping the Cultural Life of Crises and Catastrophes', in Carsten Meiner and Kristin Veel (eds), *The Cultural Life of Catastrophes and Crises. Facts, Forms, Fantasies* (Berlin: De Gruyter, 2012), 59–88.

19 Dorothea Redepenning, 'Musik und Katastrophe – Katastrophenmusik', in Gerrit Jasper Schenk, Monica Juneja, Alfried Wieczorek, and Christoph Lind (eds), *Mensch, Natur, Katastrophe von Atlantis bis heute* (Regensburg: Schnell & Steiner, 2014), 129–33, at 133.

20 See for emotive responses to disaster images and texts: Susan Sontag, *Regarding the Pain of Others* (New York: Picador/Farrar, Straus & Giroux, 2003); Thomas Labbé and Gerrit Jasper Schenk (eds), *Une histoire du sensible. La perception des victimes de catastrophe du XIIe au XVIIIe siècle* (Turnhout: Brepols, 2018).

creating a representation may have personally experienced the disaster they are describing or depicting, making them both the perceiving and transmitting agents. But even if they learn about a disaster from personal experience, they combine this knowledge with other sources of information. What people are transmitting is never informed merely by individual experience but also by historical documents and cultural representations. In other words, they combine perception with reception. In other instances, recipients of a cultural representation – who are never passive but rather add their own layer of meaning to a disaster text or image – may become communicating agents. In the end, the cultural representations are always the results of the closely intertwined stages of perception, transmission, and reception.

After a while, some disasters may become part of collective memory, while others do not. This difference in outcome is the result of many variables. Scholars of memory culture have analysed the different elements that increase an event's chances of being remembered. A connection with stories, images, gestures, rituals, and places has a decisive effect, sometimes more than facts and figures.[21] You could say that narratives and images sell a disaster.[22] However, the opposite is also possible: once a disaster has become part of collective memory, it may become a unique selling point. Illustratively, in 2021 a so-called States Bible in very bad condition was put up for sale on a Dutch online sourcing and selling platform.[23] The seller described the item as 'an 1846 States Bible with water damage from the North Sea Flood of 1953'. With this description, they presented the mangled book as an artefact of a disastrous flood that has become part of collective memory for many Dutch citizens.[24] Because 'memory installs remembrance within the sacred', to quote Nora, the objects connected to disaster memories become relics.[25]

In a joint article of the research group *Dealing with Disasters*, we have proposed using the concepts of 'appropriation', 'representation', and

21 Nora, 'Between Memory and History', 9. For an example of St Elisabeth Flood, see Pollmann, 'Of Living Legends', 108–15.

22 For the connection between images of disasters and sales figures, see Paul Hockings, 'Disasters Drawn. *The Illustrated London News* in the Mid-19th Century', *Visual Anthropology* 28 (2015), 21–50.

23 'Statenbijbel 1846 met waterschade door overstroming van 1953', advertisement on the online selling platform *Marktplaats*, published online 21 December 2021, https://www.marktplaats.nl/ (accessed 9 February 2022).

24 Lotte Jensen, 'Floods as Shapers of Dutch Cultural Identity. Media, Theories and Practices', *Water History* 13: 2 (2021), 217–33, at 221.

25 Nora, 'Between Memory and History', 9.

'identification' as main analytical categories in studying the cultural responses to disasters.[26] We argue that people appropriate disasters through two simultaneous and interdependent acts: representation and identification. Representation substitutes a disastrous event with a cultural artefact, a monument, a text, an image, or a combination of these. In this creative process, an event is categorised as disastrous and framed within certain narrative patterns: it is transformed into a story with a logical order (logical at least to the producers and theoretically to their intended audience as well) and an ending or consequence, for instance punishment for those who will not improve their sinful way of living which was believed to have caused the disaster. The narrative enables producers, reproducers, and consumers to relate to the events and identify themselves with (parts of) the representation. This perhaps motivates them to adjust behaviour accordingly or to feel strengthened in their already chosen approach.

Historical analyses of representations of disasters should therefore consider four aspects: voice (of performer, author, or artist), medium (including generic conventions and intended audience), discourse (narrative and visual structures), and community. As the above-mentioned example of Borculo illustrates, all these layers are at work at the same time. To grasp the full meaning of the cover, information about the producers, their intended consumers, and medium is necessary. By 'discourse' we mean that representations have a narrative or visual structure. While crafting representations of disasters, producers work with existing tropes, narratives, and lines of thought. In other words, existing vocabularies and images influence new representations. In fact, it is nearly impossible, and quite undesirable, to invent entirely new vocabularies or images of disasters; in order to relate to horrific events, an audience needs to recognise these as such. Therefore, intertextuality – and intervisuality – are very important to cultural artefacts of disasters. They can hardly survive without them.

Besides voice, medium, and discourse, scholars must scrutinise the communities that producers of images and texts are cultivating, criticising, or ignoring. Sociologists have demonstrated that catastrophes have been used in the media to increase community building, because, for example, people reach out to each other seeking comfort, solidarity, and support.[27] Disasters

26 Jensen et al., 'Appropriating Disasters', 34–42.

27 John Drury, Chris Cocking, and Steve Reicher, 'Everyone for Themselves? A Comparative Study of Crowd Solidarity Among Emergency Survivors', *British Journal of Social Psychology* 48 (2009) 487–506; Anthony Oliver-Smith, 'The Brotherhood of Pain. Theoretical and Applied Perspectives on Post-Disaster Solidarity', in Oliver-Smith and Hoffman, *The Angry Earth*, 156–72; Allen H. Barton, *Communities in Disaster. A Sociological Analysis of Collective Stress Situations*

have time and again turned out to be occasions for artists to strengthen communal ties on religious, social, or political grounds and on the local, regional, and even transnational levels. Yet, at the same time, cultural media could also set groups apart: by stressing communality between people from a specific group, others were, sometimes consciously and explicitly, excluded or ignored. By studying cultural representations of disasters, it is possible to scrutinise such processes of in- and exclusion, both implicit and explicit, and to gain more understanding of how people relate to different communities, what binds them and what sets them apart. They develop different identities, strengthening communal bonds and accentuating social divides in the process.

Frames of Interpretation

The volume *Dealing with Disasters* attempts to show how communities responded to disasters. The authors investigate how people perceived, transmitted, and received disasters using cultural representations, i.e. texts and images, as their primary source material. Many of the contributors focus on one type of medium, such as pamphlets (van Egeraat), the illustrated press (van Os), chronicles (Dekker), newspapers (Koopmans), or commemorative books (Duiveman). Others discuss multiple genres over a longer period (Meijer, Jensen, Wieser, Jacoby, and Rohr). Susan Broomhall analyses natural violence in the written works and drawings of one artist. Finally, other contributions include case studies focusing on one disaster representation, such as a printed image (van Asperen), a poem (Hulsenboom), or a novel (Scholten). Looking at disasters from these diverse angles, this volume investigates what roles cultural representations play when the events are mediated and moulded according to different wishes and desires.

The fourteen contributions have been divided over three thematic sections: 'Disaster and Emotions', 'Disaster and Blame', and 'Disaster and Time'. Each of these play a crucial role in cultural responses to disasters. The first of the three sections is dedicated to emotions which help artists to make sense of catastrophic events. Moreover, emotions tie people together, because they reflect shared feelings or even prescribe appropriate responses to people's misery. Another common type of immediate reaction

(New York: Doubleday, 1969); Anthony F.C. Wallace, *Tornado in Worcester. An Exploratory Study of Individual and Community Behavior in an Extreme Situation* (Washington, DC: National Academy of Sciences – National Research Council, 1956).

to disasters is to assign blame: after large-scale catastrophes people often look for scapegoats who are presumed to have provoked or aggravated the crisis. Finally, time is an important factor. Some authors and artists place recent disasters in a long-term perspective, relating them to events in the more distant past. Additionally, stories and images about the one and the same disaster irrevocably change after time has passed. Furthermore, responses are always time-bound: every period has its own time-specific values and predominant ways of coping. Nevertheless, comparing cultural responses from different periods inevitably reveals patterns of continuity. Although different times ask for different emotional responses and different scapegoats, cultural media have always been indispensable when trying to increase societal resilience. They have continually offered an outlet for emotions, provided consolation, propagated solidarity and charity, or accentuated social divides. In other words, emotions, blame, and time are used here not to categorise cultural responses and divide these into separated groups, but rather as thematic lenses to help scholars understand how contemporaries conceptualised and appropriated disasters.

The first part of this volume, 'Disaster and Emotions', is devoted to the visualisation of feelings in, and affective responses to, representations of disasters. The first chapter focuses on one artist's 'voice': that of the early modern *uomo universale* Leonardo da Vinci. Susan Broomhall investigates how the creative mind tried to capture essential elements of natural violence and to explore its meanings in various works. Perception of natural violence – disasters from the human perspective – operated through fear and other emotions, which Da Vinci tried to visualise on multiple occasions. However, in his approach natural violence was not just destructive but also generative. Nature could destroy human creation but at the same time transform and regenerate herself, offering Da Vinci an artistic framework in which natural violence is a conceptually female counterpart to the artist's genius, which was – traditionally – conceived as male.

The next chapters highlight the workings of emotions, elaborating on the theoretical framework of historian and medievalist Barbara Rosenwein, who recognised and investigated the formation and effects of 'emotional communities' in society.[28] In his contribution, Paul Hulsenboom demonstrates how a seventeenth-century author forged an emotional bond between literates across nations (Chapter 2). Because of economic and

28 Barbara H. Rosenwein, *Generations of Feeling. A History of Emotions, 600–1700* (Cambridge: Cambridge University Press, 2016), 3; and Barbara H. Rosenwein, *Emotional Communities in the Early Middle Ages* (New York and London: Cornell University Press, 2006).

cultural relations, there was a lively exchange of news about disasters that happened in foreign countries. These accounts inspired cultural expressions that promoted solidarity among the international scholarly community. To demonstrate these transnational connections, Hulsenboom focuses on the prolific Silesian-Polish author Joachim Pastorius, who wrote a Latin poem about a 1654 gunpowder disaster in Delft in the Dutch Republic. Written in Gdańsk in 1657, this Latin poem stimulated feelings of solidarity between educated people and shaped an emotional community which transgressed state borders.

Next, Hanneke van Asperen shows how a printed image could trigger emotional engagement within the Dutch Republic by presenting a geographic map as a focus of emotional attention (Chapter 3). To create an efficacious visual discourse, the Dutch printmaker Simon Fokke consulted traditional Catholic images of pity focusing on the suffering body of Christ, which had served to stimulate compassion and co-suffering for centuries. Fokke addressed the predominantly Protestant readers of the *Nederlandsche jaerboeken* ('Dutch Yearbooks') with the image, thus introducing the iconography of the suffering body to a larger audience. Reaching an audience across the Dutch Republic, the image also shows the budding wish to tie the relatively independent provinces together.

In the fourth chapter, Fons Meijer investigates nineteenth-century effects of national identification. For the concept of 'emotional communities', Rosenwein built on Benedict Anderson's notion of 'imagined communities', which he outlined in his study on the invention of national identities.[29] In their footsteps, Meijer uses occasional poetry and eyewitness reports in the periodical press to demonstrate how, in the aftermath of disasters, representations could be used to generate transregional affection for people in other, distant parts of the country. The author argues that authors did not simply evoke feelings of compassion but that they helped to shape this emotion using a typical 'disaster discourse', creating a culture of compassion that focused on national or 'Dutch' disasters.

In the final entry in this section, Lotte Jensen focuses on cholera outbreaks (Chapter 5). She shows how religious explanations could compete with or exist parallel to scientific and institutional interpretations. Most nineteenth-century writers urged Dutch citizens to repent their sins and pray, although there were some contemporary critics, including a physician, who claimed that the outbreaks had nothing to do with sinful behaviour

29 Benedict Anderson, *Imagined Communities. Reflections on the Origin and Spread of Nationalism* (London: Verso, 1991).

and could be attributed to poor hygiene and poverty. By collectively raising money for the sufferers, some also aimed at fostering a sense of the nation as a moral community. The narrative framework of blame, sin, and repentance increased people's sense of a national community and of the responsibility to help others. Jensen argues that these cultural responses help to enhance societal resilience. Together, the chapters of 'Disaster and Emotions' highlight how disasters were appropriated to bolster feelings of community.

The second part, entitled 'Disaster and Blame', is centred around the search for explanations by looking for scapegoats after a disaster had occurred. After all, people often appropriated disasters to draw or accentuate social divides. Through the ages religion has been at the heart of cultural responses; the providential explanation – with its defining characteristic of recognising divine intervention – often went hand in hand with placing blame on certain groups of people. This could be the given community, which had to repent their sins, or competing religious, denominational, social, or political circles. The contributions demonstrate how the focus on solidarity and community in one group could often lead to the exclusion of another.

The first two chapters show how producers of representations (authors, artists, and publishers) carefully considered the way they portrayed catastrophes. Marieke van Egeraat focuses on explanations of disasters in sixteenth-century pamphlets (Chapter 6). She observes that Dutch pamphlet sellers downplayed apocalyptic interpretations, while their German counterparts highlighted an eschatological view. Subsequently, van Egeraat argues that pamphlet sellers in the Low Countries preferred the less polemic idea of divine punishment as an explanation, because it fitted the diverse religious landscape and helped to increase sales figures.

Florian Wieser investigates the ways both Spanish invaders and indigenous elite made sense of the epidemic depopulation in sixteenth-century Mesoamerica (Chapter 7). Members of the clergy, *conquistadores* and their successors, and Mesoamerican intellectuals offered religious explanations. All of them stated that sinful behaviour provoked the diseases, but each blamed different groups depending on their own position in society. Both Wieser and van Egeraat emphasize the importance of identifying the voice behind the medium.

Sophie van Os focuses on the depictions of the Great Irish Famine in the Victorian illustrated press (Chapter 8) and demonstrates how visual depictions belong to a larger nineteenth-century cultural discourse of poverty, hunger, and famine. Representations of the famine combined elements from established visual repertoires and in their turn influenced how later disasters were depicted and interpreted. Van Os observes how techniques and pictorial elements were employed to create a distance between viewer and the famine victims, reaffirming social divides.

Blame, exclusion, and scapegoating are at the heart of the novels analysed by Anneloek Scholten (Chapter 9). Her sources engage with group dynamics in small rural communities during an outbreak of rinderpest. Depicting fictive situations against the historical background of the cattle plague, novelists describe the processes of in- and exclusion in rural communities. A close reading of Josef Cohen's *Ver van de menschen* ('Far from the People', 1910), a novel set in a rural area in the Low Countries, reveals how the author challenges and scrutinises dynamics of exclusion and notions of belonging by encouraging the audience to imagine grief as a shared experience across nations and generations.

The third and final section, 'Disasters and Time', deals with aspects of temporality, memory, and (dis)continuity. Each of the essays examines a different type, or types, of medium – chronicles, newspapers, prints, and/or commemoration books – over a longer period. These enable the contributors to identify patterns and transformations in discourses. The sources often bring past and present together, sometimes even looking forward into the future. The contributions thus relate to the work of historian Michael Kempe, who defined a phase in which collective memory crystallises into a 'culture of disaster'.[30] In these amphibian societies, cultural representations reminded the audience that floods might happen again so that they could adapt their attitudes and living environment accordingly.

First, Theo Dekker reflects on issues of continuity and discontinuity regarding fatal epidemics between 1500 and 1850 (Chapter 10). His main sources are Dutch chronicles, in which predominantly male, white literates describe their personal experiences. Dekker demonstrates that while natural explanations became more detailed and complex over time, they remained supplementary or subordinate to divine explanations. Religious and scientific explanations were not incompatible discourses. On the contrary, they could easily exist side by side and even complement each other. Although the role of God also underwent significant changes, from a punishing to a more benevolent being, in desperate circumstances both Catholics and Protestants turned to traditional religious practices.

30 Michael Kempe, '"Mind the Next Flood!" Memories of Natural Disaster in Northern Germany from the Sixteenth Century to the Present', *The Medieval History Journal* 10: 1–2 (2007), 327–54. Petra J.E.M. van Dam elaborates on this idea of 'amphibian cultures' in 'An Amphibious Culture. Coping with Floods in the Netherlands', in Peter Coates, David Moon, and Paul Warde (eds), *Local Places, Global Processes. Histories of Environmental Change in Britain and Beyond* (Oxford: Oxbow Books, 2017), 78–93.

Joop W. Koopmans focuses on the discourse of early modern Dutch newspapers reporting about earthquakes and tremors before the 1755 Lisbon Earthquake (Chapter 11). According to a widely accepted religious idea, earthquakes were punishments from God, and people should repent their sins to prevent more and worse disasters from happening. This thought was neither denied nor questioned, nor openly proclaimed in the reports, but they were implicitly taken for granted.

Julia Mariko Jacoby then analyses remarkably consistent cultural representations of Japanese earthquakes in different media in early modern and modern times (Chapter 12). The almost mythical creature of the catfish is at the heart of many Japanese images in which the trembling of the earth was paralleled with Western power: both could entail business opportunities and pose dangers at the same time. Single-page newsprints appeared anonymously and were tolerated only as long as the content was not offensive to the Tokugawa Shogunate. Jacoby also describes the discourses in commemoration books that assessed and mapped damage; narrated tragedies, heroism, and absurdities that emerged from disaster; and concluded on a positive note with first reports on reconstruction. These gave cosmological or scientific explanations for the disaster and contextualised it in the long history of similar events.

Unlike chronicles and newspapers, commemoration books often situate a catastrophe in longer historical perspective: Their authors deliberately reflected upon the repetitive occurrences of floods and other disasters to provide moral and practical lessons for the future, as Adriaan Duiveman shows (Chapter 13). In his analysis of the memorial functions of four commemoration books, he demonstrates that the authors emphasised a cyclical recurrence of floods in the Netherlands and that they provided their audience with technological and religious advice to prevent future ones, thus counteracting the decline narrative that was also prominent in eighteenth-century Dutch thought.

This book ends with some valuable reflections on the present. The fourteenth and final chapter is devoted to disaster memory in the nineteenth and twentieth centuries and in our own time. Looking at floods, avalanches, storms, and earthquakes in Europe and the United States of America, Christian Rohr argues that societies that do not actively remember disasters are less prepared for unexpected extreme events and therefore more vulnerable. Rohr thus promotes a new culture of disaster memory, making use of new media and new forms of knowledge transfer between people whose families have lived in vulnerable areas for centuries, on the one side, and newcomers, on the other.

Combined, the three thematic sections on emotions, blame, and time offer an overview of the various ways to study cultural representations of disasters across academic disciplines from three complementing and often overlapping perspectives. Together, the contributions demonstrate who appropriated disasters, i.e. whose voices we hear when we investigate cultural representations. Furthermore, they show for what reasons people appropriated disasters and which media and discourses people used to achieve their goals.

Future Paths of Investigation

As scholars have done in the past, some contributors to this volume mention the somewhat paradoxical trope of human incapability to describe pain and destruction on a large scale.[31] Despite this often-used disclaimer, many authors and artists in the past and present have tried to capture suffering and loss in their work. One important goal of visual art and literature focusing on disasters is the aesthetics, which was the subject of the exhibition 'Entfesselte Natur' ('Nature Unchained') in Hamburg in 2018.[32] People enjoy – and have enjoyed – looking at and contemplating images that show the violent and destructive side of nature. Additionally, art in different shapes and forms has given humans a tool to come to grips with destructive forces that are beyond immediate comprehension.

One of the explicit goals of this volume is to demonstrate how representations in different media work to achieve goals of identification and appropriation. The chapters demonstrate how people in different times produced and distributed cultural representations of disaster in different media to cope with crises, come to terms with them, and manipulate them. In other words, this book investigates how people mould disasters into visuals and narratives to appropriate them and make them manageable or even functional.

Nevertheless, *Dealing with Disasters* leaves unanswered questions. To start, this volume focuses on tangible and surviving sources, e.g. printed texts, printed images, manuscripts, and drawings. After time has passed, we are only able to read about the songs people sang, the rituals they enacted, and the stories they told in materialised, solidified, and crystallised form.

31 e.g. van Asperen (Chapter 3), Jensen (Chapter 5), and van Os (Chapter 8) in this book.

32 Markus Bertsch and Jörg Trempler (eds), *Entfesselte Natur. Das Bild der Katastrophe seit 1600*, exh. Hamburger Kunsthalle, 29 June–14 October 2018 (Petersberg: Imhof, 2018).

These performances, however, are flexible and dynamic.[33] Moreover, voices do not fall silent once the stories, songs, and ceremonies are put to paper, which leaves the question of how the surviving material sources relate to oral culture, as both their informant and their percipient.

Secondly, it would be worth investigating more of the blind spots and prejudices in the surviving source material. Historians mostly read and learn of those people who left material sources for us to study. The perspectives are therefore often elite, white, and male. Even when they acknowledge this, scholars still struggle to overcome this bias. Furthermore, it is difficult to investigate those who are not (or hardly) mentioned in the sources and who did not write, draw, or paint themselves. This does not mean, however, that it is entirely impossible to study their experiences and memories. Sometimes we do have sources that were produced by marginalised people, but these may take different shape and form. A few examples are textiles, votive images, or additional handwritten notes.[34] A case in point is presented in a manuscript obituary book (*obituarium*) from the convent of Mariënburg, a religious community of Tertiaries in 's-Hertogenbosch, situated in the Southern Low Countries (the present-day Netherlands).[35] The manuscript book contains a calendar with the names of the sisters and the convent's major benefactors who had died and were to be commemorated on specific days during the liturgical year (Fig. 0.2).

An anonymous author, closely connected to the convent and probably living there, added three short notes to the book's flyleaf. The third and final one is a recollection of the 'big fire' ('den groten brant') in 1463. In that year, a conflagration destroyed a substantial part of the city of 's-Hertogenbosch. The note comes after two others. The first mentions the date of the convent's confirmation 'on the day of St John's birth', i.e. 24 June, in 1459.[36] The

33 Paul Post, 'Introduction. Some Conceptual and Historiographical Explorations on Ritual, Disaster, and Disaster Ritual', in Martin Hoondert, Paul Post, Mirella Klomp, and Marcel Barnard (eds), *Handbook of Disaster Ritual. Multidisciplinary Perspectives, Cases and Themes* (Leuven, Paris, and Bristol, CT: Peeters, 2021), 1–48.

34 For examples of votive gifts as representations of disasters, see Giovanni Gugg, 'The Missing Ex-Voto. Anthropology and Approach to Devotional Practices During the 1631 Eruption of the Vesuvius', in Domenico Cecere, Chiara de Caprio, Lorenza Gianfrancesco, and Pasquale Palmieri (eds), *Disaster Narratives in Early Modern Naples. Politics, Communication and Culture* (Rome: Viella, 2018), 221–36.

35 The anonymous sister wrote: 'Item den groten brant ten boesch was a° m cccc lxiij', which translates as: 'The same, the great fire in 's-Hertogenbosch was in the year 1463' in the *liber obituum*, or *obituarium*, of the convent Mariënburg, kept in Brabant-Collectie of Tilburg University library, Ms KHS 6.

36 This year 1459 might be an error, because the sisters took up the rule of St Francis only in 1469. There may be another explanation: the book itself was probably not written until the first quarter of the sixteenth century, so the notes were most likely copied. Perhaps the author

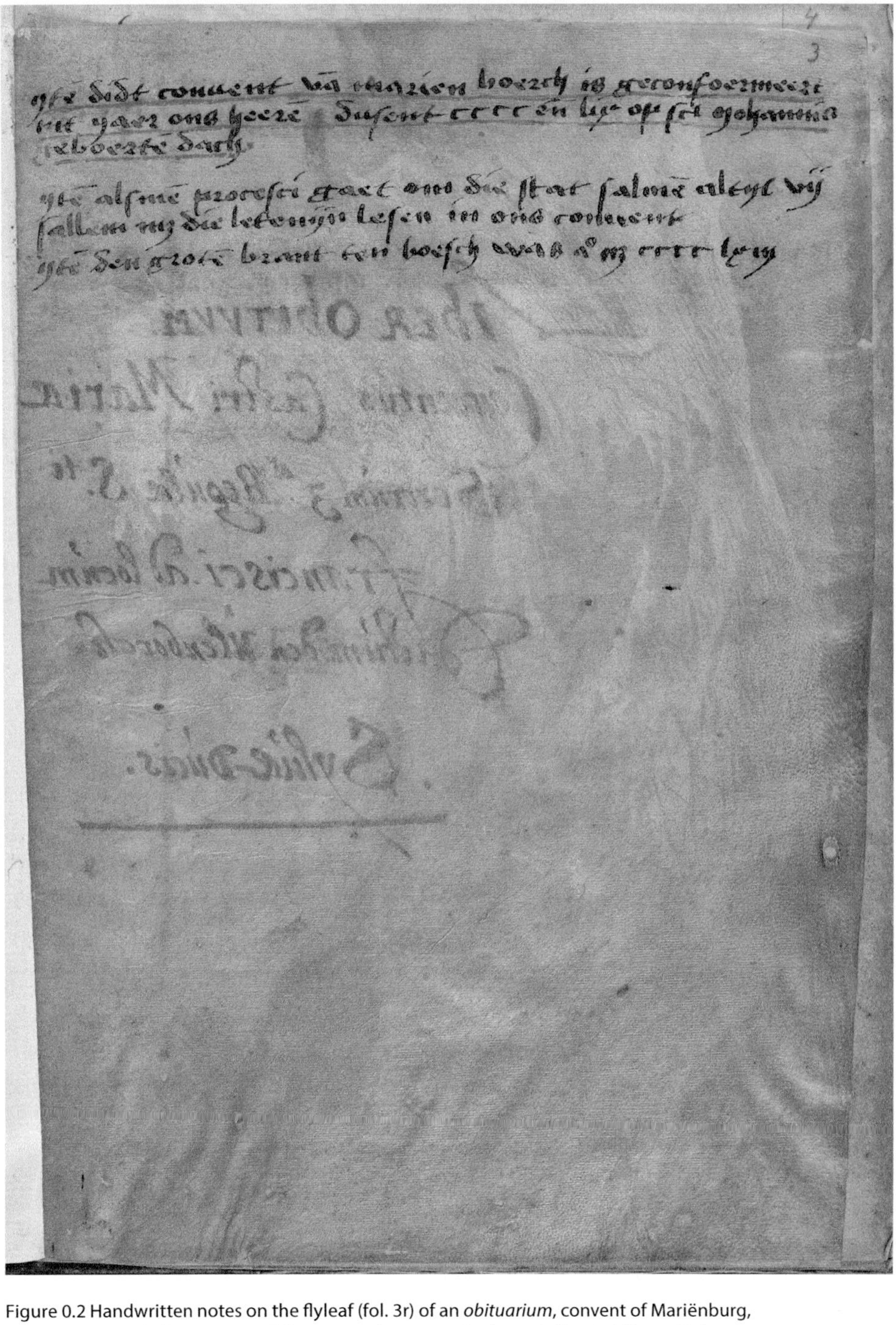

Figure 0.2 Handwritten notes on the flyleaf (fol. 3r) of an *obituarium*, convent of Mariënburg, 's-Hertogenbosch, first quarter of the sixteenth century. Brabant-Collectie, Tilburg University Library, KHS 6

second note is a reminder to read the psalms and litany on the day of the procession of 's-Hertogenbosch, which was always held on the first Sunday after 24 June. This type of evidence is important for our understanding of disaster experiences in the early modern period. First, the note identifies communities of religious women as bearers of memory.[37] Additionally, it indicates how significant dates for the sisters and the larger local community were combined with rituals and how disasters were embedded in an existing calendar of commemoration.[38]

Along with a quest for additional source material, we must keep on carefully scrutinising the sources that we already have and realise that there are always people who are forgotten, ignored, or stigmatised. We should ask ourselves, who is *not* addressed? Throughout the centuries, cultural representations – be they sermons, poems, paintings, songs, prints, handwritten chronicles, or commemoration books – offered consolation and explanation but also reflected and affirmed hierarchy, authority, and social status. These cultural products have been used as instruments to enhance resilience because they aimed to forge connections between people across time and space. However, inclusion always entails exclusion, and scholars benefit from an awareness of the exclusion processes. Furthermore, there is a lesson for society to learn as well, particularly for creative industries and (government) communications offices. The manifold links between past and present-day situations underline the value of the historical discipline for modern debates about cultural and creative strategies in response to disasters.

About the Authors

Hanneke van Asperen is an art historian affiliated with the Centre for Art Historical Documentation (CKD) at the Radboud University in Nijmegen. As postdoctoral researcher, she is part of the NWO-funded project *Dealing with Disasters. The Shaping of Local and National Identities in the Netherlands (1421–1890)*. She has expertise in visual images of charity and nature-induced

also copied the note from an older source. On the procession of 's-Hertogenbosch, see Anton Schuttelaars, 'De Heeren van de Raad. Bestuurlijke elite van 's-Hertogenbosch in de stedelijke samenleving, 1500–1580', PhD dissertation, Radboud University Nijmegen, 1998, 116–17.

37 These women helped to create the necessary *lieux de mémoire* to remember the fire. On *lieux de mémoire*, see Nora, 'Between Memory and History', 12.

38 For another example of research into rituals and disasters, see Adriaan Duiveman, 'Praying for (the) Community: Disasters, Ritual and Solidarity in the Eighteenth-Century Dutch Republic', *Cultural and Social History* 16: 5 (2019), 543–60.

disasters in premodern times. Additionally, she has an interest in medieval pilgrimage and religious badges. Her recent book *Silver Saints. Prayers and Badges in Late Medieval Books* (Brepols 2021) focuses on the practice of adding religious badges to devotional manuscripts.

Lotte Jensen is Professor of Dutch Cultural and Literary History at Radboud University, Nijmegen and Principal Investigator of the NWO-funded Vici project *Dealing with Disasters. The Shaping of Local and National Identities in the Netherlands (1421–1890)*, which investigates the impact of natural disasters on Dutch local and national identities from a cultural-historical perspective. She has published books on Napoleon's legacy in the Netherlands, Dutch literary history and the emergence of Dutch identity. Her most recent book is a cultural history of Dutch floods: *Wij en het water. Een Nederlandse geschiedenis* (De Bezige Bij 2022).

Part 1

Disaster and Emotions

1 Temporality, Emotion, and Gender in Leonardo da Vinci's Conceptualisation of Natural Violence

Susan Broomhall

Abstract
This chapter explores how the Italian engineer and artist Leonardo da Vinci conceptualised forms of violence in nature, to nature, or by nature. In using the term 'natural violence', this chapter aims to capture Da Vinci's broad-ranging consideration of such violence related to the natural world upon which he reflected across his work and to which he gave varied and ongoing responses over the course of his life. It argues that his perception of temporality, emotion, and gender were important aspects that helped Da Vinci make sense of natural violence. In doing so, the chapter suggests that while Da Vinci may have been radical in some aspects of thinking, in others he was representative of his era, and that investigating his conceptualisation of natural violence brings these distinctions into sharper focus.

Keywords
Leonardo da Vinci – nature – violence – gender – temporality – emotion

Throughout his life, the Italian engineer and artist Leonardo da Vinci (1452–1519) reflected upon violence in nature, to nature, or by nature, which could wreak destruction upon human and non-human species alike. These complex forms that I will consider under the term 'natural violence' were phenomena that he sought to understand, and he also worked to understand the relationships between them. In his work, Da Vinci theorised natural violence, responded to it, and aimed to guard against it. He explored natural

Asperen, van, Hanneke and Lotte Jensen (eds): *Dealing with Disasters from Early Modern to Modern Times. Cultural Responses to Catastrophes.* Taylor & Francis Group, 2023
DOI: 10.5117/9789463725798_CH01

violence in different forms and multiple sites – sketches, engineering designs, artworks, and writings across scattered pages of notes and notebooks. Central to his conceptualisations was an understanding of particular dimensions that underpinned natural world systems. Scholars of natural sciences in particular have tended to express these dimensions in terms of laws that Da Vinci sought to articulate through experimental investigation and upon which he drew within his engineering and artistic practices. As a consequence, other elements of his thinking – about dimensions such as time and emotion – have been less often explored in this scholarship. In this chapter, I suggest that analysis of Da Vinci's perception of temporality, emotion, and also gender is important to help to make sense of his conceptualisation of natural violence.

Early modern thinking about a world in transition has long formed the backdrop for modern scholars' consideration of the Anthropocene age. Temporality is of course central to this conceptualisation of our world, but so too is a certain historical vision of the early modern. When historian Dipesh Chakrabarty speaks of how anthropogenic explanations of climate change collapse a humanist distinction between natural and human histories, he is thinking of a mode of history widely attributed to philosopher Giambattista Vico, 'that we, humans could have proper knowledge of only civil and political institutions because we made them, while nature remains God's work and ultimately inscrutable to man'.[1] Sociologist and philosopher Bruno Latour similarly articulates a 'European prescientific vision of the Earth ... as a cesspool of decay, death, and corruption from which our ancestors, their eyes fixed toward the incorruptible spheres of suns, stars, and God, had a tiny chance of escaping solely through prayer, contemplation, and knowledge', while philosopher and historian Michel Serres considers how 'nature acted as a reference point for ancient law and for modern science *because it had no subject*'.[2] As modern scholars continue to reflect on the relationship between human agency and the world around us, further study of early modern conceptualisations of this relationship is also fruitful. Da Vinci may have been radical in some aspects of his thinking, as has long been claimed, but in others he was representative of his era. Investigating

1 Research for this chapter was supported by an Australian Research Council Discovery Project (DP210100104). My thanks to the editors for their insightful feedback on an earlier draft of this chapter.
Dipesh Chakrabarty, 'The Climate of History. Four Theses', *Critical Inquiry* 35: 2 (2009), 197–222, at 201.

2 Bruno Latour, 'Agency at the Time of the Anthropocene', *New Literary History* 45 (2014), 1–18, at 4, and citing Michel Serres, *The Natural Contract*, trans. Elizabeth Macarthur and William Paulson (Ann Arbor: University of Michigan Press, 1995), 86.

his conceptualisation of natural violence brings these distinctions into sharper focus.

Knowledge production about natural violence occurred in a wide range of genres in the early modern period, including fables, sketches, engineering, poetry, paintings, essays, and journal accounts. Although they produced very different forms of knowledge, collectively, they suggest that perception of such violence as disaster might only be one position, or perhaps juncture, temporal and emotional, on the interpretive spectrum of such destructive events.[3] Moreover, as Gerrit Jasper Schenk has explored, terms such as 'disaster' and 'catastrophe' held different valency in different early modern linguistic and cultural traditions.[4] Disaster as much as nature is thus a concept that finds meaning in specific cultural contexts.[5] Scholars are increasingly demonstrating how early moderns narrated events in which the natural world produced catastrophic consequences for human populations through historically specific cultural frames but did not always conceptualise such devastation as disastrous.[6] For some authors and diarists, damage that the natural world created in the lives of humans was witnessed for posterity in enumerations that appeared to operate as part of an eschatological framework pointing to end-times. For others, these could be affirmative of specific community identities and faith positions, since they could potentially demonstrate God's mercy towards those who survived to attest to the experience. Additionally, there were other early moderns who emphasised the relativity of human perceptions of disaster. Thus, the studied philosophical response of essayist Michel de Montaigne could utilise a dramatic language of feeling for devastating natural world

3 See Susan Broomhall, 'Devastated Nature. The Emotions of Natural World Catastrophe in Sixteenth-Century France', in Erin Peters and Cynthia Richards (eds), *Early Modern Trauma. Europe and the Atlantic World* (Lincoln: University of Nebraska Press, 2021), 31–53.

4 Gerrit Jasper Schenk, 'Historical Disaster Experiences. First Steps Toward a Comparative and Transcultural History of Disasters Across Asia and Europe in the Preindustrial Era', in Gerrit Jasper Schenk (ed.), *Historical Disaster Experiences, Transcultural Research – Heidelberg Studies on Asia and Europe in a Global Context* (Cham: Springer, 2017), 3–44, at 15–23.

5 On nature, see Schenk, 'Historical Disaster Experiences', 4, 13–14.

6 A phenomenon explored in detail by Jean Céard, *La nature et les prodiges* (1977; 2nd ed., Geneva: Droz, 1996) and further nuanced by Michael Kempe 'Noah's Flood. The Genesis Story and Natural Disasters in Early Modern Times', *Environment and History* 9: 2 (2003), 151–71; Philip M. Soergel, 'Portents, Disaster, and Adaption in Sixteenth-Century Germany', *The Medieval History Journal* 10: 1–2 (2007), 303–26; Elaine Fulton and Penny Roberts, 'The Wrath of God. Explanations of Crisis and Natural Disaster in Pre-Modern Europe', in Penny Roberts with Mark Levene and Rob Johnson (eds), *History at the End of the World? History, Climate Change and the Possibility of Closure* (Penrith, UK: Humanities-Ebooks, 2010), 67–79; Raingard Esser and Marijke Meijer Drees, 'Coping with Crisis. An Introduction', *Dutch Crossing* 40: 2 (2016), 93–96.

violence that he recognised as life-changing for some specific communities and not others, without providing explanation for activities that he perceived as beyond human understanding.[7]

Da Vinci's understanding of natural violence does not fit neatly within modern scholarly analysis of environmental events perceived by early moderns as disaster.[8] Schenk distinguishes between 'extreme natural events, which can take place, so to speak, in the absence of human society' from disasters that were 'always society-related'.[9] In using the term 'natural violence', I seek to capture Da Vinci's broad-ranging consideration of violence wrought in nature, to nature, and by nature upon which he reflected across his work and to which he gave varied and ongoing responses over his life. I employ this term to open up rather than foreclose the diverse avenues of his thought, in order to express the numerous relationships between violence and nature that he articulated, the different causative agents for such violence in different contexts, and the possibilities of such violence to produce creative as well as destructive consequences. Da Vinci's thought offered reflections upon natural violence and its relationship to humans and other species, considering its potential to be both generative and devastating and the human capacity to express affective and imaginative (both artistic and engineered) responses to it.[10]

Feelings about Destruction

Natural violence forms a special place in Da Vinci's works, his thoughts produced across written, visual, and material texts. He articulated, in the sense of conceptualising and interpreting, acts of violent intensity for which he saw evidence past and present in the world around him. Understanding natural violence mattered to Da Vinci. He explored it in drawings and paintings, in experiments and sketches, in designs for engineering works, and in those writings termed the literary works, which include riddles, fables, prophecies, and what may be experimental fiction. This suggests the

7 Broomhall, 'Devastated Nature', 39–40.

8 Monica Juneja and Frank Mauelshagen, 'Disasters and Pre-Industrial Societies. Historiographic Trends and Comparative Perspectives', *The Medieval History Journal* 10: 1–2 (2007), 1–31, at 13–16; Schenk, 'Historical Disaster Experiences', 3–44.

9 Schenk, 'Historical Disaster Experiences', 24.

10 Raingard Esser has explored the importance of engineering responses to early modern disaster responses in her '"Ofter gheen water op en hadde gheweest". Narratives of Resilience on the Dutch Coast in the Seventeenth Century', *Dutch Crossing* 40: 2 (2016), 97–107.

strongly interconnected nature of Da Vinci's work as part of his thinking process. The exploration of nature's destruction in these diverse genres, text types, and sites also suggests its affective power to consume and drive him. Indeed, in a first-person narration among his notes, Da Vinci, seemingly reversing the allegory known as Plato's Cave, described powerful emotions that underpinned his search to delve deeper into nature's crevices:

> Unable to resist my eager desire and wanting to see the great ... of the various and strange shapes made by formative nature ... I came to the entrance of a great cavern ... two contrary emotions arose in me, fear and desire – fear of the threatening dark cavern, desire to see whether there were any marvellous thing within it.[11]

We might even say that Da Vinci found natural violence good to think with.

Da Vinci was intensely interested in human emotions, both the value of their experience and their expression, and he frequently utilised an explicitly affective vocabulary himself. In what was perhaps a work of experimental fiction, Da Vinci penned a letter describing a powerful weather event in the Taurus Mountains of Armenia to which the narrator appeared to be a witness. Scholars have debated whether Da Vinci could have been present at such an event, but the view of art historian Daniel Arasse is now widely shared, that it numbered among other literary writings as a site in which Da Vinci 'allowed his imagination to run loose on subjects he was passionately interested in: the creative and destructive omnipotence of nature and the infinite mutation of shapes'.[12] The text provides an insightful example of how Da Vinci's thinking about natural violence is articulated through explicit attention to responses of human feeling. He expresses natural violence as a deeply affecting experience for humans:

> during the last few days I have been in so much trouble, fear, peril, and loss, besides the miseries of the people here, that we have been envious of the dead; and certainly I do not believe that since the elements by their

11 'E tirato dalla mia bramosa voglia, vago di uedere la gran cō... delle varie e strane forme fatte dalla artifiziosa natura, ... pervenni all'ēntrata d'una grā caverna ... si destarono in me 2 cose, pavra e desiderio; paura · per la minacciosa oscura spilonca, desidero per vedere se là ētro fusse alcuna miracolosa cosa'. *The Literary Works of Leonardo da Vinci*, ed. and trans. Jean-Paul Richter, 2 vols (New York: Dover, 1970), vol. 2, 395, no. 1339.

12 Daniel Arasse, *Leonardo Da Vinci. The Rhythm of the World*, trans. Rosetta Translations (New York: Konecky & Konecky, 1998), 45. See also the commentary by Carlo Pedretti on *The Literary Works of Leonardo da Vinci*, 2 vols (New York: Phaidon, 1977), vol. 2, 294.

> separation reduced the vast chaos to order, they have ever combined their force and fury to do so much mischief to man.[13]

Da Vinci's descriptions of emotional responses to natural violence could also extend to expressions of feeling about the cruelty of nature and of nature's creations to each other, not least of which was the cruelty of humankind itself: 'such a cruel and horrible monster'.[14] Thus, in a brief musing within his notebooks that considered humans' violent activities to extract metals from the earth, Da Vinci laments: 'O monstrous creature! ... For this the vast forests will be devastated of their trees; for this endless animals will lose their lives'.[15] 'Why did nature not ordain that one animal should not live by the death of another?' he asks in another reflection in his notes on the cruelty of nature.[16] These expressions of and about feeling deserve attention as we think about his articulations of natural violence.

Analysis of the power and meaning of emotional expression has been an important feature of recent scholarly literature exploring early modern disaster.[17] This literature has emphasised the importance of setting these affective productions, displays, or performances in different cultural forms within early modern social and rhetorical conventions that made sense to their authors and to their readers, audiences, or viewers. Culturally specific memory and emotion templates for acts and texts could strengthen or dissipate feelings

13 'ne'giorni · passati · sono stato · in tāti affanni, pavre, pericoli e danno · insieme con questi miseri paesani, che avevano d'avere invidia ai morti, e cierto · io nō credo · che, poichè gli elemēti con lor separatione · disfeciono · il grā caos, che essi riunissino · lor forza, anzi rabbia ·, a fare tanto nocimēto alli omini'. Richter, *The Literary Works*, vol. 2, 392–93, no. 1337.

14 'si crudele e spietato mōstro!' Richter, *The Literary Works*, vol. 2, 365, no. 1296.

15 'o animal mostruoso! ... per costui rimarrā diserte le grā selue delle lor piāte; per costui infiniti animali perderanno la uita'. Richter, *The Literary Works*, vol. 2, 360, no. 1295.

16 'Perchè la natura non ordinò che l'uno animale nō uivesse dalla morte dell altro?' Richter, *The Literary Works*, vol. 2, 310, no. 1219. Pedretti notes that this text and no. 1162 may be two parts of a dialogue, *The Literary Works*, vol. 2, 240 and 257.

17 Susan Broomhall, 'Narrating Experiences and Emotions of Distressing Events in the French Wars of Religion', and Andreas Bähr, 'Remembering Fear. The Fear of Violence and the Violence of Fear in Seventeenth-Century War Memories', in Erika Kuijpers, Judith Pollmann, Johannes Müller, and Jasper van der Steen (eds), *Memory before Modernity. Practices of Memory in Early Modern Europe*, (Leiden: Brill, 2013), 253–67 and 269–82; Susan Broomhall, 'Disorder in the Natural World. The Perspective of a Sixteenth-Century French Convent', in Jennifer Spinks and Dagmar Eichberger (eds), *Religion, the Supernatural and Visual Culture in Early Modern Europe* (Leiden: Brill, 2015), 240–59; Jennifer Spinks and Charles Zika (eds), *Disaster, Death and the Emotions in the Shadow of the Apocalypse, 1400–1700* (London: Palgrave Macmillan, 2016); Marijke Meijer Drees, '"Providential Discourse" Reconsidered. The Case of the Delft Thunderclap (1654)', *Dutch Crossing* 40: 2 (2016), 108–21; Peters and Richards, *Early Modern Trauma*.

of traumatic events at individual and collective levels.[18] So too is temporality critical to early modern ideas and emotions about disaster. Descriptions of fear, for example, within a particular moment of devastating experience could be rhetorically resolved within the work's overarching narrative that attested to the continuation of the author, and community, in the long term. Narrating such an event then became not a story of a disaster but one of survival.

That Da Vinci articulated feelings throughout his work about such violence has not been ignored by scholars. Arasse suggests that Da Vinci's texts 'reveal certain collective anxieties and sometimes ... personal fantasies and pessimism', and that his 'consciousness of these obscure, violent pulses' informed his expression of the self in his art.[19] Literary scholar Michael Jeanneret argues that it is in regard to 'natural disasters' that Da Vinci is inspired to develop some of his most personal works, in which he allowed imagination to prevail over analysis.[20] I suggest here, by contrast, that Da Vinci's analysis of emotions and the expression of emotions were fundamental alongside temporality to the way he engaged with natural violence and that these dimensions formed part of his intellectual process. This practice is somewhat obscured, however, as Da Vinci returned to this topic repeatedly throughout his life, leaving his thought to be reconstructed only imperfectly from scattered texts as they were subsequently arranged by others.[21]

18 Erika Kuijpers, 'The Creation and Development of Social Memoires of Traumatic Events. The Oudewater Massacre of 1575', in Michael Linden and Krzysztof Rutkowski (eds), *Hurting Memories. Remembering as a Pathogenic Process in Individuals and Societies* (London: Elsevier, 2013), 191–201; Kuijpers, '"O, Lord, save us from shame". Narratives of Emotions in Convent Chronicles by Female Authors During the Dutch Revolt, 1566–1635', in Susan Broomhall (ed.), *Destroying Order, Structuring Disorder. Gender and Emotions in Medieval and Early Modern Europe* (Aldershot: Ashgate, 2015), 127–46; Erika Kuijpers, 'Fear, Indignation, Grief and Relief. Emotional Narratives in War Chronicles from the Netherlands (1568–1648)', and Susan Broomhall, 'Divine, Deadly or Disastrous? Diarists' Emotional Responses to Printed News in Sixteenth-Century France', in Spinks and Zika, *Disaster, Death and the Emotions*, 93–111 and 321–39; Erika Kuijpers and Judith Pollmann, 'Turning Sacrilege into Victory. Catholic Memories of Iconoclasm, 1566–1700', in Éva Guillorel, David Hopkin, and William G. Pooley (eds), *Rhythms of Revolt. European Traditions and Memories of Social Conflict in Oral Culture* (London: Routledge, 2018), 155–70. See also Allie Terry-Fritsch and Erin Felicia Labbie (eds), *Beholding Violence in Medieval and Early Modern Europe* (Aldershot: Ashgate, 2012); Judith Pollmann, 'Remembering Violence. Trauma, Atrocities and Cosmopolitan Memories', in her *Memory in Early Modern Europe, 1500–1800* (Oxford: Oxford University Press, 2017), 159–85; Domenico Cecere, Chiara De Caprio, Lorenza Gianfrancesco, and Pasquale Palmieri (eds), *Disaster Narratives in Early Modern Naples. Politics, Communication and Culture* (Rome: Viella, 2018).

19 Arasse, *Leonardo*, 232, 479–80.

20 'désastres naturels'. Michel Jeanneret, *Perpetuum mobile. Métamorphoses des corps et des oeuvres de Vinci à Montaigne* (Paris: Macula, 1997), 73.

21 See Augusto Marinoni, 'I manoscritti di Leonardo da Vinci e le loro edizioni', in A. Marazza (ed.), *Leonardo. Saggi e Ricerche* (Rome: Inst. Poligr. Stato, 1954), 229–74; Pedretti, *Literary Works*, 393–402.

Destructive Thinking

Da Vinci's thinking about natural violence appears to be animated by a conceptualisation of nature in constant flux as matter moves through different forms and is created and destroyed, and as nature interacts with time. Da Vinci is hardly alone in his deep and abiding fascination with the metamorphosis of matter; creative forces of swirling energy are seen in myriad art forms at this period.[22] For Da Vinci, such metamorphosis was commonly narrated through elemental forces, as he characterised fire in a brief passage of his notes:

> One shall be born from small beginnings which will rapidly become vast. This will respect no created thing, rather will it, by its power, transform almost every thing from its own nature into another.[23]

Nature in flux was inherently destructive even to itself, he suggested in another passage that seemed to imagine the end of life on earth, which would 'end with the element of fire … its surface will be left burnt up to cinder and this will be the end of all terrestrial nature'.[24] Destruction, in these articulations, was embedded in nature's dynamic and transformative processes of the movement of matter from one form to another.

In his thinking about matter's transformation, Da Vinci echoed ancient ideas encountered through extensive reading and vibrant interactions with his contemporaries.[25] The influence of the pre-Socratics appears strong, especially the Ionian philosophers of the sixth century, Heraclitus, Anaxagoras, and Democritus among them. Heraclitus, for example, conceptualised a process of continual flux in aphorisms such as 'Fire lives the death of earth and air lives the death of fire, water lives the death of air, earth that of water'.[26] Heraclitus' ideas were well known to the early modern community of intellectuals and artists of which Da Vinci was a

22 Jeanneret, *Perpetuum mobile*.

23 'Nascierà di piccolo principio, chi si farà cō prestezza granda; questo non stimerà alcuna creata cosa, anzi colla sua potētia quasi il tutto avrà in potentia di transformare di suo essere in vn altro'. Richter, *The Literary Works*, vol. 2, 360, no. 1295.

24 'terminare collo elemēto del fuoco; allora la sua superfice rimarrà in riarsa cienere, e questo fia il termine della terrestre natura'. Richter, *The Literary Works*, vol. 2, 310, no. 1218.

25 For his reading, see Romain Descendre, 'La biblioteca di Leonardo', in Sergio Luzzatto and Gabriele Pedullà (eds), *Atlante della letteratura italiana* (Turin: Einaudi, 2010), vol. 1, 592–95.

26 Maximus of Tyre, 41.4, cited in Diogenes Laertius, *Lives of the Philosophers*, DK 22 B76, in Patrick Lee Miller and C.D.C. Reeve (eds), *Introductory Readings in Ancient Greek and Roman Philosophy* (Indianapolis, IN: Hackett Publishing, 2015), 12.

part. The Italian painter Raffaello Sanzio da Urbino depicted the ancient philosopher with the features of fellow Italian artist Michelangelo di Lodovico Buonarroti Simoni in his fresco *The School of Athens* around 1510.[27] Da Vinci's own attachment to Heraclitan ideas is suggested by another, earlier artwork by Donato Bramante around 1486, in which the philosopher, paired with Democritus, is presented with features reminiscent of Da Vinci himself.[28] The intellectual connections between Da Vinci and Bramante, recently explored by art historian Jill Pederson, leave no doubt that the latter knew much of Da Vinci's developing thought.[29] Anaxagoras is mentioned by name in a short commentary among Da Vinci's notes, providing a similar conceptualisation of dynamic flux to that of Heraclitus:

> Anaxagoras. Every thing proceeds from every thing, and every thing becomes every thing, and every thing can be turned into every thing else, because that which exists in the elements is composed of those elements.[30]

Roman authors Ovid and Plutarch provided further influences upon Da Vinci's metamorphic vision of nature.[31] In relation to such works as well as his own experiences, Da Vinci developed a conceptualisation of the dynamic flux of all nature, in which destruction as part of transformation appeared an inherent and inevitable process.

This transformation of nature's matter was informed by temporality in Da Vinci's work. He considered time's transformative influence at the individual human scale, as one red chalk work on paper portraying old and young men together suggests.[32] His notebook reflections pondered the same question at the scale of humankind as a species. In one exploratory passage, he considered different scales of time in which humans, as part of nature, were embedded:

27 In the Apostolic Palace, Vatican City.

28 In the Pinacoteca di Brera, Milan.

29 Jill M. Pederson, *Leonardo, Bramante, and the Academia. Art and Friendship in Fifteenth-Century Milan* (Turnhout: Brepols, 2020).

30 'Anasagora; ogni cose viē da ogni cosa, – ed ogni cosa si fa ogni cosa, e ogni cosa torna in ogni cosa; perchè ciò ch'è nelle elemēti è fatto da essi elemēti'. Richter, *The Literary Works*, vol. 2, 445, no. 1473.

31 André Chastel, 'Léonard et la culture', in his *Fables, formes, figures*, 2 vols (Paris: Flammarion, 1978), vol. 2, 251–63.

32 See *Heads of an Old Man and a Young Man*, c. 1495, red chalk on paper, 208 x 150 mm, Galleria degli Uffizi, Florence, Gabinetto Disegni e Stampe, 423 E r.

> O time, swift robber of all created things, how many kings, how many nations hast thou undone, and how many changes of states and of various events have happened since the wondrous forms of this fish perished here in this cavernous and winding recess. Now destroyed by time thou liest patiently in this confined space with bones stripped and bare; serving as a support and prop for the superimposed mountain.[33]

In another, his musings on the end of life on earth, Da Vinci suggested that human destruction was simply one small part of what will succumb to the inevitable flux, a process that was conceptualised and narrated as both past and future.[34]

> The watery element was left enclosed between the raised banks of the rivers, and the sea was seen between the uplifted earth and the surrounding air which has to envelope and enclose the complicated machine of the earth, and whose mass, standing between the water and the element of fire, remained much restricted and deprived of its indispensable moisture; the rivers will be deprived of their waters, the fruitful earth will put forth no more her light verdure; the fields will no more be decked with waving corn; all the animals, finding no fresh grass for pasture, will die and food will then be lacking to the lions and wolves and other beasts of prey, and to men who after many efforts will be compelled to abandon their life, and the human race will die out.[35]

33 'O tēpo, velocie predatore delle create cose, quāti rè, quāti popoli ài tu disfatti, e quāte mutazioni di stati e vari casi sono seguite dopo che la maravigliosa forma di questo pescie qui morì per le caverne e ritorte interiora; ora disfatto dal tēpo patiēte giacci ī questo chiuso loco; colle spolpate e ignivde ossa ài fatto armadura e sostegnio al sopra posto mote'. Richter, *The Literary Works*, vol. 2, 310, no. 1217.

34 Paolo Galluzzi, 'Leonardo da Vinci's Concept of "Nature". "More Cruel Stepmother than Mother"', in Marco Beretta, Karl Grandin, and Svante Lindqvist (eds), *Aurora Torealis. Studies in the History of Science and Ideas in Honor of Tore Frängsmyr* (Sagamore Beach: Science History Publications, 2008), 13–29, at 15–16. Leonardo's use of shifting tense structures is also noted by Jeanneret, *Perpetuum mobile*, 78.

35 'Rimase lo elemēto dell'acqua rīchiuso īfra li crescivti argini de' fiumi, e si vede 'l mare jfra la crescivta terra e la circunatricie aria, avēdo a fasciare e circonscrivere la moltificata machina della terra, e la sua grossezza, che staua fra l'acqua e lo elemēto del fuoco, rimāga molto ristretta e private dalla bisogniosa acqua, la fertile terra nō māderà piv leggieri frōde, nō fieno piv i cāpi adornití dale ricascāti piāte; tutti li animali nō trovādo da pasciere le fresche erbe, morranno, e mācherà il cibo ai rapaci lioni e lupi e altri animali che vivono di ratto, e agli omini dopo molti ripari cōverrà abādonare la loro vita, e mācherà la gienerazione vmana'. Richter, *The Literary Works*, vol. 2, 310, no. 1218.

Figure 1.1 Leonardo da Vinci, *A Deluge*, c. 1517–18, black chalk, pen and ink, wash, 16.2 × 20.3 cm (sheet of paper). Royal Collection Trust, RCIN 912326. © His Majesty King Charles III 2023

Existential destruction of humankind by nature was part of a large process of transformative flux that in Da Vinci's texts appeared inevitable.

A number of Da Vinci's drawings depicted what seem to be such cycles of transformation innate in nature. Ten such drawings are similarly sized and mostly in black chalk, suggesting their possible relationship as a kind of thought experiment and visual theorisation (Fig. 1.1).[36] Such images were not observations of nature, although they may have been informed by observation: they were natural philosophical *and* artistic articulations.[37]

36 On the Windsor Deluge drawings, Arasse, *Leonardo*, 112; Martin Kemp, *Leonardo* (Oxford: Oxford University Press, 2004), 158–64; Martin Kemp, *Leonardo da Vinci. The Marvellous Works of Nature and Man* (1981; revised ed., Oxford: Oxford University Press, 2007), 311–20.

37 Ernst Gombrich, 'The Form of Movement in Water and Air', in C.D. O'Malley (ed.), *Leonardo's Legacy* (Berkeley: University of California Press, 1969), 171–204; Martin Kemp, 'Leonardo and the Idea of Naturalism. Leonardo's Hypernaturalism', in Andrea Bayer (ed.), *Painters of Reality. The Legacy of Leonardo and Caravaggio in Lombardy* (New York: Metropolitan Museum of Art, 2004), 65–73; Enzo O. Macagno, 'Leonardo's Methodology in his Fluid Mechanical Investigations', in Ching Jen Chen (ed.), *Turbulence Measuring and Flow Modelling* (New York: Hemisphere, 1987), 833–94.

These depictions of natural violence communicated repeated swirls and spirals that churned all in their wake. The overwhelming, existential nature of the disaster is barely contained within the frame of paper on which Da Vinci sought to articulate it. There was nothing beyond, nothing that remained outside the path of such violence.

Nature as Mistress and Mother and Masculine Generation

If nature's cycle of destructive and generative transformation was aided by time in Da Vinci's work, so too was his conceptualisation shaped by the gender ideologies of his era. Gender mattered to the complicated pedagogical and power relationship between nature, normatively female in Da Vinci's works, and the artist, normatively male. When he reflected upon the art of painting in his notes, Da Vinci conceived of painting in a filial relationship with nature, its mother: 'for painting is born of nature – or, to speak more correctly, we will say it is the grandchild of nature; for all visible things are produced by nature'.[38] Elsewhere, engaging with a long-standing debate within ancient and contemporary artistic discourse about imitation, emulation, and invention, Da Vinci argued, in his writings on the history of artistic achievements, that only by direct engagement with nature, 'the mistress of all masters', could painting as an artform be renewed and developed.[39]

Nature's creations could even be surpassed, Da Vinci suggested, for painting

> is of such excellence that it keeps alive the harmony of those proportionate parts which nature, for all her powers, cannot manage to preserve. How many paintings have preserved the image of a divine beauty, which in its natural manifestation has been rapidly overtaken by time or death. Thus the work of the painter is nobler than that of nature, its mistress.[40]

38 'perchè la pittura · è partorita da · essa natura·; ma per dire piv corretto dirremo nipote di natura·, perchè tutte le cose evidēti sono state partorite dalla natura'. Richter, *The Literary Works*, vol. 1, 326, no. 652.

39 'natura, maestra dei maestri'. Richter, *The Literary Works*, vol. 1, 332, no. 660. On these debates, see George W. Pigman III, 'Versions of Imitation in the Renaissance', *Renaissance Quarterly* 33: 1 (1980), 1–32; Andrea Bolland, 'Art and Humanism in Early Renaissance Padua. Cennini, Vergerio and Petrarch on Imitation', *Renaissance Quarterly* 49: 3 (1996), 469–87; and beyond Italy, see Anton W.A. Boschloo (ed.), *Aemulatio. Imitation, Emulation and Invention in Netherlandish Art from 1500 to 1800. Essays in Honor of Eric Jan Sluijter* (Zwolle: Waanders, 2011).

40 Leonardo da Vinci, *Treatise on Painting* (*Codex Urbinas Latinus 1270*), trans. A. Philip McMahon, (Princeton, NJ: Princeton University Press, 1956), no. 30, cited in Galluzzi, 'Leonardo

Here time assisted the male painter's copy to assert supremacy over female nature's original and imperfect creation.[41] Nature's maternity could be rivalled by the male artist's act of generation. Moreover, the artist created not only through the production of a painted simulacrum. If the painter did more than draw 'merely by practice and by eye, without any reason ... like a mirror', he could participate in the creation of 'true knowledge'.[42] Nature, simultaneously mistress and mother, Da Vinci suggested, could and should be controlled, subordinated by a certain kind of skilled individual who could disclose her inner workings at a time of his choosing.

Nature's Desire

Da Vinci feminised the cycle of destructive violence in the natural world. Nature was of course commonly presented as female, both in the ancient sources and among his contemporaries. Da Vinci may thus have been following a standard view. Historian of science Marco Beretta proposes that his terminology suggests a conceptualisation of natural processes in the model of Epicureanism that was advanced by the Roman poet and philosopher Lucretius. Da Vinci's vision of progressive decay of fertile nature to a state of barrenness within a single earth-scaled transformative cycle echoed the path of Lucretius's industrious Venus Daedala and Genetrix to her inevitable desolate end over the course of his *De Rerum Natura* ('On the Nature of Things').[43] Da Vinci's characterisation of nature as female was not unexpected within his cultural worldview, just as within that of Lucretius, but we should not take it for granted. Firstly, Da Vinci may have been among the most radical of the era's thinkers about movement and metamorphosis, as Jeanneret suggests, but gender ideology appears one area in which his ability to think outside contemporary socio-cultural systems was more limited.[44] Secondly and importantly, the consequences of this particular characterisation shape Da Vinci's thinking about natural violence.

Da Vinci's nature as both mistress and mother would display particular feeling characteristics consistent with contemporary gender ideologies, which were fundamental to his consideration of natural violence. The

da Vinci's Concept of "Nature"', 25.

41 A claim by no means unique to Leonardo, see Jeanneret, *Perpetuum mobile*.

42 'Il pittore che ritrae per practica e givditio d'ochio, sanza ragione è come lo spechio', 'veramēte questa è sciētia', Richter, *The Literary Works*, vol. 1, 18, no. 20, and 326, no. 652.

43 Marco Beretta, 'Leonardo and Lucretius', *Rinascimento* 49 (2009), 341–72, at 363–64.

44 Jeanneret, *Perpetuum mobile*, 59.

affective dimensions of nature's maternity appeared to confound, perhaps even confront, Da Vinci. In a brief reflection in his notes about the cruelty of humans towards asses, he expounded: 'O Nature! Wherefore art thou so partial; being to some of thy children a tender and benign mother, and to others a most cruel and pitiless stepmother?'[45] Da Vinci's depiction of nature's forms of motherhood has attracted scholarly attention because of the strong feeling with which he expresses his perception of nature's ambivalence towards its creations and the violence rendered unto them. In another notebook, a single sentence jotting observed: 'nature appears with many animals to have been rather a cruel stepmother than a mother, and with others not a stepmother, but a most tender mother'.[46] Da Vinci's repeated connection between particular affective dispositions of different forms of female parenting emphasised normative expectations of maternal affections and, furthermore, the assumption that these were forged by direct blood relations that distinguished them from the care of a stepmother. It may be tempting to draw reference to Da Vinci's own experiences in such relationships, for he was not the child of any of his father's three wives. Da Vinci likely intended that humans held no privileged position in relation to nature, as historian of science Paolo Galluzzi argues.[47] Yet Da Vinci's framing of this idea is distinctly emotive. A rich literature has examined how early moderns voiced strong expectations of women as mothers, especially where a breach in performing culturally normative affective displays or acts of care led to the abandonment or death of a child and could incur severe punishment for the women concerned.[48] Da Vinci's powerful emotive exclamations and questions about nature's actions for her creations are deeply embedded in cultural expectations about maternal care.

45 'O natura, perchè ti sei fatto partiale, facciēdoti ai tua figli d'alcuna pietosa e benignia madre, ad'altri crudelissima e spietata matrignia?' Richter, *The Literary Works*, vol. 2, 354, no. 1293.

46 'La natura pare qui in molti o di molti animale stata più presto crudele matrignia che madre, e d'alcuni nō matrignia ma pietosa madre'. Richter, *The Literary Works*, vol. 2, 131, no. 846.

47 Galluzzi, 'Leonardo da Vinci's Concept of "Nature"', 14.

48 For recent examples of this literature, focused particularly on emotional dimensions, see Susan Broomhall, 'Understanding Household Limitation Strategies among the Sixteenth-Century Urban Poor in France', *French History* 20: 2 (2006), 121–37; Garthine Walker, 'Child-Killing and Emotion in Early Modern England and Wales', in Katie Barclay, Kimberley Reynolds, and Ciara Rawnsley (eds), *Death, Emotion and Childhood in Premodern Europe* (Basingstoke: Palgrave Macmillan, 2017), 151–72; Sally Holloway, 'Materializing Maternal Emotions. Birth, Celebration and Renunciation in England, c. 1688–1830', in Stephanie Downes, Sally Holloway, and Sarah Randles (eds), *Feeling Things. Objects and Emotions through History* (Oxford: Oxford University Press, 2018), 154–71.

Contemporary gender ideologies that assumed women's inherent desire for maternity and constructed how that maternity should be *affected* were thus reflected in Da Vinci's works via the emotions with which he accredited his mistress/mother nature. Da Vinci's ideas about human women and their reproductive organs are suggestively complex in their simplicity. As cultural historian Sander Gilman observes, his depiction of a woman and man in coitus figures her as 'but womb and breasts'.[49] Just as women were widely perceived to be driven to procreate by the urges of the wandering animal inside them, a widespread early modern (mis) understanding of Greek philosopher Plato's description of the uterus, so too did nature desire generation.[50] However, like a woman, nature's generative potential would eventually be exhausted, as time ensured that her fertility was inevitably followed by sterility. Nature's time-limited capacity for creative production, subsequent natural world change, and, ultimately, destruction were within and indeed central to natural world processes. The feeling force that drove feminised nature was, in Da Vinci's mind, hugely powerful. As a sustained speculative commentary among his notes records, against time's inevitability to destroy, nature's urge to create was, for Da Vinci, ferocious:

> taking pleasure in creating and making constantly new lives and forms, because she knows that her terrestrial materials become thereby augmented, is more ready and more swift in her creating, than time in his destruction.[51]

Nature's desire for continual diversity of forms involved all kinds of her production, even where one of her creations endangered others among them.

> Nay, this not satisfying her desire, to the same end she frequently sends forth certain poisonous and pestilential vapours upon the vast increase

49 Sander L. Gilman, 'Leonardo Sees Him Self. Reading Leonardo's First Representation of Human Sexuality', *Social Research* 54:1 (1987), 149–71, at 160. Gilman also produced an intriguing account of Leonardo's depictions of the uterus outside the body as the sign of the bestial, at 169.

50 Ian MacLean, *The Renaissance Notion of Woman. A Study in the Fortunes of Scholasticism and Medical Science in European Intellectual Life* (Cambridge: Cambridge University Press, 1980); Mark J. Adair, 'Plato's View of the "Wandering Uterus"', *The Classical Journal* 91: 2 (1995–96), 153–63.

51 'pigliãdo piacere del creare e fare cõtinue vite e forme, perchè cognioscie che sono accrescimẽto della sua terrestre materia, è volonterosa e piv presta col suo creare che 'l tẽpo col cõsumare'. Richter, *The Literary Works*, vol. 2, 310–11, no. 1219.

> and congregation of animals; and most of all upon men, who increase vastly because other animals do not feed upon them.[52]

Thus, nature's desire to generate delivered its own forms of violence. And 'so she has ordained that many animals shall be food for others', the speculation continued, as ever-new forms of life upon the earth were forced into confrontation and contest with each other.[53]

Da Vinci referred here to nature as 'inconstant'.[54] However, what he articulated was the predictable outcome of a state of desiring generation, perceived to be specific to women, that was inherently irrational. This would produce in nature, as in women, affects and their consequences that would appear unpredictable at a local scale. Nature's intense emotions, her desire for maternity that pushed her even to neglect the creations she produced, might be irrational, but these feelings were logical and consistent with the expectations of her feminine characterisation. It was thus not only time that brought about nature's violence to herself and her creations as her generative capacity waned, but the consequences of her generative desire that would destroy them too: 'This earth therefore seeks to lose its life, desiring only continual reproduction'.[55]

For Da Vinci, emotion, interpreted through contemporary gender templates that produced specific affects in women, informed his understanding of nature's processes and particularly its violence. Feminised irrational emotion, as Da Vinci constructed it, provided a logical key to nature's acts of both generation and destruction within a cycle of transformation. Nature's powerful maternal desires were as critical to understanding natural world systems as was time, which contributed to pushing the generative desire of nature and, ultimately, encouraged nature to destroy even herself.

Nature's Affective Legacy: Feeling as Logos

Da Vinci appeared to identify both humans and other species as having capacity for emotions; after all, they were all nature's children. One set of

52 'e nō soddisfaciēdo questo a simile desiderio, e' spesso māda fuora cierti avelenati e pestilēti vapori sopra la grā moltiplicazioni e cōgregazioni d'animali, e massime sopra gli omini, chef anno grade accrescimēto, perchè altri animali nō si cibano di loro'. Richter, *The Literary Works,* vol. 2, 311.

53 'e pero à ordinato che molti animali sieno cibo l'uno dell'altro'. Ibid.

54 'vaga'. Richter, *The Literary Works*, vol. 2, 310, no. 1219.

55 'adūque questa terra cierca di mācare di sua vita, desiderādo la continva moltiplicazione'. Richter, *The Literary Works*, vol. 2, 311, no. 1219.

Figure 1.2 Leonardo da Vinci, *A Rearing Horse, and Heads of Horses, a Lion, and a Man* (recto), c. 1503–4, pen and ink, wash, a little red chalk, 19.6 cm × 30.8 cm (sheet of paper). Royal Collection Trust, RCIN 912380. © His Majesty King Charles III 2023

drawings on a sheet of paper placed expressions of intense emotion displayed on the bodies and heads of horses, a lion, and a man, side by side, seemingly in comparative analysis.[56] For Da Vinci, humans were not above nature but a deeply integrated part of it. He sometimes foregrounded the capacities of humans above other of nature's creations to shape the world around them; indeed his own engineering designs functioned to enhance human experience of the natural world. Yet his was a highly ambivalent position.[57] In one long set of prophetic reflections in his notes, the topic to which he gave most attention was the cruelty of humankind. Here, he opined that, despite hopes and aspirations to be more than earthly, humans were pulled back to the earth: 'from their immoderate pride they will desire to rise towards heaven, but the too great weight of their limbs will keep them down'.[58]

Moreover, Da Vinci suggested that humans fully participated in natural world processes with the affective inheritance that they had been given from nature. Humans were not only subject to nature's desires and processes, but they reproduced them in their turn. Da Vinci depicted humans driven

56 Jeanneret, *Perpetuum mobile*, 63.

57 Galluzzi, 'Leonardo da Vinci's Concept of "Nature"', 15.

58 'e per la loro smisurata superbia questi si vorranno leuare inverso il cielo, ma la superchia gravezza delle lor membra gli porrà in basso'. Richter, *The Literary Works*, vol. 2, 365, no. 1296.

to acts of violence, sharing nature's inherent desire for both creation and destruction. The same prophetic musing on the cruelty of humankind in his notes thus continued:

> there will be no end to their malignity; by their strong limbs we shall see a great portion of the trees of the vast forests laid low throughout the universe ... Nothing will remain on earth, or under the earth or in the waters which will not be persecuted, disturbed, and spoiled.[59]

Plants and animals consumed by humans became matter once more transformed, 'their bodies will become the sepulture and means of transit of all they have killed'.[60] Da Vinci suggested that the forces driving humans' violent tendencies were expressed as strong feeling processes, for 'the satisfaction of their desires will be to deal death and grief and labour and wars and fury to every living thing'.[61] Human destruction also involved humans as well; humankind was driven to self-destruction. They were, Da Vinci opined, no more than animals 'who will always be fighting against each other with the greatest loss and frequent deaths on each side'.[62] Emotion, for Da Vinci, was then not simply the recorded witness to destruction, to human existence and the transformation of nature. In a musing among his notes concerned with morality and the meaning to life, Da Vinci considered how emotions were integral to violence itself, because it was in human nature to desire it.

> Now you see that the hope and the desire of returning home and to one's former state is like the moth to the light, and that the man who with constant longing awaits with joy each new spring time, each new summer, each new month and new year – deeming that the things he longs for are ever too late in coming – does not perceive that he is longing for his own destruction.[63]

59 'questi non avrā termine nelle lor malignità; per le fiere mēbra di questi uerranno a terra grā parte deli alberi delle gran selue dell'universo ... nulla cosa resterà sopra la terra e l'acqua che nō sia perseguitata ·, remossa o guasta'. Ibid.

60 ''l corpo di questi si farà sepultura e transito di tutti i già da lor morti corpi animati'. Ibid.

61 'il nutrimēto de' loro desideri sarà, di dar morte e affanno e fatiche e guerre e furie a qualūche cosa animata'. Ibid.

62 'I quali senpre conbatterranno infra loro e con danni grandissimi e spesso morte di ciascuna delle parti'. Ibid.

63 'Or vedi la sperāza e'l desiderio del ripatriarsi e ritornare nel primo caso fa a similitudine della farfalle al lume, e l'uomo che cō cōtinvi desideria sēpre cō festa aspetta la nvova primavera, sempre la nvova state, sempre e nvovi mesi, e nvovi anni, parēdogli che le desiderate cose, venēdo, sieno trope tarde, E' non s'avede che desidera la sua disfazione'. Richter, *The Literary Works*, vol. 2, 291, no. 1162.

Nature had bestowed her propensity for both feeling and violence upon humankind.

Da Vinci saw how humans perceived violence in and of nature through intense emotional frameworks. Whether or not he ever witnessed the Armenian event narrated in the draft letter composition, he attested to it as an emotional experience: 'those few who remain unhurt are in such dejection and such terror that they hardly have courage to speak to each other, as if they were stunned'.[64] This was beyond words, perception through feeling. Just as he asserted painting could operate to discover natural world processes, feeling became for Da Vinci an alternative mode of knowledge in relation to natural violence. Emotions held power in Da Vinci's work because they helped him make sense of the status of humankind. Nature could silence not only human tools of language but also human records of expression, as he reflected in a passage concerned with forms of authority for geographical knowledge:

> Since things are much more ancient than letters, it is no marvel if, in our day, no records exist of these seas having covered so many countries; and if, moreover, some records had existed, war and conflagrations, the deluge of waters, the changes of languages and of laws have consumed every thing ancient.[65]

Time would erase even the expression of these emotions. But *the experience of feeling* here and now reminded humans, at least Da Vinci, that they too were part of a larger inevitable process. As he wrote in a musing about life,

> this desire is the very quintessence, the spirit of the elements, which finding itself imprisoned with the soul is ever longing to return from the human body to its giver. And you must know that this same longing is that quintessence, inseparable from nature, and that man is the image of the world.[66]

64 'E que' pochi, che siamo restati, siamo rimasti · cō tanto sbigottimēto e tata pavra che appena come balordi abbiamo ardire di parlare · l'uno coll'altro'. Richter, *The Literary Works*, vol. 2, 393, no. 1337.

65 'Perchè molto sō piv antiche le cose che le lettere, non è maraviglia, se alli nostri giorni non apparisce scrittura delli predetti mari essere occupatori di tanti paesi; e se pure alcuna scrittura apparia, le guerre, l'incēdi, li diluvi dell'acque le mutationi delle lingue e delle leggi ànno cōsumato ogni antichità'. Richter, *The Literary Works*, vol. 2, 207, no. 984.

66 'E uo'che sapi che questo medesimo desiderio è quella quītessēza, cōpagnia della natura, e l'uomo è modello dello mōdo'. Richter, *The Literary Works*, vol. 2, 291, no. 1162.

Understanding that humans' nature was to desire violence held its own significance, as it became a new emotional, experiential knowledge. The reflection above is matter of fact, consequential and knowledge-forming: 'now you see' flows to 'and you must know'. Da Vinci presents his thought as ultimately a path towards existential enlightenment, from discovery to acceptance of humanity's inclusion within nature and her emotional cycle that would involve inevitable annihilation.

Conclusions

Leonardo da Vinci articulated natural violence – that is, violence to, within, and by nature, in creative and striking terms, which themselves formed a rich and complex set of relations and which implied diverse consequences and responses for human and non-human species alike. Natural violence was not systematically conceptualised as disaster, even when it could elicit intense and negative emotional responses. It was importantly shaped by temporal and affective dimensions. More conventionally but no less consequentially, natural violence was for Da Vinci further informed by assumptions arising from contemporary gender ideologies. Human relationships to nature were affectively complicated by 'her' identity as both a mistress and mother. Yet the experience of intense emotions was also a reminder of humankind's inheritance from nature, a legacy that subjected humans to natural world systems both as they spurred humans to produce violence upon nature's creations and to accept future existential destruction in turn. Da Vinci's conceptualisation of natural violence across timescales that were both human and geological invites comparison with Chakrabarty's call that 'the Anthropocene requires us to think on the two vastly different scales of time that Earth history and world history respectively involve', scales of time 'tinged with different kinds of affect'.[67] As we consider our agency to arrest the course of climate change, emotions have thus returned to our discussions. Or, as the title of scientist James Lovelock's influential work *The Revenge of Gaia. Earth's Climate Crisis and the Fate of Humanity* suggests, perhaps the emotions of a feminised nature have always been entrenched in our conceptualisations.[68] What we embed, what we allow, and how we respond as a result of seeing nature and human relations in gendered and emotional terms deserve closer attention not only as a matter of historical interest but as one of urgency in our own historical moment.

67 Dipesh Chakrabarty, 'Anthropocene Time', *History and Theory* 57: 1 (2018), 5–32, at 6, 13.
68 (New York: Basic Books, 2006).

About the Author

Susan Broomhall is Director of the Gender and Women's History Research Centre at the Australian Catholic University. She is a historian of the early modern world, with a particular focus on women and gender. Her most recent book is *The Identities of Catherine de' Medici* (Brill 2021). She is currently Chief Investigator in an Australian Research Council Discovery project developing a cultural history of early modern natural resource management, which includes analysis of women's roles in early modern agripreneurship and gendered ideologies shaping forest and riverways as resources to be managed.

2 Early Modern Community Formation Across Northern Europe

How and Why a Poet in Poland Engaged with the Delft Thunderclap of 1654

Paul Hulsenboom

Abstract

The Delft Thunderclap, a massive gunpowder explosion which in 1654 devastated the Dutch city of Delft, elicited numerous responses by authors and artists from the Northern and Southern Netherlands. This chapter analyses another reaction, printed outside the Low Countries: a hitherto unknown Latin poem by the Silesian-Polish polymath Joachim Pastorius. His poem was published in 1657 in the Polish cosmopolitan city of Gdańsk, which maintained strong commercial and cultural ties with the Northern Netherlands. Pastorius had many Dutch contacts and likely based his verses on Dutch sources, specifically a famous poem by Joost van den Vondel. Pastorius applied a well-known rhetorical framework to inspire solidarity with the mourning Dutch. The disastrous Delft Thunderclap thus provided Pastorius with the opportunity to shape a transnational emotional community of learned readers and writers across Northern Europe, bridging Poland and the Dutch Republic.

Keywords

Delft Thunderclap – Dutch Republic – Poland – Gdańsk – literary relations – emotional communities

On 12 October 1654, nothing of note happened in the Baltic port city of Gdańsk (Danzig). At least, nothing which was deemed significant enough to be passed down through the ages and which is remembered to the present day. Almost one thousand kilometres away, however, a tragedy occurred

Asperen, van, Hanneke and Lotte Jensen (eds): *Dealing with Disasters from Early Modern to Modern Times. Cultural Responses to Catastrophes.* Taylor & Francis Group, 2023

DOI: 10.5117/9789463725798_CH02

Figure 2.1 Herman Saftleven II, *View of Delft after the Explosion of the Gunpowder Arsenal on 12 October 1654*, second half of the seventeenth century, black chalk, pen and brown ink, brush and brown wash on two sheets of paper, 24.9 × 74.9 cm. The Metropolitan Museum of Art, New York, Purchase, Bequest of Helen Hay Whitney, by exchange, and The Mnuchin Foundation, Mr. and Mrs. David M. Tobey and Werner H. Kramarsky Gifts, 1995, 1995.197

which quite literally shook the earth. At approximately 10:30 a.m., disaster struck in the Dutch city of Delft. A quarter of the town was wiped away by an explosion in the gunpowder magazine of Holland, which was located in or near a former monastery and which probably contained over 90,000 lb of powder. The cause of the explosion has never been established, but the story goes that a clerk entered the magazine carrying a burning lantern, sparks of which may have set fire to the highly flammable stash of gunpowder inside. The disaster that followed is known as the Delft Thunderclap.[1] Hundreds

1 I would like to thank Dr Aleks Koutny-Jones and the editors of this volume for their helpful comments on previous versions of this chapter.

On the material aspects of the disaster, as well as possible explanations of its cause, see J. Weerheijm, W.P.M. Mercx, and H.J. Pasman, 'De ontploffing van het kruitmagazijn van Holland te Delft op 12 oktober 1654', in H.L. Houtzager, G.G. Kunz, H.W. van Leeuwen, M. van Noort, and M. Tienstra (eds), *Kruit en krijg. Delft als bakermat van het Prins Maurits Laboratorium TNO* (Amsterdam: Rodopi, 1988), 31–42; G.G. Kunz, '"Alsof het hemelgewelf barstte en de aardbodem openscheurde …"', in Houtzager et al., *Kruit en krijg*, 43–51.

of houses were laid to waste, including that of the city's mayor, and an unknown but no doubt large number of human and animal casualties were either killed or severely injured: Contemporary estimates ranged from one hundred to one thousand fatalities. The Old and New Churches survived, but the blast destroyed their precious stained-glass windows. The gunpowder magazine itself was utterly eradicated, leaving only a crater filled with water and measuring some 4.5 m in depth. Allegedly, the explosion could even be heard on Texel – an island located over one hundred kilometres from Delft.

The calamity elicited a variety of responses. Drawings, paintings, and prints disseminated poignant visual images of the explosion, as well as of the material and human losses of the tragedy (Fig. 2.1).[2] In addition, Dutch pamphlets, sermons, poems, and historiographies discussed the disaster, interpreting it in various ways. This chapter analyses a unique and hitherto unknown reaction to the Delft Thunderclap: a Latin poem by the prolific

2 On visual representations of the disaster, see, for example, Carolyn Logan, 'Recording the News. Herman Saftleven's *View of Delft after the Explosion of the Gunpowder Arsenal in 1654*', *Metropolitan Museum Journal* 31 (1996), 203–10; Markus Bertsch and Jörg Trempler (eds), *Entfesselte Natur. Das Bild der Katastrophe seit 1600*, exh. Hamburger Kunsthalle, 29 June–14 October 2018 (Petersberg: Imhof, 2018), 176–77.

Silesian-Polish polymath Joachim Pastorius (1611–1681), who wrote numerous works in verse and prose and who for many years formed a central figure in the cultural, academic, and political life of Royal Prussia – an integral part of the Polish-Lithuanian Commonwealth, subject to the Polish king. In addition, Pastorius had a close, personal attachment to the United Provinces and maintained friendly relations with several Dutchmen. His poem about the Delft Thunderclap was first published in Gdańsk in 1657. It forms a short yet exceptional testimony, as to my knowledge it is the only literary reflection on the disaster written and printed outside the Low Countries.

After an introduction of Pastorius's life and works, I consider the intricate literary ties between seventeenth-century Gdańsk and the Northern Netherlands. This provides context for the following section, in which I analyse the poem about the Delft Thunderclap in comparison to other contemporary publications on the event. Lastly, I discuss the poem's main goals and social potential. I argue that Pastorius's verses acted as a tool for evoking feelings of transnational solidarity. Literary historian Marijke Meijer Drees referred to the works of medievalist Barbara Rosenwein to assert that Dutch seventeenth-century authors who responded to the Delft Thunderclap applied a so-called rhetoric of emotions as a social coping strategy.[3] Rosenwein coined the concept of 'emotional communities': 'groups – usually, but not always social groups – that have their own particular values, modes of feeling, and ways to express those feelings'.[4] She has disputed that these communities are the same as Brian Stock's textual communities, which revolve around an authoritative text.[5] However, Meijer Drees has followed historian Jan Plamper in arguing that an emotional community can also have textual components when people are emotionally connected through written media.[6] Pastorius's reaction to the Delft Thunderclap exemplifies this type of emotional community, fuelled by a disastrous event and constructed via written words. Importantly, moreover, as Pastorius poetically engaged

3 Marijke Meijer Drees, '"Providential Discourse" Reconsidered. The Case of the Delft Thunderclap (1654)', *Dutch Crossing* 40: 2 (2016), 108–21, at 118.

4 Barbara H. Rosenwein, *Generations of Feeling. A History of Emotions, 600–1700* (Cambridge: Cambridge University Press, 2016), 3. Rosenwein has studied various emotional communities, focusing predominantly on medieval written sources. See, for example, *Emotional Communities in the Early Middle Ages* (New York and London: Cornell University Press, 2006) and 'Problems and Methods in the History of Emotions', *Passions in Context* 1: 1 (2010), 1–32. For an introduction to Rosenwein's work, see Jan Plamper, *The History of Emotions. An Introduction*, trans. Keith Tribe (Oxford: Oxford University Press, 2015), 67–71.

5 Barbara H. Rosenwein, 'Worrying about Emotions in History', *The American Historical Review* 107: 3 (2002), 821–45, at 842, n. 76.

6 Meijer Drees, '"Providential Discourse" Reconsidered', 118.

with a calamity which occurred in another country, this chapter reveals that such emotional communities could also be transnational, connecting people over vast distances and across political, cultural, and linguistic borders. In addition, this chapter contributes to the study of the poetical culture of early modern Gdańsk and Dutch–Polish literary relations.

From Silesia via the Dutch Republic to Gdańsk

The author of the poem at the heart of this chapter was Joachim ab Hirtenberg Pastorius. It is uncertain where exactly he was in the fall of 1654: he worked in Gdańsk from at least 1655 onwards, and it has been suggested that he settled there in December the previous year, but there appears to be no hard evidence to support this claim.[7] In any case, he must have written his poem about the Delft Thunderclap somewhere along the Polish Baltic coast. It is likely – though uncertain – that he disseminated it in the direct aftermath of the explosion, after which he published it in Gdańsk in 1657. In order to contextualise his verses, this section introduces Pastorius's life and works.

Pastorius was born in Głogów (Glogau), in Silesia, in a Protestant household.[8] He studied in Leipzig and Halle and from 1631/32 until 1641 travelled around Europe as tutor of multiple Socinian/Arian and Calvinist Polish youths. Keeping pace with the wider European trend of the time, Polish noblemen of means as a rule went on extended educational journeys across the continent.[9] Pastorius visited the Dutch Republic at least thrice

7 See Kazimierz Kubik, *Joachim Pastorius. Gdański pedagog XVII wieku* (Gdańsk: Gdańskie Towarzystwo Naukowe, 1970), 19–20.

8 For more information about Pastorius's life and works, see A. Birch-Hirschfeld, 'Autobiografia Joachima Pastoriusa', *Reformacja w Polsce* 9–10 (1937–39), 473–77; Kubik, *Joachim Pastorius*; Stanisław Salmonowicz, 'Kazimierz Kubik: *Joachim Pastorius gdański pedagog XVII wieku. Gdańsk 1970 …*', *Kwartalnik Historii Nauki i Techniki* 17: 1 (1972), 135–39; Andrzej Skrobacki, 'Testament sekretarza królewskiego, historiografa, lekarza i kanonika warmińskiego Joachima Pastoriusa', *Komunikaty Mazursko-Warmińskie* 1–2 (1973), 73–92; Lech Mokrzecki, 'Joachim Pastorius – Dyrektor elbląskiego Gimnazjum Akademickiego', in Mokrzecki (ed.), *Wokół staropolskiej nauki i oświaty. Gdańsk-Prusy Królewskie-Rzeczpospolita* (Gdańsk: Wydawnictwo Gdańskie, 2001); Elżbieta Starek and Grzegorz Kotłowski, *Łacińskie inskrypcje w kościołach Warmii. Katedra we Fromborku* (Pelplin: Wydawnictwo Bernardinum, 2017), 230–33; Paul Hulsenboom and Alan Moss, 'Tracing the Sites of Learned Men. Places and Objects of Knowledge on the Dutch and Polish Grand Tour', in Koen Scholten, Dirk van Miert, and Karl A.E. Enenkel (eds), *Memory and Identity in the Learned World. Community Formation in the Early Modern World of Learning and Science* (Leiden and Boston, MA: Brill, 2022), 257–306, at 282–90.

9 Literature on the subject is extensive. See, for example, Hans-Jürgen Bömelburg, 'Adelige Mobilität und Grand Tour im polnischen und litauischen Adel (1500–1700)', in Rainder Babel

Figure 2.2 *Joachim Pastorius*, painted portrait after a print by Johann Alexander Böner, c. 1700, part of Pastorius's epitaph in Frombork Cathedral. Source: Stanisława Czajkowska

and also sojourned to London, Oxford, Orléans, and Paris. While he was a tutor, he also followed classes himself, for example at the Amsterdam *Athenaeum Illustre* and the University of Leiden,[10] where he obtained a doctorate in medicine. He befriended several Dutch scholars, moreover, such as Gerardus Joannes Vossius (1577–1649), Hugo Grotius (1583–1645), and Marcus Zuerius Boxhornius (1612–1653). Back in Poland, Pastorius worked as a doctor and – from 1651 onwards – as a professor of history, first at the *Gymnasium* in Elbląg (Elbing) and then in Gdańsk. Pastorius was appointed court historian in 1649, and he also served as a Polish diplomat, for example participating in the peace negotiations between the Swedes and Poles in 1660. Two years later he received a Polish nobleman's title. Pastorius changed his confession several times, alternating between Arianism, Calvinism, and

and Werner Paravicini (eds), *Grand Tour. Adeliges Reisen und Europäische Kultur vom 14. bis zum 18. Jahrhundert* (Ostfildern: Jan Thorbecke Verlag, 2005), 309–40; Dorota Żołądź-Strzelczyk, *O przedsięwzięciu peregrynacyjej. Edukacyjne wojaże szlachty z Rzeczypospolitej w świetle instrukcji podróżnych* (Warsaw: Muzeum Pałacu Króla Jana III w Wilanowie, 2020), 17–57.

10 Pastorius's name features twice in Leiden's *album studiosorum*, in 1636 and 1641. See *Album Studiosorum Academiae Lugduno Batavae MDLXXV–MDCCCLXXV* (The Hague: Apud Martinum Nijhoff, 1875), 280, 321.

Lutheranism but ultimately converting to Catholicism. Towards the end of his life, Pastorius held various ecclesiastical positions. He lies buried in Frombork Cathedral, Poland (Fig. 2.2).[11] A street in Gdańsk is named after Pastorius, and his native Głogów honoured him with a statue in 2019.

Pastorius's career path, which led him from Silesia via the Dutch Republic to Gdańsk, was typical at the time. It has long been established that seventeenth-century Silesian-born authors had close relations with the Northern Netherlands and were heavily indebted to Dutch examples. Multiple writers from Silesia, such as Martin Opitz (1597–1639) and Andreas Gryphius (1616–1664), were influenced by the works of their Dutch acquaintances, particularly Daniel Heinsius (1580–1655).[12] Silesia was at that time part of the Habsburg Monarchy, but the cultural ties with neighbouring Poland were strong.[13] A number of Silesian authors studied and/or settled in Gdańsk, bringing their Dutch-inspired verses with them. However, research concerning this type of literary reception has been dedicated almost exclusively to German plays and poems from the first half of the seventeenth century, mostly leaving out later material and largely disregarding works written and published in other languages, particularly Latin.[14] Joachim Pastorius has therefore thus far been excluded from studies regarding Silesian–Dutch literary relations.

11 My thanks go out to Stanisława Czajkowska of Frombork Cathedral for supplying me with valuable secondary material about Pastorius's epitaph, as well as for allowing me to use a photograph of the epitaph in this chapter. On the epitaph, see Tadeusz Piaskowski and Henryk Szkop, *Zabytki Fromborka* (Frombork: Muzeum Mikołaja Kopernika we Fromborku, 2003), 70; Starek and Kotłowski, *Łacińskie inskrypcje*, 234–37.

12 See, for example, Ulrich Bornemann, *Anlehnung und Abgrenzung. Untersuchungen zur Rezeption der niederländischen Literatur in der deutschen Dichtungsreform des siebzehnten Jahrhunderts* (Assen and Amsterdam: Van Gorcum, 1976); Ferdinand van Ingen, *Holländisch-deutsche Wechselbeziehungen in der Literatur des 17. Jahrhunderts* (Bonn: Presse- und Kulturabteilung der Kgl. Niederländische Botschaft, 1981); Stefan Kiedroń, 'De Silezische dichterscholen en de Nederlanden', *Neerlandica Extra Muros* 54 (1990), 63–69; Evgeny Kazartsev, 'Frühe deutsche Jamben und ihre niederländischen Vorbilder', *Neerlandica Wratislaviensia* XVIII (2009), 23–39; Evgeny Kazartsev, 'Niederländische Quellen von Martin Opitz' Versrhythmik', *Zeitschrift für Germanistik*, n.s. XXIII: 3 (2013), 499–509.

13 See Mirosława Czarnocka, 'Polnisch-deutsche Wechselbeziehungen in Kultur und Kommunikation im Barock', in Jan Papiór (ed.), *Polnisch-Deutsche Wechselbeziehungen im zweiten Millennium* (Bydgoszcz: Akademia Bydgoska, 2001), 130–53.

14 A few Latin exceptions are discussed in Stefan Kiedroń, 'Neerlandica uit de kring van Martin Opitz (1597-1639)', *Colloquium Neerlandicum. Nederlands in culturele context. Handelingen twaalfde Colloquium Neerlandicum* 12 (1994), 153–64; Ümmü Yüksel, 'Daniel Heinsius als Leitfigur auf dem Wege zur deutschen Kulturnation im Spannungsfeld von Latein und Landessprachen', in Tom Deneire (ed.), *Dynamics of Neo-Latin and the Vernacular. Language and Poetics, Translation and Transfer* (Leiden and Boston, MA: Brill, 2014). These contributions do not consider literature from Gdańsk, however.

Pastorius left a vast oeuvre, the majority of which has yet to be examined.[15] His most influential work, *Florus Polonicus* ('The Polish Florus'), a Latin study of Polish history, was first published in 1641 in Leiden by Franciscus Hegerus (born c. 1602).[16] Besides this and other works in prose, such as pedagogical treatises and speeches, Pastorius wrote countless Latin poems. While some scholarly attention has been aimed at Pastorius the pedagogue and historiographer,[17] his Latin poetical oeuvre is still largely unexplored.[18] Many of his poems were occasional and separately printed, but Pastorius also produced several larger poetical collections. His compositions cover a wide range of subjects, from European travel destinations to Christian saints, and from foreign politics to Polish royalty. Much like his poem about the Delft Thunderclap, most of these works were published in Gdańsk. The following section therefore considers the poem's place within the contemporary literary world of Gdańsk, which maintained strong economic and cultural relations with the United Provinces.

Early Modern Literary Relations Between Gdańsk and the United Provinces

The seventeenth-century Dutch Republic and Gdańsk both prospered immensely from their mutual trade relations, which stretched back to the

15 For an extensive yet incomplete overview of his works, see Kubik, *Joachim Pastorius*, 199–210.

16 Joachim Pastorius, *Florus Polonicus Seu Poloniae Historiae Epitome Nova* (Leiden: Apud Franciscum Hegerum, 1641). On the *Florus Polonicus*, see Ignacy Lewandowski, '"Florus Polonicus" Joachima Pastoriusa', *Meander* 23: 11–12 (1968), 522–29 and Lewandowski, *Florus w Polsce* (Wrocław, Warsaw, and Kraków: Zakład Narodowy Imienia Ossolińskich and Wydawnictwo Polskiej Akademii Nauk, 1970), 29–46. The book was reissued in Leiden in 1642, in Gdańsk in 1651, in Amsterdam in 1664, and in Gdańsk-Frankfurt in 1679. Kubik, *Joachim Pastorius*, 204 furthermore mentions editions printed in Leiden in 1648 and Gdańsk in 1669, but I have been unable to trace these versions. Moreover, Pastorius had at least one other book published in the United Provinces: Eustache de Refuge, *Aulicus Inculpatus*, trans. Joachim Pastorius (Amsterdam: Apud Lud. Elzevirium, 1644). A second edition of this Latin translation of an originally French courtier book was issued in 1649. A work on politics, also translated by Pastorius from French to Latin, has a title page hinting at an edition from Amsterdam, but its place of publication has been disputed: *Differentiae inter politicen genuinam ac diabolicam*, trans. Joachim Pastorius (Juxta Exemplar Amsterodami, 1659). Kubik, *Joachim Pastorius*, 208, 210 mentions two other titles published in the United Provinces, which I have not been able to verify.

17 Kubik, *Joachim Pastorius*, 115–87; Hans-Jürgen Bömelburg, *Frühneuzeitliche Nationen im östlichen Europa. Das polnische Geschichtsdenken und die Reichweite einer humanistischen Nationalgeschichte (1500–1700)* (Wiesbaden: Harrassowitz, 2006), 207–11.

18 Pastorius's poems are discussed globally in Kubik, *Joachim Pastorius*, 46–55; Edmund Kotarski, *Gdańska poezja okolicznościowa XVII wieku* (Gdańsk: Instytut Bałtycki, 1993), passim.

medieval Hanseatic League. From c. 1590 onwards, Baltic commerce was dominated by Dutch merchants, particularly by tradesmen from Amsterdam. Gdańsk at that time formed the central commercial hub of the Baltic, as well as the largest and wealthiest city in Royal Prussia. It was also a highly cosmopolitan city, drawing in Germans, Poles, Silesians, Scandinavians, Dutchmen, and many others. The main commodities which the Dutch imported via Gdańsk were grain and timber. Conversely, traders in Gdańsk chiefly bought luxury wares from the Dutch, such as wines, fine linens, exotic fruits, and spices.[19] In the second half of the seventeenth century, the Baltic grain trade went into decline, but relations between the Dutch Republic and Gdańsk remained strong. The United Provinces on several occasions worked hard to secure their trade with Gdańsk, most ostensibly so when they sent a war fleet to repel a Swedish force which threatened the port in 1656.[20] Symbolising the dependency of the young Dutch Republic on Baltic commerce – and hence especially on Gdańsk – is the phrase 'the mother trade', which the Grand Pensionary of Holland Johan de Witt (1625–1672) used to underline its importance in 1671.[21]

The close commercial ties between the United Provinces and Gdańsk, and the presence of Dutchmen in the city, went hand in hand with a flourishing cultural exchange, which lasted hundreds of years. The historical centre of Gdańsk has often been compared with Amsterdam, for example, due to the many houses with 'Dutch' façades. In fact, a number of architects and artists from the Low Countries designed and decorated several eye-catching buildings, such as the city's Town Hall (Fig. 2.3). Various studies – mostly art historical – have focused on Dutch architects, engineers, painters, and

19 Literature on the Dutch Baltic trade is extensive. Historian Maria Bogucka has devoted numerous publications to the topic, with an emphasis on the commercial ties between Gdańsk and the United Provinces, particularly Amsterdam. See, for example, Maria Bogucka, 'Amsterdam and the Baltic in the First Half of the Seventeenth Century', *The Economic History Review* 2nd ser., 26: 3 (1973), 433–47; Maria Bogucka, 'Dutch Merchants' Activities in Gdansk [sic] in the First Half of the Seventeenth Century', in J.Ph.S. Lemmink and J.S.A.M. van Koningsbrugge (eds), *Baltic Affairs. Relations between the Netherlands and North-Eastern Europe, 1500–1800. Essays* (Nijmegen: INOS, 1990), 19–32.

20 Nicolaas Frans Noordam, *De Republiek en de Noordse Oorlog 1655–1660* (Assen: Van Gorcum, 1940), 25–43. On the presence of the Dutch in the Baltic during the (long) seventeenth century more broadly, see Francis J. Bowman, 'Dutch Diplomacy and the Baltic Grain Trade, 1600–1660', *Pacific Historical Review* 5: 4 (1936), 337–48; Anja Tjaden, 'The Dutch in the Baltic, 1544–1721', in Göran Rystad, Klaus-R. Böhme, and Wilhelm M. Carlgren (eds), *In Quest of Trade and Security. The Baltic in Power Politics, 1500–1990* (Lund: Lund University Press, 1994), 61–136.

21 Milja van Tielhof, *The 'Mother of all Trades'. The Baltic Grain Trade in Amsterdam from the Late 16th to the Early 19th Century* (Leiden, Cologne, and Boston, MA: Brill, 2002), 5.

Figure 2.3 *The Town Hall in Gdańsk*, engraving in Jan Lodewijk Schuer, *Beknopte Beschryving Van de Stadt Dantzig* (Amsterdam: Janssoons van Waesberge, 1735). Warsaw University Library, S.8.152

engravers who were active in Gdańsk.[22] Other areas of cultural interaction, for instance those relating to book production or diplomatic practices, constitute equally promising yet largely untouched fields of research. Similarly, scholars have barely scratched the surface of the early modern literary relations between Gdańsk and the United Provinces.

Seventeenth-century Gdańsk was a thriving centre of literature and learning. The city boasted a *Gymnasium Academicum*, founded in 1558, as well as several printing companies. Moreover, the City Council was an active patron of poetry.[23] Throughout the century, therefore, local authors wrote an abundance of (mostly occasional) poems. The richness of their literary productivity can be viewed in the PAN Biblioteka Gdańska (Polish Academy of Sciences Library in Gdańsk), which houses a wealth of occasional compositions, both printed and handwritten. Older German studies of this material have mainly focused on its presumed 'Germanness', while Polish scholarship has brought to the fore works with a strong Polish connection.[24] However, the collection in the PAN Biblioteka Gdańska contains numerous poems on international relations and foreign topics, such as royal marriages, the deaths of monarchs, or wars and peace treaties. This rich poetical heritage should invite scholars to examine the cosmopolitan outlook of seventeenth-century Gdańsk.[25]

The city's ties with the Northern Netherlands are particularly well-represented: a comparatively high number of poems in the PAN Biblioteka Gdańska – mostly written in Latin, but also German and Dutch – relate to the United Provinces. For example, local authors eulogised Dutch diplomats

22 See, for example, Teresa Sulerzyska, 'Ryciny Cornelisa van Dalen w siedemnastowiecznych drukach gdańskich', *Biuletyn Historii Sztuki* 39: 4 (1977), 325–55; Teresa Grzybkowska, 'Niderlandyzm w sztuce gdańskiej', in Teresa Hrankowska (ed.), *Niderlandyzm w sztuce polskiej* (Warsaw: Wydawnictwo Naukowe PWN, 1995), 11–18; Juliette Roding, 'Dutch Architects and Engineers in Danzig and the Southern Baltic in the 16th and 17th centuries', in J.J. van Baak (ed.), *The Baltic. Languages and Cultures in Interaction. Proceedings NOMES-Conference, 19 & 20 May, 1994* (= *Tijdschrift voor Skandinavistiek* 16: 2 (1995)), 223–35.

23 See Edmund Kizik, 'Remuneration for Works Dedicated or Donated to the City Council of Gdańsk in the 17th century', *Polish Libraries* 2 (2014), 167–79.

24 The principal German example, which was plainly guided by nationalist sympathies, is Heinz Kindermann, *Danziger Barockdichtung* (Leipzig: Philipp Reclam, 1939). The seminal Polish work on seventeenth-century occasional poetry from Gdańsk is Kotarski, *Gdańska poezja*. Kotarski's study mostly deals with Latin and German poems in honour of Polish royalty, as well as local magistrates, soldiers, and scholars.

25 For a helpful overview of Gdańsk's seventeenth-century literary culture, see Astrid Dröse, *Georg Greflinger und das weltliche Lied im 17. Jahrhundert* (Berlin: De Gruyter, 2015), 50–77. Dröse acknowledges that a full examination of the matter is still lacking. A valuable case study is provided by Dick van Stekelenburg, *Michael Albinus 'Dantiscanus' (1610–1653)* (Amsterdam: Rodopi, 1987).

and admirals, bewailed the deaths of Dutch soldiers, and responded in horror to the so-called 'Disaster Year' of 1672. Many of these verses were original to Gdańsk, while others were republished or copied by hand from other sources. All this testifies to a considerable local interest in and literary engagement with the Northern Netherlands. Pastorius's composition about the Delft Thunderclap not only forms part of the history of Silesian-Dutch literary relations, but it also fits into a rich context of internationally oriented and specifically Dutch-themed poetry from Gdańsk.

Such poems could have diverse functions. For instance, if they were recited during banquets or handed to foreign ambassadors as tokens of friendship, they acted as diplomatic tools stressing mutual relations.[26] This chapter reveals how poetry from Gdańsk could serve yet another purpose: to express feelings of international solidarity and thus shape an emotional community which transgressed state borders, specifically those between the Dutch Republic and Poland. Pastorius was a central figure in this development. Indeed, his poem about the Delft Thunderclap is not the only one in his oeuvre to have a Dutch connection, as he wrote numerous verses which directly concern the United Provinces.

Pastorius regularly engaged with Dutch topics throughout his literary career. To begin with, he authored a series of flattering (and at times humorous) Latin compositions about interesting sights and travel destinations in the Northern Netherlands, based on his own sojourns in the 1630s, as well as poems in honour of several Dutch men of letters and Anna Maria van Schurman (1607–1678), one of the most renowned female scholars of her time. These verses were first published in 1644 by Daniel Vetter (1592–1669) in Leszno (Lissa), a stronghold of Polish Protestantism.[27] Subsequently, Pastorius composed a number of Latin poems about the Second Anglo-Dutch War (1665–67), which were published in Gdańsk. They lament the death of the Dutch admiral Jacob van Wassenaer Obdam (1610–1665), for instance, and celebrate the eventual peace.[28] Chronologically speaking, the piece

26 On the role played by poetry in diplomatic encounters between the Dutch Republic and Gdańsk, see Paul Hulsenboom, 'Diplomats as Poets, Poets as Diplomats. Poetic Gifts and Literary Reflections on the Dutch Mediations between Poland-Lithuania and Sweden in the First Half of the Seventeenth Century', *Legatio* 3 (2019), 63–110, at 74–76, 97–100.

27 Joachim Pastorius, *Heroes Sacri, Musa Peregrinans, Flos Poloniae, & Epigrammata Varia* (Leszno: Ex Typogr. Danielis Vetteri, 1644). The poems in this edition were partly republished in Gdańsk in 1653. Several of these compositions are discussed in Hulsenboom and Moss, 'Tracing the Sites', 285–90. The poems also feature prominently in my forthcoming dissertation.

28 These are preserved in Gdańsk: Joachim Pastorius, *In luctuosum Illustris Viri, Baronis Wassenarii Dni. de Opdam Prothalassiarchi Batavi fatum* [1665], kept in the PAN Biblioteka

about the Delft Thunderclap is nestled between these two clusters of verses, as it dates from after Pastorius's international travels but before the Second Anglo-Dutch War. Thematically, too, the poem obviously differs from the rest. It is a relatively stand-alone occasional composition, but it ties in neatly with the literary interest Pastorius displayed for the United Provinces from the 1630s onwards.[29] In short, the poem forms a potent example of the literary side of 'the mother trade', which has thus far remained unstudied and in which Pastorius held a key position.

Published in Poland, Inspired by Dutch Sources

The poem can be found in Pastorius's *Sylvarum Pars Secunda*, published in Gdańsk in 1657. This is the second part of the author's *Sylvae*, a two-volume collection of Latin verses on all manner of subjects, divided into seven 'books'. According to the title page of part one, which is dated 1656, the tomes were printed at the author's own expense. The piece on the Delft Thunderclap features in book four, which mainly contains poems dedicated to Polish nobles and dignitaries from Gdańsk. It also comprises a number of compositions in honour of the author's Dutch acquaintances, which had previously been published in the 1644 edition from Leszno. The verses about the disaster in Delft thus align with book four's partly Dutch theme and further strengthen Pastorius's relationship with the United Provinces. This section offers a close reading of the poem and considers its possible sources of inspiration.

The piece is entitled 'On the City of Delft, which has been sadly scorched by gunpowder':

> The Citizen who shone most beautifully amongst the Batavian Cities,
> Nourished by a calm bay of the tranquil sea –
> Oh! – now lies scattered by the flames of a sulphur storm,
> And burnt, she searches for herself in her own ruins.
> And she hardly finds herself remaining in any part,
> [But] – alas! – she has become her own burial, funeral pyre, and pile.

Gdańska, Oe 13, 216; Joachim Pastorius, *In Pacem Bredanam inter Magna Britanniae Regem et Foederati Belgii Ordines* [1667], kept in the PAN Biblioteka Gdańska, Oe 7, 245.

29 Another example is a Latin poem by Pastorius in honour of the Dutch diplomat and classicist Nicolaas Heinsius. See Nicolaas Heinsius, *Poematum Nova Editio, Prioribus Longe Auctior* (Amsterdam: Apud Danielem Elzevirium, 1666), 'Adoptivorum carminum libri duo', 37–38.

The tower-bearing monuments groaned under the polluted sky,
And the approaching flame terrified the Gods themselves.
Houses rushed against houses, bodies which were entirely
Parched by murderous alkali flew through the air.
Together with a host of students, the afflicted Lyceums burned,
And the entire scorched City already bewails her sons.
And while she laments that her people's sweet hope has been cut down,
She says that she has perished here twice on one funeral pile.
Learn, mortals, to confide in arms more carefully,
And may your own weapons inspire you with fear.
For if fate will so decide, that with which we used to strike our enemy,
Will itself become our death-dealing foe.[30]

As Pastorius personified the city, the poem constitutes a fitting parallel with the book's verses addressed to specific Dutchmen. The composition has a clear structure. First, Pastorius highlights the supposedly exceptional beauty and calmness of Delft prior to the explosion. Although the city does not actually border the sea, it does lie relatively close to it, leading Pastorius to imagine Delft in a peaceful bay. This tranquil scene is immediately shattered, however, as the poem suddenly describes the ruinous state of the city after the catastrophe. Indeed, the detonation even appears to have erased Delft off the map in its entirety. This contrast between serene beauty and total devastation intensifies the sense of tragedy. Next, Pastorius discusses the explosion itself, highlighting the threat posed to the 'tower-bearing monuments' (i.e. the Old and New Churches), the fires which scorched the city, and the blast which demolished houses and sent corpses flying. The poem then concentrates on the destruction of local schools and the death of young students – facts which are represented as particularly lamentable. Furthermore, the verse about the repeated perishing of Delft 'on one funeral pile' probably refers to the fire which consumed the city

30 '*De Urbe Delphensi pulvere pyrio miserè ustulatâ,* / Quae Batavas inter pulcherrima splenduit Urbes / Tranquilli placido Civis Alumna sinu, / Nunc ah sulphureae facibus jacet icta procellae, / Seq[ue] in ruderibus quaerit adusta suis. / Vixq[ue] aliquâ sese reperit sibi parte relictam, / Funus, & heu! bustum facta rogusq[ue] sibi. / Turrigerae vapido crepuêre sub aethere moles, / Attonuitq[ue] ipsos flamma propinqua Deos. / Concurrêre domus domibus: per inane volarunt / Corpora, fulmineo torrida tota nitro. / Cum grege discentum correpta arsêre Lycea, / Totaq[ue] jam natos Urbs gemit usta suos. / Spemq[ue] sibi dulcem generis cecidisse quiritans, / Sese, ait, hîc uno bis periisse rogo. / Discite mortales confidere cautiùs armis, / Incutiantq[ue] animis propria tela metum. / Nam si fata volent, quo pellere suevimus hostem, / Hoc ipsum nobis lethifer hostis erit.' Joachim Pastorius, *Sylvarum Pars Secunda* (Gdańsk: 1657), 33–34. Translation from Latin is mine. I thank Michiel Sauter for his valuable comments.

in 1536. This accentuates the author's knowledge of Delft's tragic history. Finally, Pastorius uses the Dutch calamity to formulate a lesson aimed at 'mortals' in general, urging them to consider the fickleness of fate and take care with their own weaponry. On the one hand, therefore, the poem bewails the doom of Delft specifically, while on the other, it bears a more universal message.

Pastorius did not fashion his response to the Delft Thunderclap out of nowhere. Instead, Dutch sources – which could reach Poland in a matter of weeks – guided his imagination and his pen. Considering his longstanding ties with and interest in the Northern Netherlands, it is safe to assume that Pastorius had mastered the Dutch language. Perhaps he received word of the disaster via one or more of his Dutch contacts. In fact, I would argue that it is particularly likely that Pastorius was sent news about the Thunderclap by one of his closest Dutch friends: Joannes Naeranus (1608–1679). Naeranus was a Dutch Remonstrant and publisher active in Rotterdam, who had lived and travelled in Poland for several years and corresponded with various members of the Polish Brethren (also known as Socinians/Arians), to which Pastorius himself also for some time belonged.[31]

Extant letters reveal that Pastorius and Naeranus corresponded from the 1630s until at least 1650 and regularly sent each other books.[32] In addition, Pastorius had published a Latin poem in honour of Naeranus in 1644, which he also included in the *Sylvarum Pars Secunda*, several pages after his reaction to the Delft Thunderclap.[33] Importantly, Naeranus was the publisher of a five-page Dutch letter dated 13 October 1654 – i.e. a day after the explosion.[34] The anonymous letter addresses an unnamed 'good friend' and dramatically discusses the tragedy of Delft, mourning the immense destruction of the city and loss of human life. The text also draws attention to the ruin of two local schools and the deaths of forty-eight pupils and a teacher. It is possible that this information provoked Pastorius's verse about the unfortunate students and 'afflicted Lyceums' of Delft. Considering the companionship between the two men and the letter Naeranus published

31 On the correspondence between Pastorius and Naeranus, see Sibbe Jan Visser, *Samuel Naeranus (1582–1641) en Johannes Naeranus (1608–1679). Twee remonstrantse theologen op de bres voor godsdienstige verdraagzaamheid* (Hilversum: Verloren, 2011), 171–73.

32 Visser, *Samuel Naeranus,* 173, 511–12.

33 Pastorius, *Sylvarum Pars Secunda*, 47.

34 *Brief Uyt Delft Verhalende de Gruwelijke en Verschrikkelijke Ruine van een groot gedeelte der zelver Stad, door het op-springen van het Kruyd-huys aldaar. Geschied op den 12. October 1654* (Rotterdam: By Joannes Naeranus, 1654).

shortly after the Thunderclap, it seems likely that Pastorius learned about the disaster via his Dutch friend. Perhaps Pastorius received a copy of the letter, which may have motivated him to write his poem.

However, he probably found concrete inspiration in another source. As stated in this chapter's introduction, multiple Dutch authors responded to the catastrophe, offering a variety of interpretations in both prose and poetry.[35] The best-known testimony about the Delft Thunderclap is a Dutch composition by the renowned poet Joost van den Vondel (1587–1679), in which he considered three either mythical or historical disasters, before turning to the calamity of Delft itself. The poem was published as a broadsheet, addressed to the mayor of Amsterdam. It was also included in other publications, for example in a pamphlet about the Thunderclap, issued in Amsterdam in 1654.[36] The poems by Vondel and Pastorius show a number of noteworthy similarities.

First, both pieces echo Virgil's *Aeneid*. Vondel's poem does so most clearly, carrying verse II.368 as its motto: 'plurima mortis imago' ('death in multiple guises'), which occurs in a passage narrating the destruction of ancient Troy. Pastorius's comments about the engagement of 'the Gods' and the scattering of dead bodies throughout Delft are reminiscent of the same passage. In addition, several details in both poems are similar. Vondel described the Thunderclap as follows:

In Delft, where against style and order,
  Our powder, turned into the country's enemy,
Spares neither town hall nor church,
  And delves a citizens' grave for Delft,
Dug in rouble, human flesh, and waves
  Of smouldering ash and glass.[37]

35 On Dutch literary reactions to the Delft Thunderclap, see G.C. Kunz and D.P. Oosterbaan (with G.G. Kunz), 'Buskruitramp was "straf op de zonden"', in Houtzager et al., *Kruit en krijg*, 53–64; Timna Hacquebord, 'Herinnering aan De Delftse Donderslag', unpublished student thesis, Universiteit van Groningen, 2014; Meijer Drees, '"Providential Discourse" Reconsidered'.

36 Joost van den Vondel, *Op het Onweder van 's Lants Bussekruit te Delft* (Amsterdam: Voor de weduwe van Abraham de Wees, 1654); Jan Philipsz. Schabaelje, *Historisch Verhael / Van het wonderlick en schrickelick opspringen van 't Magasyn-Huys, Voor-gevallen op den 12. October, 1654. Binnen Delft* (Amsterdam: Gedruckt by Tymon Houthaeck, 1654), fols 14v–[15r]. The poem was also reprinted in editions of Vondel's oeuvre, as well as in a 1667 description of Delft.

37 'Te Delf, daer, tegens styl en orden, / Ons kruit, 's lants vyandin geworden, / Stadthuis ontziet, noch kerckgewelf, / En delft een burgergraf voor Delf, / In puin, en mensevleesch, en golven / Van gloejende assche en glas gedolven'. Joost van den Vondel, *De werken van Vondel*, ed. J.F.M. Sterck, H.W.E. Moller, C.G.N. de Vooys et al., 10 vols, vol. 5: *1645–1656* (Amsterdam:

Much like Pastorius, Vondel depicted the exploded gunpowder as its owners' enemy.[38] Moreover, Vondel also mentioned some of the characteristic buildings of the city, and in a manner resembling Pastorius conjured up an image of Delft as a grave – something which results in a play on words in both Dutch ('delft ... Delf') and English ('delves ... Delft'). Lastly, whereas some observers represented the tragedy as God's punishment for the sinfulness of Delft's inhabitants, Vondel refrained from any references to divine retribution and ended by saying that '[one may] build for a century and [spend] Croesus's treasure, / [but] one spark, one moment destroys a city'.[39] Vondel thus voiced a universal message about the unpredictability of fate, comparable to the idea defined by Pastorius. In my opinion, all of these parallels together convincingly suggest that Pastorius knew and was directly inspired by Vondel's poem.

Disaster Poetry as a Transnational Community Builder

This leaves one final question: Why did Pastorius write a poem about the Delft Thunderclap? To begin with, the event may have struck a personal chord with Pastorius. Not only had he written about the Dutch Republic before, but he also had an interest in Delft specifically. His 1644 collection of travel verses contains a flattering epigram about the funerary monument of William (1533–1584) and Maurice of Orange (1567–1625) in the town's New Church, hinting that Pastorius had visited the city himself.[40] Delft was not a distant, abstract location for Pastorius, therefore, but rather a city with which he had a personal history and previous literary attachment. Additionally, his connection with the New Church informs the focus in the poem about the Thunderclap on the 'tower-bearing monuments'. To Pastorius, the churches were probably the most familiar sights of Delft. This may be why he gave them pride of place in his description of the explosion. Similarly, the losses suffered by the schools of Delft may have moved Pastorius in particular because of his own background as a teacher and tutor, leading him to emphasise the sad fate of the town's young population. Therefore,

De Maatschappij voor goede en goedkoope lectuur, 1931), 822. Translation from Dutch is mine.

38 The Dutch poet Jeremias de Decker applied the same motif. See his *Gedichten* (Amsterdam: Gedruckt by Jacob Colom, 1656), 74.

39 'Verbouw een eeuw, en Krezus schat: / Een vonck, een blick verwoest een stadt'. Vondel, *De werken van Vondel*, vol. 5: *1645–1656*, 823.

40 Pastorius, *Heroes Sacri*, 17.

one of the reasons Pastorius wrote about the tragedy of Delft may have been to personally cope with the disaster.

Pastorius also had more ambitious goals, however, aimed at an international readership. There is no evidence that he had his poem published in 1654, but he undoubtedly composed and disseminated it either in printed or manuscript form soon after the disaster, when it was still newsworthy. Considering his use of Latin, it is likely that Pastorius's main audience consisted of fellow men of letters, both in Poland and abroad. For example, he probably sent one or more copies of the poem to Joannes Naeranus. When he published the piece in Gdańsk in 1657, furthermore, he obviously believed that there was a market for it – though the volume's print run is unknown. The topic of the poem aligns with the aforementioned cosmopolitan character of Gdańsk and the city's strong relations with the United Provinces in particular. Pastorius must have hoped that his literary skills and knowledge of Dutch history and news would impress learned audiences in Poland and the Northern Netherlands alike. The poem about the Delft Thunderclap was one of the ways in which Pastorius participated in the transnational Republic of Letters, linking the United Provinces and Poland specifically.

While doing so, moreover, Pastorius showed a strong emotional commitment. This is evidenced by the contents of the poem itself. Marijke Meijer Drees has argued that Dutch responses to the Delft Thunderclap applied a so-called rhetoric of emotions.[41] As explained in this chapter's introduction, Meijer Drees has taken Rosenwein's concept of emotional communities and expanded it with a textual component. Following the time's literary theories, authors often used a common set of narratological tools when lamenting disasters. For example, writers elaborated on the horrors of catastrophes and filled their readers' hearts with fear for their own lives. One of their primary aims was to arouse feelings of compassion in their audiences. Pastorius did just that. He focused on several terrible details, such as scorched bodies and burned-down schools, and warned his readers to avoid sharing this tragic fate. Furthermore, he used a number of emotionally charged words and images: even the Gods were 'terrified', and the destroyed city – trying in vain to recover herself – 'bewails' and 'laments' the death of her children. The author also added two outcries of despair ('ah' and 'eheu!'), thus expressing his own anguish. Through these rhetorical devices, Pastorius engaged in emotional persuasion. As had many Dutch observers, he urged his readers to feel compassion with Delft and its population.

41 Meijer Drees, '"Providential Discourse" Reconsidered', 115–18.

Pastorius thus inspired a sense of transnational solidarity and social cohesion: an emotional community of authors and readers in the United Provinces and Poland. Herein lies the uniqueness of Pastorius's contribution: not only did he fashion an emotional community via a written text, but he also transcended political, cultural, and linguistic borders across Northern Europe. Through his poem, written nearly a thousand kilometres from the scene of tragedy, Pastorius inscribed himself into the same emotional community to which Naeranus, Vondel, and other Dutch authors belonged. Using familiar rhetorical tools, he entered their mournful narrative about Delft and joined their appeal for compassion with the victims. Indeed, he even personified the city, thereby giving his poem an additional personal level: Delft is effectively placed on the same plane as Pastorius's Dutch friends and acquaintances. His verses about the disaster in Delft constituted both a sign and a building block of a transnational emotional community of men of letters, which stretched from the Dutch Republic to Poland and vice versa. Tragic though it was, the Delft Thunderclap provided Pastorius with the ideal opportunity to reinforce this emotional community by means of his literary toolkit.

Conclusion

This chapter contributes to the study of early modern Dutch–Polish literary relations by bringing into focus the Silesian-Polish polymath Joachim Pastorius, whose career path and vast oeuvre made him a key figure in a community of intellectuals living in Holland and Gdańsk. His poem about the Delft Thunderclap offers a potent case study hereof. Even though his indebtedness to Naeranus cannot be proven beyond all doubt, it is likely that Pastorius's personal contacts with Dutch men of letters were instrumental for spreading news and literature from the United Provinces to Poland. Additionally, the parallels with Vondel almost certainly form a rare example of this Dutch literary giant's imprint on the Polish Baltic shore. In turn, Pastorius's verses demonstrate that these Dutch–Polish literary relations were not a one-way street, but could also inspire a response.

The poem by Pastorius is a unique testimony, moreover, as it reveals how the Delft Thunderclap of 1654 left a mark far beyond the Northern Netherlands. The piece exemplifies the value of expanding our perspective across the Dutch borders. Much can be gained by asking how and why people in other parts of the world engaged with disasters in the United Provinces, and what effects their reactions had, both locally and abroad. The case of

Pastorius and the Delft Thunderclap shows that large and terrible calamities carried the potential to manifest and construct transnational emotional communities via written texts. The poem was composed by an author with a powerful personal attachment to the Dutch Republic and published in a city with a similarly strong connection. Pastorius further emphasised these bonds by including his address to Delft amongst his poems to learned friends and acquaintances in the Northern Netherlands. What is more, by applying a well-known rhetorical framework, Pastorius was able to use the disaster to deepen a transnational sense of solidarity and social cohesion between the United Provinces and Poland, and between Delft and Gdańsk in particular. It seems that a dramatic event like the Delft Thunderclap was pre-eminently suited to forge an emotional community – even if its members were separated by a considerable distance. Pastorius helped shape an emotional community of intellectuals living in Northern Europe, bridging Poland and the Dutch Republic.

The case study discussed in this chapter illustrates the possibilities and payoffs of research into the literary side of 'the mother trade'. It has been made clear that a 'Dutch reading' greatly benefits our understanding of Gdańsk's early modern literary heritage. Scholarship devoted to the production, circulation, and uses of literature in seventeenth-century Gdańsk should take into account the city's cosmopolitan character and its cultural interwovenness with the Northern Netherlands in particular. Conversely, a focus on Gdańsk would enhance our knowledge of the early modern transnational impact of Dutch textual culture, especially in Poland. In addition, it would inform our awareness of Poland's place within the early modern Republic of Letters, the study of which tends to concentrate on countries in present-day Western Europe. The intricate workings of the seventeenth-century literary interaction between Holland and Gdańsk thus offer new and promising lines of inquiry within the rich field of Dutch–Polish historical relations.

About the Author

Paul Hulsenboom is researcher and lecturer at the department of Dutch Language and Culture at Radboud University, Nijmegen. He specialises in early modern Dutch–Polish relations, particularly in the fields of cultural representation and literary exchange. His interests and publications cover a variety of topics, including seventeenth-century perceptions of East and West, travel, translation, intertextuality, cultural transfer, and the interaction between poetry and diplomacy. He has a background in teaching Classics and has also worked as a lecturer at Utrecht University. In addition, he has translated books from Polish to Dutch. His PhD dissertation, which studies the ways in which the Dutch and Poles viewed one another during the long seventeenth century, is due to appear in 2023.

Figure 3.1 Simon Fokke, *Breaching of the Rhine and IJssel Dykes*, 1754, etching, 16.5 × 10.7 cm, published in *Nederlandsche jaerboeken*, vol. 4 (Amsterdam: Frans Houttuyn, 1754). Rijksmuseum, Amsterdam, RP-P-1960-400

3 Landscape as Wounded Body

Emotional Engagement in Visual Images of Floods

Hanneke van Asperen

Abstract

Throughout the centuries, artists employed many models to picture inundations. The most often cited are depictions of mythical, biblical, and historical floods. In addition, artists sometimes used unexpected but equally dramatic models. In 1754 the prolific eighteenth-century draughtsman and printmaker Simon Fokke designed an image of recent dyke breaches for the triweekly periodical *Nederlandsche jaerboeken*. Prominently depicted in the foreground of Fokke's image is a map of the flooded region. This element of a map, with mourning cherubs on either side, recalls religious, specifically Catholic, imagery that centralises the dead body of Christ. Arguably, Fokke used these images of pity to depict the land as a tangible and suffering body in order to evoke feelings of compassion. Hence, the depiction encouraged viewers to extend these feelings of empathy towards a part of the country where they may not have resided themselves, in keeping with the synthesising goals of the periodical.

Keywords

art history – prints – emotion – compassion – affective piety – Passion iconography

After heavy rains in December 1753, the Rhine dykes broke near Emmerich in Germany, close to the Dutch border. In January 1754 the dykes farther along the rivers Rhine and IJssel could not hold the excess water either, leading to inundations near the Dutch cities of Arnhem, Zutphen, Deventer, Zwolle, and Kampen. The subsequent floods devastated parts of the county of Zutphen and large parts of the province of Overijssel. Farmers suffered

Asperen, van, Hanneke and Lotte Jensen (eds): *Dealing with Disasters from Early Modern to Modern Times. Cultural Responses to Catastrophes.* Taylor & Francis Group, 2023
DOI: 10.5117/9789463725798_CH03

enormously from the disaster. A cattle plague epidemic had already been punishing the Dutch Republic since the previous year, and now farmers lost even more of their livestock to the water.[1]

Shortly after the catastrophe, the publisher Maria van den Berge printed a lamentation, or *Klaag-liet*, on the tragedy.[2] The text of the unsigned poem describes how the flood hit farmers especially. 'Where there was once grass and grain, for the sustenance of cattle and people, now there is only water', the poem says.[3] Farms were destroyed, crops rotted, cattle drowned. The anonymous author did not shy away from the usual rhetorical devices to accentuate the drama. We learn, for example, that the disaster is indescribable. 'No tongue can ever explain / No pen describe the pain / No quick brush ever compose / The tragedy that there arose'.[4] No author, no artist can do justice to the pain and loss. Still, the observation that the pain and tragedy are unspeakable did not prevent authors from taking up the pen. Nor did artists refrain from picking up chalk, pencils, and brushes.

Artists tried to visualise unfathomable tragedies, such as the inundations of 1754, sometimes using seemingly unlikely models. In this chapter I will focus on an outstanding example: an etching by the Amsterdam-based printmaker Simon Fokke (1712–1784).[5] Among many other subjects, this prolific artist represented different catastrophes over the years, which makes him an interesting subject of disaster research. In this study I demonstrate that artists had other dramatic models besides the usual depictions of mythical, biblical, and historical floods to visualise recent

1 The research for this article was funded by the Dutch Research Council NWO and carried out with a Vici grant for the research project *Dealing with Disasters. The Shaping of Local and National Identities in the Netherlands, (1421–1890)*. I would like to thank Adriaan Duiveman, Lotte Jensen, and Fons Meijer especially for their close reading of earlier versions of this article. Furthermore, I would like to thank the participants of the conference 'Dealing with Disasters. Cultural Representations of Catastrophes, c. 1500–1900' (2021) for their inquisitive questions and valuable comments when I first presented the preparatory research at the base of this article.

Jan Buisman, *Duizend jaar weer, wind en water in de Lage Landen*, ed. A.F.V. van Engelen, 8 vols, vol. 6: *1751–1800* (Franeker: Van Wijnen, 2015), 38–44, 85–90.

2 *Klaag-liet Uitgeboesemt over de inbreuke en overstroming der vreeslijke wateren omtrent Zwolle, Kampen en Zutphen, veroorsaakt door het Doorbreken van den IJssel-dijk, en Snipler-dijk* (s.l.: Maria van den Berge, 1754).

3 'Daar eertijts Gras en Koren waren / Tot onderhout van Mensch en vee / Daar ziet men niet als woeste baren / Rontom gelijk een baare Zee'. *Klaag-liet*, 3. Translation from Dutch is mine.

4 'Geen Tonge kan het ooit verhalen / Geen Pen beschrijven dit verdriet / Geen rap Penceel kan ooit afmalen / De droefheid die daar is geschiet'. *Klaag-liet*, 4. Translation from Dutch is mine.

5 *Allgemeines Künstler Lexikon* (AKL), vol. 41 (Berlin: De Gruyter, 2004), 519; Leontine Buijnsters-Smets, 'Simon Fokke (1712–1784) als boekillustrator', in *Achttiende-eeuwse kunst in de Nederlanden* (= *Leids kunsthistorisch jaarboek*, 1987), 127–46.

inundations. I propose that Fokke modelled his image of the 1754 flood after depictions of Christ's suffering. Inevitably, this observation impacts the way we should interpret his image. Arguably, Fokke referred to these highly empathic images of pity to achieve emotional engagement from his audience. In his choice of models, I show, he was most likely motivated by the type of periodical in which his print was published: the *Nederlandsche jaerboeken* ('Dutch Yearbooks'). With its focus on recent events throughout the Dutch Republic, this periodical explicitly aimed to address its readers as members of an emotional community that spanned the entire country.

State of Research

In the eighteenth-century Dutch Republic, disaster prints were produced in large quantities. Although images of historic and recent floods were hardly a new phenomenon, their number had steadily increased. This rise coincided with 'the invention of catastrophe', which historians have situated in the eighteenth century.[6] During that time, catastrophes were increasingly aestheticised. As objects of beauty and awe, they became beloved subjects of art, and artists incessantly invented new formats to visualise catastrophes and their social and political impact. In line with this development, catastrophe was a recurring theme in Fokke's oeuvre.[7]

Historians have recognised the importance of printed images. These could be incorporated in publications but were also sold separately and disseminated in great numbers. This is also true for the images in news periodicals.[8] Therefore, their audience was often larger than that of the publications in which they appeared. As part of news periodicals and historical works, prints highlighted described events, as they attracted

6 Anne Marie Mercier-Faivre and Chantal Thomas (eds), *L'invention de la catastrophe au XVIIIe siècle. Du châtiment divin au désastre naturel* (Geneva: Droz, 2008), 8.

7 Fokke depicted both recent and historic floods. An example of the latter are his depictions of St Elisabeth Flood of 1421 for different works of historian Jan Wagenaar. See, for example, my article 'Dordrecht aan de horizon. Veranderingen in de visuele verbeelding van de Sint-Elisabethsvloed', in Hanneke van Asperen, Marianne Eekhout, and Lotte Jensen (eds), *De grote en vreeselike vloed. De Sint-Elisabethsvloed 1421–2021* (Amsterdam: De Bezige Bij, 2021), 27–47.

8 Joop W. Koopmans, *Early Modern Media and the News in Europe. Perspectives from the Dutch Angle* (Leiden: Brill, 2018), 79–80; Joop W. Koopmans, *Het nieuws verbeeld. Oorlog en vrede in de titelprenten van de Europische Mercurius (1690–1750)* (Hilversum: Verloren, 2021), 25–31, at 26. See also Lotte Jensen, Hanneke van Asperen, Adriaan Duiveman, Marieke van Egeraat, Fons Meijer, and Lilian Nijhuis, 'Appropriating Disasters. A Framework for Cultural Historical Research on Catastrophes in Europe, 1500–1900', *Journal of Historical Geography* 76 (2022), 34–41.

the attention of readers and aroused curiosity. With images, publishers could thus accentuate specific passages and episodes. Moreover, an image offered other opportunities than text. Depictions visualised aspects that are difficult to capture in writing or could provide a visual summary of the written argument and thus add to text. In other words, the choice of publishers and authors to incorporate images in text is always meaningful and therefore deserves investigation.

Research into printed disaster images fits the larger field of the history of emotions. Medievalist and historian Barbara Rosenwein has shown that people lived (and live) in emotional communities and that different emotional communities co-existed simultaneously.[9] In other words, one person was part of different, partially overlapping emotional communities at the same time. Using a shared discourse, emotional communities could be activated at different times, for example during crises.[10] It is important to realise that their discourses also included images that tapped into a different vocabulary than written words.

In catastrophic times, images were extremely efficient instruments to express and generate affective responses, and artists had various methods and models to evoke a wide range of emotions. Because cultural and social contexts shape emotions, I believe that the field would benefit from case studies looking into (printed) disaster images in different periods and regions. Importantly, we should not only look at the images themselves to find out about artistic modes of expressions but also at the functional contexts of these images in order to define which groups of people were addressed. Subsequently, different cases demonstrate the various ways in which artists could achieve emotional engagement from their audiences and why. Here I focus on one such case, namely a print by Simon Fokke of the flood of 1754.

9 Barbara H. Rosenwein, *Emotional Communities in the Early Middle Ages* (New York and London: Cornell University Press, 2006). In line with the time frame of this article, see also William M. Reddy, *The Navigation of Feeling. A Framework for the History of Emotions* (Cambridge: Cambridge University Press, 2001), focusing on the eighteenth and early nineteenth centuries.

10 In addition to Rosenwein and Reddy, consider Jennifer Spinks and Charles Zika (eds), *Disaster, Death and the Emotions in the Shadow of the Apocalypse, 1400–1700* (London: Palgrave Macmillan, 2016); Monica Juneja and Gerrit Jasper Schenk (eds), *Disaster as Image. Iconographies and Media Strategies Across Europe and Asia* (Regensburg: Schnell & Steiner, 2014). Fons Meijer focuses specifically on crises as moments of national community building: *Verbonden door rampspoed. Rampen en natievorming in negentiende-eeuws Nederland* (Hilversum: Verloren, 2022).

Floods and Images

Simon Fokke was extremely productive as an etcher and engraver. Via publications aimed at a national market, his images – often after his own designs – found a broad audience. He produced representations of wide-ranging events, including historical ones for Jan Wagenaar's book series on Dutch history, *Vaderlandsche Historie* ('History of the Fatherland').[11] In his turn, Wagenaar honoured Fokke's memory in his history of Amsterdam (1788), writing that 'the innumerable number of book prints are witness to his extraordinary diligence and his productive genius'.[12] Fokke's productivity guaranteed his versions of historic events were omnipresent. Reproduced and copied many times, his visualisations, of both the recent past and long ago, have shaped later views of Dutch history in ways that have not yet been sufficiently acknowledged and studied.

Fokke also designed many prints for the *Nederlandsche jaerboeken*, published in Amsterdam from 1747 to 1765 and continued as *Nieuwe Nederlandsche jaerboeken* ('New Dutch Yearbooks') until 1806.[13] Contrary to what their title suggests, the *Nederlandsche jaerboeken* (hereafter *Jaerboeken*) were published as a periodical more or less triweekly. In his monograph on the subject, historian Donald Haks elaborates on the role of the *Jaerboeken* in the Dutch media landscape of the second half of the eighteenth century.[14] He characterises them as a medium combining the synthesising views of news periodicals – which reported on the news with more distance than

11 Jan Wagenaar and Petrus Loosjes Azn, *Vaderlandsche historie, vervattende de geschiedenissen der Vereenigde Nederlanden, inzonderheid die van Holland, van de vroegste tyden af: uit de geloofwaardigste Schryvers en egte Gedenkstukken samengesteld. Met konstplaaten opgehelderd*, 24 vols (Amsterdam: Allart, 1749–89). See also Buijnsters-Smets, 'Simon Fokke (1712–1784)', 132.

12 'De ontelbaare menigte Boekprenten, door hem vervaardigd, kan, van zyne ongemeene arbeidzaamheid, en van zyn vrugtbaar vernuft getuigen'. Jan Wagenaar, *Amsterdam in zyne Geschiedenissen, Voorregten, Koophandel, Gebouwen, Kerkenstaat, Schoolen, Schutterye, Gilden en Regeeringe beschreeven om te dienen ten vervolge op het werk van Jan Wagenaar, historieschryver der Stad* (Amsterdam: Conradi, 1788–90), vol. 4 (1788), 562. Translation from Dutch is mine.

13 Rietje van Vliet, 'Nederlandsche Jaarboeken (1747–1765)', in *Encyclopedia Nederlandse Tijdschriften: Nederlandstalige periodieken tot aanvang Koninkrijk der Nederlanden (tot 1815)*, https://www.ent1815.nl/ (latest update 22 July 2021; accessed 3 April 2023); K.L. Sprunger, 'Frans Houttuyn, Amsterdams boekverkoper. Preken, uitgeven en doopsgezinde Verlichting', *Doopsgezinde bijdragen* 31 (2005), 183–204.

14 Donald Haks, *Journalistiek in crisistijd. De (Nieuwe) Nederlandsche Jaarboeken 1747–1822* (Hilversum: Verloren, 2017), 11–17.

daily newspapers – and the opinionising comments of pamphlets.[15] Publisher Frans Houttuyn explicitly related his *Jaerboeken* to historical works, writing that their correspondents preserved events, not just for personal use but also for future generations.[16]

Furthermore, the *Jaerboeken* addressed an audience throughout the Dutch Republic. Their scope was not limited to subjects from one city or one province alone. In fact, one of their explicit goals was informing readers about events from every corner of the country. To achieve this nationwide coverage, they used often anonymous correspondents throughout the Republic. Significantly, Houttuyn addressed both his readers and his correspondents in the prologues.[17] Adding to evolving ideas about the Dutch Republic as a unity, the *Jaerboeken* were explicitly marketed as a communal attempt of the citizenry throughout the Republic to co-write Dutch history.

Because the *Jaerboeken* appeared less frequently than daily newspapers, it was possible to include printed images. According to Haks, the illustrations fit their audience promoting an enlightened vision 'of a world revolving around the citizenry'.[18] The *Jaerboeken*'s intended audience were educated people, both in the urban centres and rural regions, with a wide-ranging interest in politics, religion, commerce, art, and science.[19] Consequently, their images mostly depicted people from the middle classes. The Princes of Orange, who were Stadholders, were also portrayed, but their position of authority was never debated. People from the lower strata of society, on the other hand, were hardly depicted.

When the *Jaerboeken's* correspondents from the Dutch cities of Zutphen, Arnhem, Zwolle, and Deventer wrote about the flooding in January 1754, the memory of the disaster was still fresh. Adding to these recent reports from the affected areas, Fokke produced a fascinating print depicting the various

15 Haks, *Journalistiek*, 13–15. The heirs Houttuyn regarded the *Nederlandsche jaerboeken* as a continuation of the *Europische Mercurius*, as suggested in advertisements in *Haerlemse* and *Leydse courant* where they offer their stock of old *Nederlandsche jaerboeken* for sale. See examples of these advertisements, in *Haerlemse courante* (13 October 1768), 1 and *Leydse courant* (16 November 1768), 2 (both accessed via Delpher, 4 April 2023). On the *Europische Mercurius*, see Koopmans, *Het nieuws verbeeld*.

16 *Nederlandsche jaerboeken, inhoudende een verhael van de merkwaerdigste geschiedenissen: die voorgevallen zyn binnen den omtrek der Vereenigde Provintien, sedert het begin des Jaers MDCCXLVII* (Amsterdam: Frans Houttuyn, 1754–), vol. 2 (1749), ii.

17 *Nederlandsche jaerboeken*, vol. 2 (1749), i–iii.

18 Haks, *Journalistiek*, 36.

19 Haks, *Journalistiek*, 20.

dyke breaches along the Rhine and IJssel Rivers (Fig. 3.1).[20] To visualise the disaster, the artist effectively combined two genres: allegory and narrative. Looking at the image, we are first confronted with an allegory of two mourning cherubs with a map standing in front of a parapet. Secondly, we see the narrative of the unfolding disaster behind the low wall. There, groups of people are trying to save themselves and their loved ones from the water flowing rapidly through the large gap in the dyke. Despite the tragedy unfolding in the background, the focus of attention is the large map in the foreground. It depicts the IJssel River flowing from Arnhem in the south to Kampen in the north. The map is draped across the parapet. Not only is it placed in the immediate foreground, but the cherubs also draw attention towards it by looking at it and pointing it out. One is wiping its tears with a cloth. Underneath one of the cherubs' feet is a branch copying the shape of the river. Leafless and bare, this branch adds to the aura of death and decay.

Images of Pity

Though Fokke's print is mentioned sporadically, it has been the focus of some scholarly research. Haks regards the map as an indication of the growing interest in cartography and science.[21] Environmental historian Adam Sundberg also discusses the image and points especially to the pictorial element of the map, which he calls 'a common motif in disaster prints'.[22] He continues by saying that the main purpose of the map is to centralise the scale of the disaster. Otherwise, the allegorical part of Fokke's print has never been examined, even though it contextualises the narrative part of the print. As Sundberg rightfully points out, the inclusion of a map is not new, but it is presented in a way that deserves further investigation. The map with the mourning cherubs on either side brings to mind specific types of religious, specifically Catholic, imagery that centralise the dead body of Christ.

20 *Nederlandsche jaerboeken*, vol. 4 (1754), ill. between pp. 36 and 37. Rijksmuseum also has in its collection a transfer drawing for the print in pencil and red chalk, pen and brush in grey, grilled. Rijksmuseum, Amsterdam, RP-T-00-1590.

21 Haks, *Journalistiek*, 36.

22 Adam D. Sundberg, 'Floods, Worms, and Cattle Plague. Nature-Induced Disaster at the Closing of the Dutch Golden Age, 1672–1764', PhD dissertation, University of Kansas, 2015, 23. Sundberg does not centralise the visual element of maps in his recent *Natural Disaster at the Closing of the Dutch Golden Age* (Cambridge: Cambridge University Press, 2022), although he does describe some prints with maps on them. Fokke's print of the flood of 1754 is not among these.

More specifically, the image of two mourning cherubs calls to mind images of the *Engelpieta*, which was invented in the fourteenth century and became extremely popular throughout Europe in the fifteenth, also in the Low Countries. The *Engelpieta* elaborates on the Man of Sorrows, which was the subject of a classic study by art historian Erwin Panofsky.[23] With its roots in the twelfth century, the Man of Sorrows shows Christ in isolation, half-figure, bare-chested, his side and hands pierced, often in front of the cross and/or his tomb. It combines two related but originally separate images, of Christ on the cross, not yet dead, and the same lying in his tomb. Presenting Christ's wounded body as a focus of emotional attention, these images were also called *imagines pietatis* (images of pity).

Following in Panofsky's footsteps, much scholarly work has been done on these types of Catholic images that served to evoke emotional response and facilitate contemplative absorption, or *Versenkung*.[24] The tendency to regard images as sources of empathic meditation is really of early Christian date, but in the fourteenth and fifteenth centuries it became more and more prominent. These images were not only the subject of large altarpieces, but they were also depicted on a much smaller scale on panels suitable to put up or hang on a wall inside a home. Art became an aid to achieve deep emotional experiences, not necessarily in the public space of a church but often in the privacy of one's home. Although art historians have studied these religious images extensively, their influence on other genres for other audiences remains to be investigated.[25]

Here I propose that images of pity, especially the *Engelpieta*, directly inspired Fokke. In an especially dramatic version by the Italian Renaissance painter Carlo Crivelli (Fig. 3.2), crying cherubs are depicted on either side of the dead

23 Erwin Panofsky, 'Imago Pietatis. Ein Beitrag zur Typengeschichte des Schmerzensmanns und der Maria Mediatrix', in *Festschrift für Max J. Friedländer zum 60. Geburtstage* (Leipzig: Seemann, 1927), 261–308.

24 Sixten Ringbom, *Icon to Narrative. The Rise of the Dramatic Close-up in Fifteenth-Century Devotional Painting* (2nd ed., rev. and augm., Doornspijk: Davaco, 1984); Sixten Ringbom, 'Devotional Images and Imaginative Devotions. Notes on the Place of Art in Late Medieval Private Piety', *Gazette des Beaux-Arts* 111: 73 (1969), 159–70.

25 A.A. MacDonald, H.N.B. Ridderbos, and R.M. Schlusemann (eds), *The Broken Body. Passion Devotion in Late-Medieval Culture* (Groningen: Egbert Forsten, 1998); Henk van Os, *Gebed in Schoonheid. Schatten van privé-devotie in Europa 1300–1500* (Amsterdam: Rijksmuseum, 1994); Carol M. Schuler, 'The Seven Sorrows of the Virgin. Popular Culture and Cultic Imagery in Pre-Reformation Europe', *Simiolus* 21: 1–2 (1992), 5–28; Craig Harbison, 'Visions and Meditation in Early Flemish Painting', *Simiolus* 15: 2 (1985), 87–118; James H. Marrow, *Passion Iconography in Northern European Art of the Late Middle Ages and Early Renaissance. A Study of the Transformation of Sacred Metaphor into Descriptive Narrative* (Kortrijk: Van Gemmert, 1979).

Figure 3.2 Carlo Crivelli, *Dead Christ Supported by Angels*, signed, 1470–75, tempera on wood, 72.4 × 55.2 cm. National Gallery, London, NG602. © National Gallery London

Christ. They support his body while showing his gaping wounds to the viewer. The most important elements of comparison with Fokke's print are the crying cherubs. These figures direct the viewer towards the subject of their pain which they hold in their midst. In addition, there is the element of the parapet as an extremely effective illusionistic device. Because part of the image is depicted in front of a wall, it seems as if these elements – Christ's hand in Crivelli's case, the map in Fokke's – enter the realm of the viewer. Thus, the beholder is invited to enter into a close emotional relationship with the presented body/object.

The capacity of bloody and sensory images to evoke emotions has long been recognised. This objective of emotional proximity closely corresponds with late medieval and early modern Catholic images of 'affective piety', as medievalist and historian Caroline Walker Bynum has called it, focusing on Christ's suffering and the Virgin's compassion.[26] Many art historians have studied the emotional effects of these evocative images.[27] Historian and sociologist Herman Roodenburg has described how fifteenth-century authors and artists facilitated and stimulated emotional and sensory absorption in Christ's pain through their texts and images.[28] Their production certainly did not cease after the Reformation. Aiming for feelings of empathy in their viewers, images of affective piety were still invented, sculpted, and painted in the Catholic parts of Europe. Importantly, many of these artists were admired and copied in the northern parts of the Low Countries. Their inventions were readily available in the Northern Netherlands in the seventeenth and eighteenth centuries via many painted and printed reproductions.

Wounded Bodies

To note the formal similarities between Fokke's image and the iconographical theme of the *Engelpieta* is one thing, but how can we know these similarities are not just coincidences? Fokke was a fervent art collector, and his impressive art collection is mentioned by the aforementioned historian Jan Wagenaar.[29] In his short biography of Fokke, Wagenaar even called this engraver 'one of the best connoisseurs of Dutch and Italian art'.[30] Already in the 1740s, when Fokke started designing his images for the *Jaerboeken*, he bought paintings, drawings, and prints from various estates.[31] He gathered artworks, especially printed copies, because he admired their inventors.

26 Caroline Walker Bynum, *Jesus as Mother. Studies in the Spirituality of the High Middle Ages* (Berkeley: University of California Press, 1982).
27 Marrow, *Passion Iconography*.
28 Herman W. Roodenburg, 'Empathy in the Making. Crafting the Believer's Emotions in the Late Medieval Low Countries', *BMGN – Low Countries Historical Review* 129: 2 (2014), 42–62.
29 Wagenaar, *Amsterdam*, vol. 4 (1788), 562.
30 'Behalven zyne bekwaamheid in het graveeren, was Fokke een der beste kenners van italiaansche en Nederlandsche Tekeningen; van welke hy ook een aanzienlyke Verzameling heeft nagelaaten, by zyn overlyden'. Ibid.
31 Buijnsters-Smets, 'Simon Fokke (1712–1784)', 132.

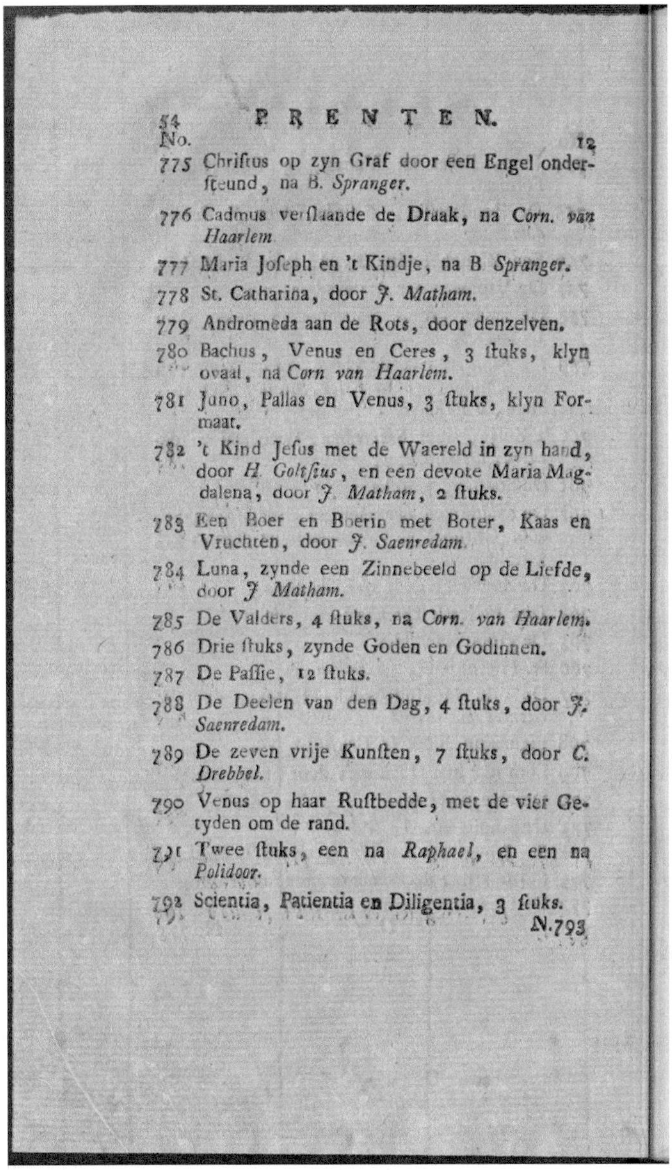

54 PRENTEN.

No. 12

775 Chriſtus op zyn Graf door een Engel onderſteund, na *B. Spranger.*

776 Cadmus verſlaande de Draak, na *Corn. van Haarlem.*

777 Maria Joſeph en 't Kindje, na *B. Spranger.*

778 St. Catharina, door *J. Matham.*

779 Andromeda aan de Rots, door denzelven.

780 Bachus, Venus en Ceres, 3 ſtuks, klyn ovaal, na *Corn van Haarlem.*

781 Juno, Pallas en Venus, 3 ſtuks, klyn Formaat.

782 't Kind Jeſus met de Waereld in zyn hand, door *H. Goltſius*, en een devote Maria Magdalena, door *J. Matham*, 2 ſtuks.

783 Een Boer en Boerin met Boter, Kaas en Vruchten, door *J. Saenredam.*

784 Luna, zynde een Zinnebeeld op de Liefde, door *J. Matham.*

785 De Valders, 4 ſtuks, na *Corn. van Haarlem.*

786 Drie ſtuks, zynde Goden en Godinnen.

787 De Paſſie, 12 ſtuks.

788 De Deelen van den Dag, 4 ſtuks, door *J. Saenredam.*

789 De zeven vrije Kunſten, 7 ſtuks, door *C. Drebbel.*

790 Venus op haar Ruſtbedde, met de vier Getyden om de rand.

791 Twee ſtuks, een na *Raphael*, en een na *Polidoor.*

792 Scientia, Patientia en Diligentia, 3 ſtuks.

N.793

Figure 3.3 Page from *Konstboek* 12, the catalogue of art works, assembled and bequeathed by the late Simon Fokke, Amsterdam, 6 December 1784, 2 vols (Amsterdam: P. Yver, 1784), vol. 2. KB – National Library of the Netherlands, The Hague, KW 528 G 25. Source: Delpher (accessed 22 April 2022)

He subsequently used prints of favourite works to inspire him and help him formulate new ones.

Fokke's art collection was put up for auction after his death in 1784. Fortunately, we still have a printed two-volume (!) catalogue of his impressive collection of artworks, shortly stating themes, inventors, and engravers.[32] From the catalogue we know that Fokke had several prints in his possession depicting the dead Christ. For example, lot no. 162 in the sale was a print of a dead Christ in Mary's lap accompanied by an angel holding his hand after the Venetian Baroque painter Paolo Veronese. Lot no. 287 was a print of Christ's dead body supported by an angel, by the Bolognese artist Agostino Carracci.

The second volume of the auction catalogue is entirely dedicated to prints subdivided among various portfolios and books. One of these – *Konstboek* no. 12 – is dedicated to the most important prints ('de voornaamste Prenten') by the Haarlem artist Hendrick Goltzius. In this book, we find a 'Christ on his tomb, supported by an angel, after Spranger' (Fig. 3.3).

This 'Spranger' can be no other than the famous Antwerp painter Bartolomeus Spranger, who became court painter to Emperor Rudolph II in Prague in 1581. Spranger's work had considerable impact on artists in the Northern Netherlands, especially via Goltzius.[33] In his turn, Goltzius had gotten acquainted with Spranger's work via the Flemish painter Karel van Mander. Van Mander wrote that he 'showed [Goltzius] some drawings by Spranger which sparked great interest in him'.[34] After this introduction, Goltzius must have personally contacted Rudolph II's court painter, because he regularly started making engravings after Spranger's designs, for which he received international praise.[35]

Among his many works, Spranger painted an *Engelpieta* depicting Christ on his tomb with angels (Fig. 3.4). Goltzius subsequently produced a print

32 *Catalogus van een zeer fraay kabinet prenten enz. Alles met veel kunde in meer dan vyftig jaaren byeen verzameld en nagelaaten door wylen den Heer Simon Fokke, in leven beroemd kunstgraveerder* ... (6 December 1784), vol. 2, 54, no. 778. Art dealer P. Yver announced the auction and the two-volume catalogue in the newspaper *Nederlandsche courant* (29 November 1784), 2 (accessed via Delpher, 3 April 2023).

33 Nadine M. Orenstein, 'Eindelijk Spranger. Prenten en prentontwerpen 1586–1590', in Huigen Leeflang and Ger Luijten (eds), *Hendrick Goltzius (1558–1617). Tekeningen, prenten en schilderijen*, exh. Rijksmuseum, Amsterdam, the Metropolitan Museum of Art, New York, and the Toledo Museum of Art (Zwolle: Waanders, 2003), 81–114; Huigen Leeflang, 'Post uit Praag. Over een teruggevonden tekening van Bartholomeus Spranger', *Bulletin van het Rijksmuseum* 56: 1–2 (2008), 114–27.

34 '... liet hem enige tekeningen van Spranger zien, waar hij groot belang in stelde'. Karel van Mander, *Het schilder-boeck* (Haarlem: 1604; repr., Utrecht: 1969), fol. 284r.

35 Leeflang, 'Post uit Praag', 115.

Figure 3.4 Bartholomeus Spranger, *Dead Christ Supported by Angels*, 1587, oil on copper, 33.7 × 26.6 cm. Rijksmuseum, Amsterdam, Gift of B.P. Haboldt, in memory of the victims of COVID-19

Figure 3.5 Hendrick Goltzius, after Bartholomeus Spranger, *Dead Christ Supported by Angels*, 1587, engraving, 35 × 25.6 cm. Rijksmuseum, Amsterdam, RP-P-OB-10.376

after this painting (Fig. 3.5). It was undoubtedly this engraving that Fokke considered to be among his most important prints by Goltzius. The parallels between the scene of mourning and emotional response after impactful disasters did not go unnoticed by Fokke and his contemporaries. When Fokke was asked to illustrate the dyke breaches of 1754, he chose to translate Spranger's image of suffering into a stirring image of a recent disaster. Goltzius's print after Spranger's *Engelpieta* provided the formal model for Fokke's depiction, copying its outward form. However, I suspect that Fokke may have had more substantive reasons to refer back to Spranger: the image's capacity to evoke feelings of compassion.

The depiction of a disaster modelled after an image of mourning showed viewers how they should respond to the catastrophic event. The composition of Fokke's image prescribes that the viewer focus not on details but on the afflicted area in its entirety. To make it easier for the viewer to see the disrupted land as a unity, Fokke focuses on the IJssel River, which is central to the map. The main river course and all of its distributaries pervade the land. The waterways interconnect the depicted cities and villages. The artist prominently depicts the river and its branches, almost as an artery with all its smaller blood vessels.

Furthermore, the artist clearly indicates the sites of the dyke breaches, probably based on news reports. There are no less than seven places showing breaches. Fokke does not just point out these different sites – all mentioned in the *Jaerboeken*'s reports of the events – but he actually depicts the water gushing onto the land. In other words, he adds narrative detail to the map. It is almost as if the rivers have started to bleed. We can identify the areas where the dykes break and water is flowing, like blood from a wound. The cherub on the left points to one of these 'wounds' while patting its eyes with a cloth. Thus, multiple elements add to the idea of the land as a wounded body. These visual elements help the viewer to identify the land as something to feel compassion with.

Emotional Engagement

Arguably, the map helped viewers to focus feelings of compassion onto an abstract idea of a unified land. The *Jaerboeken*'s readership will have been familiar with maps in allegorical prints, also in images of floods. The use of maps to identify the region where empathy should be directed was not uncommon. A map was also used, for example, in a flood print of 1702, published by Pieter van den Berge (c. 1665–1737) in Amsterdam probably not long after the disaster occurred (Fig. 3.6).[36] The region around Amsterdam had been affected by a heavy storm, which had caused dykes to break. Here, instead of cherubs, the Maiden of Holland holds the map of the devastated area. Sitting on her knees in the water, the woman personifying the struck province of Holland points out a map to the viewer and raises her other arm to the sky in a gesture of despair.[37] For this detail, the artist must have found inspiration in images of the Virgin mourning the dead body of Christ, for example a painting by Flemish artist Anthony van Dijck, which circulated widely in print.[38] The intention of both prints was the same. Using a map, the artists transformed a geographic region into an object of compassion.

Van den Berge's audience differed from Fokke's. The former was active in Amsterdam until 1737.[39] His image depicts the area around Amsterdam which likely coincided with the living area of his customers. Fokke designed his image of the IJssel region for an audience that extended the depicted area. After all, the *Jaerboeken*'s readership were citizens from the middle classes throughout the Dutch Republic. Importantly, Houttuyn articulated a sense of social cohesion among his readers. In the prologue to the 1754 edition, he explained that humans are at heart social creatures. It is argued that all humans know they need help to remedy their own shortcomings and, at the same time, have a deep desire to share what is theirs. They would be 'the unhappiest creatures on Earth' if they did not have the means to satisfy each of these tendencies.[40] Fortunately, humans

36 The text on the print reads: 'Verbeelding van de ongemeene hooge / WAATERVLOED / In Holland, en Polders en Dorpen van dien, en de groote schade / daar door geschiet, tusschen den 5 en 6 April A°. 1702. / Zynde dezelve booven de hoogste Vloeden in een Eeuw herwaards, naar aan- / teekening op de gemeene Pyl van de Zeedyken hooger gevonden 4 duym. / De Maagd van Holland zit hier in het treurgewaad, / Beweenende, vol smart, haar jammerlyke staat; / Dewyl een Zee-Orcaan, rammeyend tegens Dyken / En Dammen, doorbreekt, en met haar Landsdouw gaat stryken. / De Noord' en Westen-wind, twee blaasbalgen, heel verwoed / Uit hunne ketenen gespat, den Waater-vloed / Doen swellen, en met Huys en Hof, en Landgenooten, / En Vee, en Stolp en Stal, om 't jammer te vergrooten, / Verbrooken en gesloopt, heen dryven, en dien schat, / Haar welvaart, smooren in 't onstuymig peekelnat. / Wat Landplaag, welk elend, kan by dees droefheid haalen! / Wie zal, als God alleen, haar groot verlies bepaalen'.

37 The posture of the woman personifying the Dutch Maiden, kneeling in the water, refers back to Romeyn de Hooghe's print depicting the Dutch state of ruination in 1675. See my 'Disaster and Discord. Romeyn de Hooghe and the Dutch State of Ruination in 1675', *Dutch Crossing* 45: 3 (2020), 241–62. The element of the map as a focus of attention is a new element, however.

38 See, for example, *The New Hollstein Dutch & Flemish Etchings, Engravings & Woodcuts 1450–1700* (Ouderkerk aan den IJssel: Sound & Vision), 536-1(2).

39 Ernst W. Moes, in *Nieuw Nederlandsch biografisch woordenboek*, vol. 2 of 10 vols (1912), 133–34.

40 'Zoo waer is het dus een Mensch een der ongelukkigste Schepselen op aerde zoude zyn, indien hy, aen de éène zyde, de noodzakelykheit der Hulpe van anderen ter vervullinge zyner Gebreklykheden

Figure 3.6 *The Flood of 1702*, 1702, etching, part of a broadsheet, published by Pieter van den Berge in Amsterdam, 40 × 22.5 cm. Rijksmuseum, Amsterdam, RP-P-OB-83.122

are not only created with the inclination for 'curative companionship', but they have the capacity to fulfil it. Significantly, the publisher stated that this 'noble characteristic' should not be limited to a circle of friends but should be extended towards all regions of the Dutch Republic.

The moralising overtone fits the reports in the *Jaerboeken*. As Haks observes, the *Jaerboeken* gladly mention technological innovations in the prevention and handling of dyke breaches. The love of science and technological progress is reflected in the lists of factual information about water levels, temperatures, precipitation, and the condition of the dykes.[41] However, the reports were not limited to this kind of factual information and technological innovations. The report of January 1754 described the inundations but also the hardship and efforts to alleviate suffering, in line with the goal of the *Jaerboeken* to stimulate 'curative companionship' in its readers.[42] The reports often use terms as misery (*ellende*), unfortunate (*ongelukkig*), miserable (*jammerlijk*), and mourn (*bejammeren*). With this choice of words, the texts also encouraged readers to feel for the victims.

The correspondents also discussed acts to assist the victims. When writing about the misery and damage among the county of Zutphen, for example, the correspondent recorded that the city magistrate sent a ship with provisions at the city's expense to alleviate the victims, 'according to their commendable habit'.[43] The activities of the magistrate included the distribution of food, clothing, and peat. Besides the magistrates, the *Nederlandsche jaerboeken* also mention charitable acts of citizens. For example, a correspondent in Zwolle wrote that 'the flood victims found refuge in churches, government buildings, and private residences because of the compassion of the citizenry'.[44]

kennende, en aen de andere zyde teffens de zucht om anderen van het zyne mede te deelen, in zich gewaer wordende, zich versteken [sic] zag van de middelen om beide die geneigdheden te voldoen. Dan de hemelsche Voorzienigheit zy gedankt, die de proefstukken harer Goedheit en Almacht willende toonen, den Menschen niet slechts een trek tot de heilzame Gezelligheit ingeschapen; maer teffens met Vermogens beschonken heeft om daer aen, op de gevoeglykste wyze, te kunnen voldoen'. *Nederlandsche jaerboeken*, vol. 4 (1754), ii–iii. Translations from Dutch are mine.

41 Haks, *Journalistiek*, 58.

42 On the importance of catastrophes as moments of national community building, see Meijer, *Verbonden door rampspoed*.

43 'volgens derzelver loffelyke gewoonte'. *Nederlandsche jaerboeken*, vol. 4 (1754), 9, also 3–4, 7, 34–35, 37–38.

44 'Doch dezen ... maekten niet veel haest om te vluchten voor den 2. dezer Louwmaend, wanneer ... zy in menigte met hun Vee herwaert de wyk namen, en door de bestelling van den Magistraet, en ontferming van velen uit de Burgerye dezer Stad, in eenige Kerken en andere Stads Gebouwen of Huizen van particulieren Huisvesting en Stallinge vonden'. *Nederlandsche jaerboeken*, vol. 4 (1754), 36–37. Translation from Dutch is mine.

Evoking empathy with a flood's victims was often an implicit goal of disaster images.[45] Fokke carefully chose the visual format to fit the educated readership of the *Jaerboeken*. These were well informed and knew about art. Some of them may have even been print collectors, since sales of art works including those by Goltzius and Spranger were advertised in the same newspapers as the *Jaerboeken* and the historical works by Wagenaar, for example.[46] Evidently, these circulating images were part of a shared visual discourse that artists could tap into. Therefore, it is fair to suggest that part of the *Jaerboeken*'s readership will have recognised the visual parallels with these images of piety, still well-known and appreciated in the second half of the eighteenth century. They will have been sensitive, therefore, to the moralising message to sympathise with the suffering.

Conclusions

As I have shown, Fokke evaluated the readership of the *Jaerboeken*, considered his models, and carefully composed an image that suited its public with the explicit goals of the periodical in mind. Referring back to Spranger's highly evocative *Engelpieta*, Fokke presented the land as a physical body, the rivers as its blood vessels, and the dyke breaches as bleeding wounds. In Fokke's decatholicised version of an *imago pietatis*, the cherubs present the wounded land as a focus of compassion, just as they present Christ in images of pity. The image thus triggered 'curative conviviality', so explicitly praised and propagated in the *Jaerboeken*. It helped readers to extend these feelings towards a part of the Dutch Republic where they may not have resided themselves, fitting the synthesising goals of the *Jaerboeken*.

At the beginning of this article, I mentioned Sundberg, who pointed out that Fokke used the map to centralise geographic scale when depicting catastrophes. Here, I would like to give more gravity to maps that were depicted so prominently. In short, I believe the map served several functions. Firstly, Fokke used the map to unify the affected areas, treated separately in consecutive *Jaerboeken* reports. Secondly, Fokke used the map to depict the land as a tangible body to evoke feelings of compassion. Finally, with

45 e.g. Lotte Jensen, Hanneke van Asperen, Adriaan Duiveman, Marieke van Egeraat, Fons Meijer, and Lilian Nijhuis, 'Omgaan met rampen in Nederland door de eeuwen heen. De rol van culturele media bij gemeenschapsvorming', *Neerlandica Wratislaviensia* 30 (2020), 43–59, at 50–51; Spinks and Zika, *Disaster, Death and the Emotions*.

46 e.g. *Oprechte Haerlemsche courant* (5 December 1765), 2; *Leydse courant* (6 May 1767), 2; *Oprechte Haerlemsche courant* (28 September 1780), 2 (all accessed via Delpher, 4 April 2023).

the visual element of the map, Fokke was able to activate an emotional community of the middle-class readership throughout the Dutch Republic because Goltzius's print of Spranger's *Engelpieta* was still part of a shared visual vocabulary. Including the lamenting cherubs, Fokke created a highly prescriptive image representing a part of the Republic as something with which the *Jaerboeken*'s readers not only could but should feel compassion. The urge to donate for relief of the victims was always implicit in remarks that good citizens feel 'a deep desire to share what is theirs'. However, the first and main goal of this highly affective image was to tie citizens of the Dutch Republic together in 'curative conviviality', as propagated by the *Jaerboeken*.

Significantly, Spranger's *Engelpieta* was in the news during another moment of crisis. An art dealer donated it to the Rijksmuseum in Amsterdam in 2020 during the COVID-19 pandemic (Fig. 3.4). In explanation of the gift, the donor stated that 'Coronavirus has affected me, in the first place emotionally. It gave me occasion to reflect on how I could make a contribution, and on how we could best memorialise this period. What is wonderful about paintings is that they are eternal and can serve as monuments to the difficult times in which we find ourselves'.[47] From this statement it is clear that Spranger's image is still regarded as a powerful and poignant representation of pity. It is this quality as icon of suffering and compassion that Fokke recognised, that must have appealed to citizens of the Dutch Republic, and that still activates emotional communities today.

About the Author

Hanneke van Asperen is an art historian affiliated with the Centre for Art Historical Documentation (CKD) at the Radboud University in Nijmegen. As postdoctoral researcher, she is part of the NWO-funded project *Dealing with Disasters. The Shaping of Local and National Identities in the Netherlands (1421–1890)*. She has expertise in visual images of charity and nature-induced disasters in premodern times. Additionally, she has an interest in medieval pilgrimage and religious badges. Her recent book *Silver Saints. Prayers and Badges in Late Medieval Books* (Brepols 2021) focuses on the practice of adding religious badges to devotional manuscripts.

47 The donor and art dealer Bob Haboldt in a press release, see 'Rijksmuseum Receives Painting by Bartholomeus Spranger', news release 27 May 2020, https://www.rijksmuseum.nl/en/press/press-releases/rijksmuseum-receives-painting-by-bartholomeus-spranger (accessed 25 June 2021).

4 Suffering Compatriots

Compassion, Catastrophe, and National Identification in the Netherlands in the Nineteenth Century

Fons Meijer

Abstract

Why do nations command such profound emotional legitimacy? This was one of the central questions of Benedict Anderson's 1983 seminal study *Imagined Communities*. Only recently have historians turned their full attention to the affective dimensions of nationalism and national identification. They argue that emotions and feelings are key to understanding the strong pull of nationalism on the lives of people. My chapter will single out one such 'national emotion': compassion for fellow countrymen. Focusing on the Netherlands of the nineteenth century, this article investigates how moments of major catastrophe were appropriated by the authors of poems and eyewitness reports to disclose the suffering of countrymen in the struck area. They invited the public to affectively identify with the victims' anguish and misery, gave shape and meaning to these feelings of compassion, and thus laid the emotional groundwork for strong ties of national unity and belonging.

Keywords

national belonging – appropriation – compassion – the Netherlands – nineteenth century

In late January 1807, Josué Teissedre l'Ange delivered a sermon in the Walloon church of Haarlem. The subject of the minister's address was the Leiden gunpowder disaster: earlier that month, a ship carrying numerous barrels of black powder had exploded in the town of Leiden, causing the death of more than one hundred fifty people and the destruction of over two hundred

---

Asperen, van, Hanneke and Lotte Jensen (eds): *Dealing with Disasters from Early Modern to Modern Times. Cultural Responses to Catastrophes.* Taylor & Francis Group, 2023
DOI: 10.5117/9789463725798_CH04

buildings in the town's city centre.[1] Teissedre l'Ange invited his audience to paint for themselves a mental picture of the 'scene of disaster' ('toneel van akeligheden'). He asked them to imagine the mortal fear the townsfolk had felt during the explosion and to hear the 'moaning of the wounded, the groans of the dying'. The listeners had to feel the pain of the inhabitants of Leiden, or so the minister made clear: 'If one member of the Fatherland suffers, the other members should suffer with him'.[2]

Making use of this expression, Teissedre l'Ange referred to a key characteristic of nation-building: the feature that the 'heart' of any nation ultimately consists of a shared sense of social and moral compassion with those individuals who allegedly belong to the same nation.[3] The cognitive understanding that compatriots are in fact part of the same national community as yourself implies a sense of mutual connectedness and affective identification, and compassion has therefore been a powerful resource for nation-building. This emotion has allowed contemporaries to affectively – rather than just rationally – identify with their compatriots and, in turn, feel ever more strongly part of the national community. Focusing on the case of the Netherlands in the nineteenth century, I argue in this chapter that major disasters – such as the Leiden gunpowder disaster, but also severe storm surges and river floods – were key moments during which a 'nationalist' culture of compassion with one's countrymen – and women, of course – was being propagated and consolidated.[4]

For a long time, nationalism scholars have shown awareness of the fact that affective experiences in general play a key role in processes of nation-building. National identification thrives on emotional experiences such as collective singing of anthems, attendance at solemn memorial services, and the passionate celebrations of national holidays.[5] As early as 1983, one

1 Arti Ponsen and Ed van der Vlist (eds), *Het fataal evenement. De buskruitramp van 1807 in Leiden* (Leiden: Ginkgo, 2007).

2 J. Teissedre l'Ange, *De verwoesting van den twaalfden Januarij, voorgesteld als eene algemeene ramp in eene leerrede over I. Corinth. XII 26* (Haarlem: J. Enschedé en zonen, 1807), 9–10.

3 Hanneke Hoekstra, *Het hart van de natie. Morele verontwaardiging en politieke verandering in Nederland 1870–1919* (Amsterdam: Wereldbibliotheek, 2005), 14–15.

4 For the role of disasters in Dutch nation-building, see Lotte Jensen, 'Floods as Shapers of Dutch Cultural Identity. Media, Theories and Practices', *Water History* 13: 2 (2021), 217–33. In my PhD dissertation I have shown that major disasters are important milestones in the history of nineteenth-century Dutch nation-building: Fons Meijer, *Verbonden door rampspoed. Rampen en natievorming in negentiende-eeuws Nederland* (Hilversum: Verloren, 2022). The following contribution is based on the research I have conducted for the fourth chapter of this book.

5 Joep Leerssen, *Nationalisme* (Amsterdam: Amsterdam University Press, 2015), 55. This is also analysed in Ann Rigney, 'Embodied Communities. Commemorating Robert Burns, 1859', *Representations* 115 (2011), 71–101.

of the pioneers of modern nationalism studies, Benedict Anderson, noted: 'To understand [nation-ness and nationalism] properly we need to consider carefully how they have come into historical being, in what ways their meanings have changed over time, and why, today, *they command such profound emotional legitimacy*'.[6] Only if we answer the latter question can we fully grasp why millions of individuals over the last two centuries have been willing to die for the sake of an abstract – or imagined – community such as the nation.

Yet only more recently, under the influence of the 'emotional' or 'affective turn' in the humanities and social sciences, have nationalism historians started to analyse more thoroughly the emotional dimensions of national identification processes, as is illustrated by the publication of *Emotions and Everyday Nationalism in Modern European History* in 2020.[7] Edited by historians Andres Stynen, Maarten Van Ginderachter, and Xosé Manoel Núñez Seixas, the volume examines the various ways in which emotions are integral to understanding the everyday pull of nationalism on ordinary people in the eighteenth, nineteenth, and twentieth centuries.[8]

This chapter focuses on the authors of two types of nineteenth-century mass media genres: occasional poetry and eyewitness reports in the periodical press. While the first genre was especially popular in the first decades of the nineteenth century, the second genre became influential towards the end of the century. I argue that the authors of these texts gave their audiences access to the suffering of compatriots in times of major disaster and contributed to Dutch nation-building by engaging their readers' affective lives. The consumers of these texts learned how to feel compassion for people with whom they had little in common but their nationality. This did not mean that authors simply *stirred* and *aroused* feelings of compassion, but they also *shaped* this emotion. They provided their readers with a vocabulary that allowed them to make sense of their affective lives.

The nineteenth-century authors' efforts to bolster a culture of compassion with unfortunate and disaster-struck compatriots can be deemed

6 Benedict Anderson, *Imagined Communities. Reflections on the Origin and Spread of Nationalism* (London and New York: Verso, 2006), 4 (my italics).

7 Not only historians but also nationalism scholars have turned their attention to the relationship between nationalism and emotions. See, for example, Jonathan Heany, 'Emotions and Nationalism. A Reappraisal', in Nicolas Demertzis (ed.), *Emotions in Politics. The Affect Dimension in Political Tension* (London: Palgrave Macmillan, 2013), 243–63; Sinem Adar, 'Emotions and Nationalism. Armenian Genocide as a Case Study', *Sociological Forum* 33 (2018), 735–56.

8 Andreas Stynen, Maarten Van Ginderachter, and Xosé M. Núñez Seixas (eds), *Emotions and Everyday Nationalism in Modern European History* (Abingdon: Routledge, 2020).

an example of what Michael Billig has called 'banal nationalism'. This concept refers to all representations of the nation that only tacitly build a collective sense of national belonging.[9] The pitiful texts studied in this chapter were not really *explicit* nation-building endeavours: they did not overtly propagate compassion as an emotion that was typically Dutch, or even as an emotion that was solely – or predominantly – felt towards one's countrymen. All the same, the *contexts* in which they were produced meant that these texts contributed to the consolidation of the Dutch nation-state as a self-evident frame of reference. As such, they implicitly contributed to the process of nation-building in the Netherlands. Indeed, they were concerned with the victims of *national* disasters and with the suffering of *Dutch* citizens. All in all, these texts silently but boldly reinforced feelings of national unity and legitimised the notion of the Netherlands as a community of solidarity.

Compassion: A Historical Phenomenon

It was not self-evident that nineteenth-century Dutchmen would be aware of the hardships their compatriots faced in times of major disaster. Most disasters occurred tens or hundreds of miles away, while, for most contemporaries, their habitat did not encompass the entire country but rather the specific sub-region where they lived. Geographically speaking, the Netherlands was an 'archipelago of regions and communities', at least before the dawn of the railways during the middle of the century. This meant that there was little chance that an ordinary citizen would regularly meet a fellow Dutchman who resided in the other side of the country.[10]

Then as now, most disasters were events that directly affected only inhabitants of one specific city or region. Compassion towards compatriots in disaster areas was not the result of people actually knowing each other but instead originated from the process of reading about – and seeing – these victims in various sorts of texts. Media made it possible to bridge the spatial divide between, on the one hand, those who suffered the consequences of disasters, and, on the other, those who did not. It was, in short, only because

9 Michael Billig, *Banal Nationalism* (London: Sage, 1995).

10 Hans Knippenberg and Ben de Pater, *De eenwording van Nederland. Schaalvergroting en integratie sinds 1800* (Nijmegen: SUN, 1990), 27–42. An example: at the dawn of the nineteenth century, it would take an inhabitant of Haarlem approximately seven hours to travel to Leiden, while the train would eventually make it possible to cover this distance in less than an hour: Carl Koopmans and Rogier Lieshout, 'Spoorwegen en groei', *ESB* 93 (2008), 301–2.

media raised awareness about catastrophes that people throughout the country came to learn about the misery of the victims.

By means of producing what historian Thomas Laqueur has called 'humanitarian narratives', the authors of these texts mobilised readers into affective identification with 'unknown' disaster victims in their country.[11] The notion of humanitarian narratives refers to texts that made visible the suffering of others, prompted a physical reaction from their readership, and generated moral indignation. Despite what the word 'narrative' seems to suggest, humanitarian narratives were not necessarily narrations with a well-defined storyline. Laqueur has rather categorised them as rhetorical forms that consisted of three recurring motifs. Firstly, these narratives relied on details as signs of truth. Authors overwhelmed the reader with specific information or minute observations about the hardships faced by individuals. As such, they engendered a 'reality effect'. Readers became aware of the fact that the misery took place in the same world they were living in. Secondly, authors of humanitarian narratives emphasised the corporality of the suffering subjects. Detailed depictions of corporal suffering affirmed the bodily commonality between reader and victim. They demanded that the reader physically identify with the described torment – *as if* it were happening to their own bodies. Thirdly, humanitarian narratives exposed causality and the possibility of human agency. They presented explicit pointers to stop the specific suffering they had described and, as such, represented intervention as a moral imperative. Their readers should not sit idly by but were instructed to take action, by signing a petition or by donating money. In turn, this realisation reinforced the readers' affective identification: aware of the fact that it was their decision as to whether the suffering was to stop, they were likely to feel even more connected to the fate of the unfortunate souls they were reading about.[12]

Historical research has shown that texts that made use of humanitarian narratives – novels, medical case studies, investigation reports – have played a critical part in several Dutch, nineteenth-century reform movements, such as the movements against slavery, vivisection, and child labour or the ethical movement in Dutch colonial politics.[13] Contemporar-

11 Thomas Laqueur, 'Bodies, Details, and the Humanitarian Narrative', in Lynn Hunt (ed.), *The New Cultural History* (Berkeley: University of California Press, 1989), 176–204.

12 Laqueur, 'Bodies, Details', 176–78.

13 Amanda Kluveld-Reijerse, *Reis door de hel der onschuldigen. De expressieve politiek van de Nederlandse anti-vivisectionisten* (Amsterdam: Amsterdam University Press, 2000); Maartje Janse, *De afschaffers. Publieke opinie, organisatie en politiek in Nederland 1840–1880* (Amsterdam: Wereldbibliotheek, 2007); Maartje Janse, 'Representing Distant Victims. The Emergence of an

ies could no longer close their eyes to the pain and misery that was being caused elsewhere and had to take a stance against the practices that made them possible, or at least this was the aim of the idealist authors of such texts.

Many types of texts that Dutch authors produced in the wake of major disasters were also meant to open readers' eyes to the hardship of others. It was with good reason that the authors of these texts often made use of theatre metaphors, just as Teissedre l'Ange had done when he spoke of the 'scene of disaster' in his sermon about the Leiden gunpowder disaster. The members of the audience had to feel as if they were actual spectators of the described suffering. Moreover, just as with the humanitarian narratives of reform movements, many disaster texts were meant to mobilise readers into taking action. As will become clear in the following paragraphs, the idea was that readers would become convinced of the societal and moral importance of charitable action and eventually would make a donation for the benefit of their compatriots in the disaster area.

Yet Laqueur does not offer any tools to analyse how humanitarian narratives actually gave shape and meaning to compassion. Ever since Laqueur first wrote about humanitarian narratives in 1989, significant changes have occurred in the ways in which historians think about emotions. Under influence of the emotional turn, historians of emotion have started arguing that emotions are not universal, timeless phenomena, but rather that emotions have a historical dimension. Each and every community and time has its own ideals, theories, and beliefs regarding specific emotions and their expression, as well as specific emotional norms and words.[14] This also applies to compassion: what this emotion meant to people, what they felt when they felt compassion, and how they expressed it differed considerably from time to time and from place to place.[15] What is more: just like other emotions, compassion had a strong cognitive component. Contemporaries, therefore, had to learn what compassion was before they could feel it.

In this article I show how the humanitarian narratives that nineteenth-century authors produced contributed to this process of learning compassion. To explain this, I make use of historian William Reddy's concept of

Ethical Movement in Dutch Colonial Politics, 1840–1880', *BMGN – Low Countries Historical Review* 128 (2013), 53–80. Hanneke Hoekstra has also argued that humanitarian narratives were the boosters of a culture of civic idealism: *Het hart van de natie*, 31–32.

14 Barbara H. Rosenwein, 'Worrying about Emotions in History', *American Historical Review* 107 (2002), 821–45

15 Aleida Assmann and Ines Detmers, 'Introduction', in Aleida Assmann and Ines Detmers (eds), *Empathy and Its Limits* (Basingstoke: Palgrave Macmillan, 2016), 1–17.

'emotive'.[16] In short, emotives are speech acts directed at interpreting an individual's affective life, yet at the same time they influence the perception of this person's affective life. Expression organises the experience: our affective lives often are 'amorphous and unintelligible ... until [they] have been shaped by mental attention'.[17] Emotives provide cognitive frameworks that allow persons to relate to their affective lives. A woman, for example, is experiencing certain bodily sensations and yells: 'I am angry!'. This statement is an effort to *describe* her affective life but simultaneously navigates the ways in which she *understands* her affective life. In other words, the statement 'I am angry' allows the individual to actually *become* angry. Historian Monique Scheer has deepened our understanding of emotives by arguing that not only speech acts but also every social interaction (she uses the notion 'emotional practice') contribute to the ways in which individuals learn a certain emotion.[18] The fact that the aforementioned woman for example *yells* that she is angry, and maybe also experiences trembling hands, contributes just as much to the way in which she learns how to experience anger as the actual content of her remark.

Many texts that consisted of humanitarian narratives about the misery of compatriots in the disaster area employed emotives. Authors addressed compassion as much as they addressed suffering: they introduced characters that explicitly felt compassion, made it known that they themselves felt compassionate towards disaster victims, or emphasised that their readers were to experience this emotion. They used various Dutch emotion words that characterised a sense of fellow feeling interchangeably: apart from compassion (*medelijden*), they used words such as mercy (*mededogen*), charity (*mensenliefde*), or pity (*deernis*).[19] At the same time, they often promoted

16 William M. Reddy, *The Navigation of Feeling. A Framework for the History of Emotions* (Cambridge: Cambridge University Press, 2001), 104–11.

17 This interpretation of Reddy's concept comes from Monique Scheer, 'Are Emotions a Kind of Practice (and Is That What Makes Them Have a History)? A Bordieuan Approach to Understanding Emotion', *History and Theory* 51 (2012), 193–220, at 212.

18 Scheer, 'Are Emotions', 212–14.

19 Writing about the eighteenth century, Dorothée Sturkenboom has argued that various words that authors used to characterise compassion were 'to a high extent ... used interchangeably': *Spectators van de hartstocht. Sekse en emotionele cultuur in de achttiende eeuw* (Hilversum: Verloren, 1998), 300. Nevertheless, it must be acknowledged that each emotion word has its own history and connotations and refers to different kinds of fellow feeling. 'Compassion', for example, has different conceptual origins than related words such as 'empathy', 'sympathy', or 'pity'. See for example: Assmann and Detmers, 'Introduction', 3–4; Rob Boddice, *The Science of Sympathy. Morality, Evolution, and Victorian Civilization* (Urbana: University of Illinois Press, 2016), 3–6.

Figure 4.1 Reinier Vinkeles, after Cornelis van Hardenbergh, *The Destruction of the Village of Beusichem after the River Flood of 1809*, 1809, etching and engraving, 25.2 × 34.1 cm, included in a commemoration book that was published after the disaster. Rijksmuseum, Amsterdam, RP-P-1944-1855

certain ideas about compassion, about the nature of this emotion and how it should be evaluated and expressed. As such, these authors consolidated an emotional norm and provided to their readers a vocabulary that allowed them to become aware of, give meaning to, and express their compassion. All in all, they bolstered a culture of compassion with suffering compatriots and encouraged their readers to participate in this culture.

Various genres played their part in boosting and bolstering this culture. Commemoration books, for example, were a type of text that, from the late eighteenth century onwards, provided extensive examples of suffering of disaster victims.[20] After major national disasters, especially flooding disasters, these books overwhelmed their readership with factual and specific information about all the forms of mayhem the disaster had caused in all the affected towns. Moreover, these books included prints that made visible these cases of suffering, all the more enhancing readers' identification with the

20 For more information about this genre, see Petra J.E.M. van Dam and Harm Pieters, 'Enlightened Ideas in Commemoration Books of the 1825 Zuiderzee Flood in the Netherlands', in Pepijn Brandon, Sabine Go, and Wybren Verstegen (eds), *Navigating History. Economy, Society, Science and Nature. Essays in Honor of Prof. Dr. C.A. Davids* (Leiden and Boston, MA: Brill, 2018), 275–97. See also the contribution by Adriaan Duiveman to the present volume.

misery of their compatriots in the disaster area (see Fig. 4.1).[21] The remainder of this contribution, however, singles out and analyses two other genres that – because of their popular appeal and widespread dissemination – made an even deeper impression with regards to boosting and shaping nationwide compassion: occasional poetry and eyewitness reports in news media.

Shaping Compassion Through Occasional Poetry

Throughout the nineteenth century, the Netherlands had a flourishing societal life. In major and medium-sized cities – both in the urbanised west and the more rural east – the infrastructure of community life was made up of one or more voluntary societies, such as scientific, cultural, and reform associations.[22] Literary scholar Marleen de Vries has shown that the ideology behind many of those associations was centred around notions of civilisation, societal virtuousness, and patriotism: members were expected to feel responsible for the national, common good.[23] Especially in the first half of the century, members would frequently give recitals during the meetings of such associations, often in the form of poetic readings.

They believed the ideal of civilisation was to be achieved more smoothly if authors of recital texts did their best to capitalise on the affective lives of the attending listeners.[24] Not surprisingly, many speakers at association meetings would make use of humanitarian narratives in recitals about catastrophes that had just happened, for this encouraged their audiences to feel responsible for the fates of their compatriots in the affected area. Afterwards, many of these poems were published and sold at modest prices on behalf of the relief of the disaster victims, which considerably increased the number of people that would eventually hear or read one of the versions of the disaster poem.[25]

21 A more exhaustive analysis of commemoration books as the boosters of a culture of compassion with suffering compatriots can be found in the fourth chapter of Meijer, *Verbonden door rampspoed*.

22 Boudien de Vries, 'Voluntary Societies in the Netherlands 1750–1900', in Graeme Morton, Boudien de Vries, and R.J. Morris (eds), *Civil Society, Associations and Urban Places: Class, Nation and Culture in Nineteenth-Century Europe* (Aldershot and Burlington: Ashgate, 2006), 103–16.

23 Marleen de Vries, *Beschaven! Letterkundige genootschappen in Nederland 1750–1800* (Nijmegen: Vantilt, 2001). For an analysis of association culture of the nineteenth century, see Willem van den Berg and Piet Couttenier, *Alles is taal geworden. Geschiedenis van de Nederlandse literatuur 1800–1900* (Amsterdam: Bert Bakker, 2009), 45–53.

24 De Vries, *Beschaven*, 118–21.

25 For an introduction to the nineteenth-century genre of charitable publications, see Marita Mathijsen, 'Ten voordele van... Liefdadigheidsuitgaven in de negentiende eeuw', *De moderne tijd* 4 (2020) 234–58.

An example of such a recital was a poem that Robert Arntzenius read during two association meetings in The Hague in February 1807, weeks after the Leiden gunpowder disaster. He dwelled extensively on all the calamities the explosion had caused. Immediately in the first verse, Arntzenius emphasised that it had been 'heart-wrenching compassion' ('hartverted'rend mededogen') that had incited him to take up his pen.[26] By using this emotive, he set the tone for the poem to come.

As was fitting for a genuine humanitarian narrative, Arntzenius furthermore explicitly emphasised the corporeal dimension of suffering when speaking about the specific cases of the misery victims of the gunpowder disaster had endured.[27] The poem, for example, contained a rather disturbing anecdote about a young, pregnant woman, who was crushed by her collapsing house, causing her to have a miscarriage. Arntzenius accentuated the corporality of both the woman and the stillborn baby. He recited how 'the neck, the chest and the head' of the woman were damaged by the walls of her house, how the baby was forced from the woman's 'intestines', and how the little girl had no 'blush on her cheeks' anymore.[28] Despite these descriptions of human suffering, Arntzenius decided to end his recital on a more positive note. He expressed the hope that Leiden would soon be rebuilt and that later generations of its inhabitants could hardly imagine that all the misery he had described had actually taken place. On the contrary, they would only feel pride in the fact that their compatriots had helped them out. 'In these times of hardship', Arntzenius remarked, 'The fatherland took care of Leiden; / And God took care of the Fatherland'.[29]

The question is, however, whether poems such as these can be categorised as humanitarian narratives. Authors often staged rather stereotypical characters, especially if their poems concerned the misery caused by flooding disasters. Over and over again, young mothers witnessed their children dying, fathers failed to save their families, old men succumbed after long lives, and farmers lost the cattle that had secured their livelihoods. While authors in their poems often relied on the body of the victims as a means to enable a physical identification between sufferer and listener, their stories still come across as sentimental clichés to modern-day readers.

26 R.H. Arntzenius, *Dichterlijk tafereel der stad Leijden, in den avond en nacht van den 12den van louwmaand, 1807* (The Hague: Gebr. Van Cleef, 1807), 5.

27 Other poems that were read in the wake of the Leiden gunpowder explosion did the same: E.J.B. Schonck, *Treurzang op het rampspoedig lot der stad Leyden, van den 12 Januarij 1807* (Nijmegen: J.C. Vieweg, 1807); Johannes Wilhelmus Bussingh, *De weldadigheid, lierzang* (Leiden: Herdingh en Du Mortier, 1807).

28 Arntzenius, *Dichterlijk tafereel*, 15.

29 Arntzenius, *Dichterlijk tafereel*, 23.

Figure 4.2 A. Lutz, after Jan Willem Pieneman, *Leiden after the Gunpowder Explosion in January 1807*, 1809, hand-coloured aquatint and etching, 66.5 × 86 cm. Rijksmuseum, Amsterdam, RP-P-OB-88.824

Obviously, these poems were inspired by sentimentalism, a genre that had been particularly popular among novelists at the end of the eighteenth century but which would also leave its rhetorical mark on other forms of communication during the nineteenth century. Sentimental rhetoric was aimed at arousing compassion and tended to focus on the feelings of its characters.[30] The sentimentalist inclination of the association members, however, did not mean that their pitiful disaster poems were mere fiction. In some cases, the described suffering was based on real events. Arntzenius's story about the miscarriage, for example, can be traced back to an incident

30 Janet Todd, *Sensibility. An Introduction* (London and New York: Methuen, 1986). For sentimentalism in the Netherlands, see, for example, Annemieke Meijer, *The Pure Language of the Heart. Sentimentalism in the Netherlands 1775–1800* (Amsterdam: Rodopi, 1998). Other forms of communication that were inspired by sentimental rhetoric were, for example, sermons and parliamentary speeches: Herman W. Roodenburg, 'Tranen op het preekgestoelte. De achttiende-eeuwse kanselwelsprekendheid tussen toneel en authenticiteit', *De achttiende eeuw* 41 (2009), 15–32; Edwina Hagen, 'Fashioning the Emotional Self. The Dutch Statesman Rutger Jan Schimmelpenninck (1761–1825) and the Cult of Sensibility', *BMGN – Low Countries Historical Review* 129 (2014), 138–62.

that had most likely actually happened in Leiden.[31] Disaster poems were perhaps not always *true*, yet they always aimed to be *truthful*. The main idea of literature of that age was that stories were not imitations of reality per se but rather compositions on the basis of authentic elements.[32] Even though there was a thin line between truthfulness and cliché – literary critics would regularly attack disaster poems for their hackneyed use of sentimental rhetoric – this postulate meant that listeners could accept that events *such as these* had taken place.[33]

Through their poetic readings, authors raised awareness about the fact that, elsewhere in the Netherlands, individuals were suffering the results of disaster and were deserving of compassion. Simultaneously, they frequently used emotive expressions and gave shape and meaning to feelings of compassion. Like Arntzenius, who had claimed that he himself was feeling 'heart-wrenched compassion', authors of disaster poems provided their audiences with a vocabulary that allowed them to articulate, understand, express, and evaluate the bodily arousals they experienced when listening to or reading about the misery of their compatriots.

Two notions played a key role in the culture of compassion with compatriots these authors propagated. First of all, authors repeatedly accentuated that people should express compassion by giving free rein to their tears. Author and medical doctor Nicolaas van der Hulst, for example, staged a potential benefactor in a poem that he wrote about the river inundations of 1809 and recited at an association meeting in Rotterdam. The staged man witnesses the destruction of a river inundation but cannot help the victims swallowed by the water. Van der Hulst wrote that the man, as a consequence, 'feels how his heart is torn apart' and 'cries, for he cannot help'.[34] The author here used the imagery of heart and tears to convey the idea that compassion, like other emotions, was felt from the *inside* and subsequently should be expressed from the *outside*.[35] The tears flowed directly from the heart, so

31 Peter van Zonneveld, 'Leiden: het Rapenburg. De buskruitramp van 12 januari 1807', in Jan Bank and Marita Mathijsen (eds), *Plaatsen van herinnering. Nederland in de negentiende eeuw* (Amsterdam: Bert Bakker, 2006), 37–47, at 38.

32 De Vries, *Beschaven*, 125–31. In the preface to a published poem about river inundations in January 1807, one author even explicitly acknowledged that it was this idea that formed the basis of his story: T. van Limburg, *De watervloed van 1809* (s.l.: s.pub., 1809), 6–7.

33 For this criticism, see, amongst others, Lotte Jensen, *Wij tegen het water. Een eeuwenoude strijd* (Nijmegen: Vantilt, 2018), 23–26; Mathijsen, 'Ten voordele van…', 253–57.

34 N. van der Hulst and W. Smits, *De overstrooming in Gelderland, dichtstuk, benevens eene bijdrage op hetzelfde onderwerp* (Rotterdam: Cornel, Krieger & Van Balen, 1809), 9.

35 This older idea played a key role in the 'cult of sensibility', which was especially popular in the eighteenth century. See, for example, Paul Goring, *The Rhetoric of Sensibility in Eighteenth-Century*

to speak. Compassion was also presented as an authentic emotion, which meant that it neither could nor should be faked. Conversely, individuals who felt compassion neither could nor should try to hold back their tears.

Secondly, many authors of disaster poems represented compassion as an emotion that should always lead to action. Historian of emotions Ute Frevert has aptly summarised: 'Feeling and acting were perceived as two sides of the same coin. He who just felt pity for a person without coming to their rescue or alleviating the other's suffering was deemed a weakling'.[36] After river inundations in 1820, Rotterdam brewer Willem Messchert wrote a poem in which he staged a personification of charity in order to characterise the relation between compassion, tears, and charitable behaviour:

> Charity opens her benevolent hands to help:
> The damage is being repaired, what has been lost is being compensated;
> And if her gold cannot stem the source of sorrow,
> She has a tear in her eyes to comfort the aching mind.[37]

Messchert represented compassion – and tears – as signs not of weakness but rather of helpfulness and willingness to take action.[38] Other authors did the same, for example by writing that both heart and hands had to be unlocked. This metaphorical expression served to communicate that compassion always had to go together with active help – the one simply could not do without the other.[39]

Some authors explicitly linked this idea to a sense of national belonging. They invited their audiences to identify as Dutchmen who had to help other Dutchmen. Case in point is, again, a verse from the poem van der Hulst wrote after the 1809 river inundations:

> Hurry, hurry, my Countrymen!
> Come and unlock your hearts and hands;

Culture (Cambridge: Cambridge University Press, 2004).

36 Ute Frevert, 'Empathy in the Theatre of Horror, or Civilising the Human Heart', in Assmann and Detmers (eds), *Empathy and Its Limits*, 79–99, at 82.

37 'Menschlievendheid ontsluit haar milde hand tot helpen: / De schade wordt hersteld, 't verlorene vergoed; / En kan haar goud de bron van 't zieleleed niet stelpen, / Zij heeft een traan tot troost van het afgepijnd gemoed'. W. Messchert, *De watersnood* (Amsterdam: M. Westerman, 1820), 28.

38 This is also argued in Sturkenboom, *Spectators*, 313–17.

39 Toussyn Woordhouder, *Het karacter van een oprecht patriot, herïnnert bij den waternood van 't Koningrijk Holland, in den Jare 1809* (Rotterdam: J. van Santen, 1809); Johannes Wilhelmus Bussingh, *Aan Nederland* (Rotterdam: J. Pippijn, 1825), 13.

Oh yes! already does charity
Happily ignite all the gifts,
While you, with raised hands
And in tears, help your brothers.[40]

The prototype that van der Hulst represented was that of the compassionate, tearful, and charitable Dutchman who felt morally responsible for the fates of individuals with whom he had little in common except his nationality. As such, he tacitly bolstered the notion of the Netherlands as a community of solidarity, made up of compatriots who affectively identified with one another and helped each other out in times of hardship.

Witnessing Suffering

Disaster poems and charitable publications had their heyday in the first half of the nineteenth century. In the second half of the century, eyewitness reports would become the leading genre with regards to bolstering a nationwide culture of compassion in times of major disaster. Who could better manage to open the eyes of audiences to the misery of others than someone who had in fact witnessed this misery themselves? Eyewitness reports were especially published in periodical press, most notably newspapers and illustrated magazines.[41] The possibilities for periodicals using eyewitness reports expanded after 1850: with the advent of trains and telegraph machines, periodicals were increasingly able to report from the disaster area within only one or two days. These were also the decades in which these media, due to a variety of factors, caught the attention of an ever-larger share of the Dutch public.[42]

As the genre of eyewitness reports published in periodicals became prominent in the second half of the nineteenth century, this article now

40 'Snelt aan, snelt aan, mijn Landgenooten! / Komt, mildlijk hart en hand ontsloten; / O ja! menschlievendheid ontsteekt / Alreeds blijmoedig de offeranden, / Terwijl met opgeheven handen / Uw broeder, snikkend, uitkomst smeekt'. Van der Hulst and Smits, *De overstrooming*, 12.

41 Sometimes they were also published in the form of individual booklets. Two examples are *Herinnering aan onzen togt naar eenige Noord-Hollandsche dorpen, overstroomd op den 4den Februarij 1825* (Amsterdam: M. Westerman, 1825) and B.S., *Twee dagen in Veenendaal gedurende den watersnood in maart 1855* (Dordrecht: J.P. Revers, 1855).

42 For an extensive overview of these developments, see the chapters by Remieg Aerts, Frank Harbers, and Thomas Smits, in Huub Wijfjes and Frank Harbers (eds), *De krant. Een cultuurgeschiedenis* (Amsterdam: Boom, 2019).

takes a leap forward in time. Moving from the early decades of the century, this last section focuses on a specific flooding disaster that happened in December 1880. The reporting of this disaster perfectly illustrates the nature and dynamics of eyewitness reports. On the penultimate night of the year of 1880, a river dyke broke, causing the Meuse to flood several polders and towns in the province of North Brabant. The town of Nieuwkuijk had been particularly strongly impacted, hence the name given to the disaster afterwards: the Disaster of Nieuwkuijk.[43]

Catholic newspaper *De tijd* took the lead. In the first weeks of 1881, this newspaper published numerous reports from various towns in and around the flooded region. Eyewitnesses recurrently described to the readers of *De tijd* what they had seen: the damage that had been done by the streaming water, inhabitants of the region who had fled to higher areas – their roofs or a nearby hill or dyke – and large numbers of refugees who were given sanctuary in nearby towns such as 's-Hertogenbosch. It is not surprising that *De tijd* was most ambitious in publishing eyewitness reports. In North Brabant, the Catholic newspaper could rely on a broad network of informants. What is more, they probably sensed their readers were especially interested to know what had happened to their fellow-believers in the Catholic south. However, *De tijd* was not the only news medium that reported on the disaster: newspapers and magazines such as *Nieuwe Rotterdamsche courant, Het nieuws van den dag*, and *Eigen haard* published one or more eyewitness reports from the affected region as well.[44] Newspaper *Algemeen handelsblad* and illustrated magazine *Katholieke illustratie* went a step further by sending their own reporters to Nieuwkuijk and giving them the task of writing about their experiences.[45]

The eyewitnesses made it clear that their ultimate goal was to make visible the situation in the flooded region. Some of them used theatre metaphors to express this: while describing specific events or providing a description of the atmosphere, they spoke, for example, of a spectacle ('schouwspel') and of

43 Francien van den Heuvel, *'s-Hertogenbosch, eiland in een onmetelijke zee. Omgangsstrategieën van de stedelijke overheid en de bevolking van 's-Hertogenbosch met hoogwater en overstromingen (1740–1995)* (Oisterwijk: Wolf Publishers, 2021), 254–59.

44 *Nieuwe Rotterdamsche courant* (2 January 1881); *Het nieuws van den dag* (4 January 1881); *Eigen haard* 7 (1881), 52–54.

45 For more information about the dawn of this genre, see Frank Harbers, 'Between Personal Experience and Detached Information. The Development of Reporting and the Reportage in Great Britain, the Netherlands and France, 1880–2005', PhD dissertation, Rijksuniversiteit Groningen, 2014.

scenes of sadness and misery ('tooneel van ellende', 'tooneel vol jammer').[46] Furthermore, the eyewitnesses emphasised their role as observers. They purposely described what they *saw* or what they *heard*. The reporter of *Algemeen handelsblad* even explicitly invited his readers to watch *along with* him in the perilous accommodation of a group of refugees: 'Oh, throw a glance with me into the hut, which has been built so hastily out of straw, in the midst of this watery pond'.[47] As such, authors allowed the readers to experience the scene via the reporter's senses, so that the readers could imagine themselves being eyewitnesses as well.

The readers of such eyewitness reports were not to count on the objective or sober stories many of us nowadays associate with good journalism. These nineteenth-century reports were very personal accounts, meant to move the readers and engage their affective lives. In order to so, they made use of the same rhetoric the authors of disaster poems had used as well: they zoomed in on the anguish and misery of the inhabitants of the disaster area. One author, for example, wrote about the situation in Nieuwkuijk in the days after the dyke breach: 'Throughout the entire town, grim cries for help could be heard. Out of windows and from roofs, one could see entire households extending their arms, and hear unfortunates begging for aid'.[48] What is more, the authors accentuated the various forms of corporeal suffering, which made these eyewitness reports genuine humanitarian narratives. They used, for example, the image of freezing bodies to pave the way for readers to physically identify with the suffering of flood victims.[49] Most nineteenth-century Dutchmen knew only too well the bodily torment that could come from the winter cold. In addition, authors frequently described the great deal of hunger they had witnessed among the various flooding victims.[50]

In addition to their senses, the authors of these reports also put their own affective lives at the readers' disposal. They described how they emotionally experienced everything they witnessed in the vicinity of Nieuwkuijk. They

46 See, amongst others, *De tijd* (4 January 1881); *Algemeen handelsblad* (4 January 1881); *Nieuwe Rotterdamsche courant* (2 January 1881); *Algemeen handelsblad* (7 January 1881); *Katholieke illustratie* 14 (1881), 147.

47 'Och, sla met mij een blik in dat van riet en stroo en palm in allerijl opgeslagen hok te midden van dien polderpoel'. *Algemeen handelsblad* (4 January 1881).

48 'Allerwege verhief zich een akelig noodgeschrei. Uit de vensters en boven op de daken zag men gehele gezinnen de armen uitsteken, en hoorde men de ongelukkigen om hulp smeeken'. *Het nieuws van den dag* (4 January 1881)

49 *Katholieke illustratie* 14 (1881), 147; *Algemeen handelsblad* (4 January 1881).

50 *Nieuwe Rotterdamsche courant* (2 January 1881); *Algemeen handelsblad* (6 January 1881).

often used heart metaphors to give substance to these emotive expressions, as a means to characterise how they were deeply touched by the suffering of the inhabitants of the region: they wrote how their hearts were either 'wrenched' or 'moved' by the various scenes they observed ('hartverscheurend', 'hartroerend').[51] And not only that: several eyewitnesses labelled their emotional response as 'compassion' or even prescribed how compassion ought to be expressed. 'Go there', one author wrote about his visit to Nieuwkuijk, 'Surely, your eye will turn wet and you will compassionately avert your eyes as soon as you set foot in the town'.[52] Readers could use these emotive expressions to make sense of their own affective lives as well, to grasp the emotions they experienced when reading about the unfortunates in Nieuwkuijk as a sign of compassion.

Like the poems analysed in the previous paragraph, the key idea of the humanitarian narratives these eyewitnesses produced was that readers would not wallow in their compassion but rather that they would convert their feelings into charitable behaviour. One author explicitly articulated this idea and linked it to the notion of the Netherlands as a community of solidarity. So far, the author from the town of Waalwijk (near Nieuwkuijk) had repeatedly reported about the misery in the flooded region. In *De tijd*'s edition of 10 January 1881, he more or less directly addressed the reader: 'Many houses have been devastated, much happiness has been put to an end, much prosperity has been destroyed – how can we repair this, even in a small degree, without all humanitarians in the Netherlands joining forces?' Here he placed compassion in a national framework, since he explicitly wrote about humanitarians *in the Netherlands*. He went on: 'That is why I want to call for the aid of our wealthy brothers, who know the word "flood" and perhaps can imagine what a flood looks like, but whose imagination still cannot match what has really happened; our fate is put into their hands, they have to help us'.[53] The author represented charitable behaviour as something that originates from imagination, from the extent to which individuals are able to

51 *De tijd* (4 January 1881); *De tijd* (6 January 1881).

52 'Gaat er thans heen … [Stellig] wordt uw oog dan vochtig en wendt ge u met deernis en innig medelijden af, zoodra ge het plaatsje betreedt'. *Eigen haard* (1881), 52.

53 'Zoovele woningen weggevaagd, zooveel geluk geknakt, zooveel welstand vernietigd – hoe dat alles, zij het dan ook slechts in geringe mate, te herstellen, zoo niet alle menschenvrienden in Nederland de handen in elkander slaan. Daarom voeg ik mijn stem bij zoo velen, om nogmaals een beroep te doen op onze bevoorrechte broeders, aan wie de watersnood niet dan bij naam bekend is, die zich ja misschien wel een watersnood kunnen verbeelden, maar wier verbeelding toch nog steeds beneden de werkelijkheid zal blijven; op hen is onze hoop gevestigd, zij moeten ons ter hulp snellen'. *De tijd* (6 January 1881).

visualise for themselves the amount of misery others have to suffer. Through his eyewitness reports, the author from Waalwijk – as well as the other authors in *De tijd* and the authors in other periodicals – aimed at providing the means for readers to form a clear notion of the situation in Nieuwkuijk so that they would start feeling compassionate towards their unfortunate compatriots and, as a consequence, take action to stop their suffering.

Concluding Remarks

The nineteenth-century Netherlands experienced many physical, social, and religious divides. Some authors hoped to bridge those divides – if only for a brief period – by producing texts that invited their readers to affectively identify with the victims of disaster. This chapter has shown that, through disaster poems and eyewitness reports, contemporaries were able to learn about the suffering that had been triggered by disasters elsewhere in the country. What is more, consumers of these texts were taught how to feel compassion with their contemporaries, or rather: what it entailed to feel compassionate towards a fellow Dutchman. The most important notion these texts conveyed was the notion that feeling and acting were two sides of the same coin. This notion was often presented with a nationalist flavour: they encouraged readers and listeners to convert their feelings into actions to help their compatriots. Feeling compassion was expected to be translated into acts of national belonging. As such, these texts laid the foundation for the notion of the Netherlands as a community of solidarity, whose members helped each other out in times of crisis and hardship. Indeed, through their mobilising appeal, these texts embedded the emotion of compassion firmly in the ways in which contemporaries identified with the nation and the national community.

About the Author

Fons Meijer is a historian. Between 2018 and 2022, he wrote his PhD thesis at Radboud University about the relationship between nationalism and processes of nation-building in the wake of major disasters in the nineteenth-century Netherlands. Previously, he studied history at Radboud University and the University of Sheffield (UK). He is specialised in the modern political and social history of Western Europe, with a specific interest in identity formation, political representation, and media culture.

5 Cultural Resilience during Nineteenth-Century Cholera Outbreaks in the Netherlands

Lotte Jensen

Abstract

The nineteenth century saw several severe outbreaks of cholera in the Netherlands. This chapter discusses how Dutch society responded to this new and frightening disease. It is argued that cultural responses played an important role in increasing societal resilience. Resilience was strengthened and shaped by cultural media, such as paintings, prose, poems, songs, prints, novels, sermons, and concerts. By offering moral and religious explanations, cultural responses helped people to make sense of the disruptive events and to cope with their fears and uncertainties. Cultural responses also aimed at fostering a sense of community by raising money for the sufferers. Citizens were called to action in the local and national newspapers, poems, songs, and sermons. This led to impressive amounts being collected in a short time.

Keywords

cholera – epidemic disease – the Netherlands – cultural resilience – relief – solidarity

In 1853 the well-known Dutch poet Hendrik Tollens published a beautifully decorated leaflet containing a short poem. By buying it for the price of 25 cents, citizens helped the cholera sufferers.[1] It was not only sold door-to-door,

1 This corresponds to approximately € 2.80 in the year 2021. See the calculator of the International Institute of Social History, https://iisg.amsterdam/nl/onderzoek/projecten/hpw/calculate.php (accessed 4 April 2023).

This chapter is based on Lotte Jensen, 'Culturele veerkracht. Lessen uit vroegere cholera-epidemieën voor de huidige coronacrisis', in Marieke Winkler, Marijke Malsch, and Jan Willem Sap (eds), *Nooit meer dansen? De veilige stad in tijden van pandemie* (Amsterdam: Boom 2021), 241–65.

Asperen, van, Hanneke and Lotte Jensen (eds): *Dealing with Disasters from Early Modern to Modern Times. Cultural Responses to Catastrophes.* Taylor & Francis Group, 2023
DOI: 10.5117/9789463725798_CH05

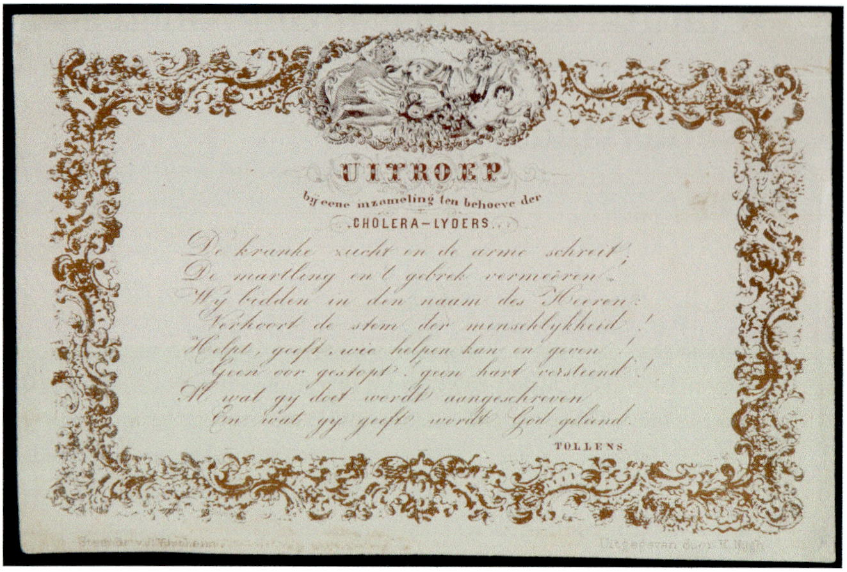

Figure 5.1 Decorated leaflet with a poem by Hendrik Tollens on behalf of the cholera sufferers, 1837–56, lithography, text in red with a golden border, published by H. Nijgh, 11.7 × 17.4 cm. Museum Rotterdam, 16316

but in several larger cities people could also leave their payments in collection boxes.[2] With his verses Tollens encouraged his countrymen to be as generous as possible: 'Help, give, who is able to give! / No ear may be deaf! No heart of stone!'[3] Tollens prayed in the name of the Lord that the voice of humanity would be heard. His campaign was quite successful: with his poem he raised 721 guilders (which corresponds to approximately € 7,870 nowadays).[4]

Tollens was not the only author who wrote about the cholera pandemic in a moving manner. Nationwide, citizens were encouraged to help the victims through songs, prose, sermons, prints, and broadsheets. In some cities, charity concerts were organised on behalf of the needy. Writers, artists, and musicians appealed to a sense of shared Dutch identity and reminded people that it was their religious duty to reach out to the distressed. In these artistic responses, there was also room for expressing emotions such as fear and grief or, as Tollens did, seeking consolation in religion.

2 *Nieuwe Rotterdamsche courant* (5 October 1853); *Algemeen handelsblad* (6 October 1953).

3 'Helpt, geeft, wie helpen kan en geven! / Geen oor gestopt! geen hart versteend'. H. Cz. Tollens, *Uitroep bij eene inzameling ten behoeve der cholera-lyders* (Rotterdam: H. Nijgh, 1853).

4 *Tollens begrafenis (25 October 1856). Uitgegeeven ten behoeve der oprigting van een gedenkteeken* (Rotterdam: H. Nijgh, 1856), 6.

In this chapter, I argue that cultural responses played an important role in increasing societal resilience in times of disaster, such as the nineteenth-century cholera outbreaks. 'Cultural resilience' is a broad concept, which derives from the social and psychological sciences, and considers 'how cultural background (i.e. culture, cultural values, language, customs, norms) helps individuals and communities overcome adversity'.[5] Here I focus on the way cultural resilience was strengthened and shaped by cultural media, such as paintings, prose, poems, songs, prints, novels, sermons, and concerts. By offering moral and religious explanations and by shaping a sense of community, cultural responses helped people to make sense of the disruptive events and to cope with their fears and uncertainties. Firstly, I will embed this research in the broader field of historical research on epidemics and offer some historical background on the cholera outbreaks in the Netherlands. Secondly, I will analyse how cultural responses were aimed at increasing societal resilience. Finally, I will tentatively draw some parallels between the past and the present.

Historical Approach

In the last two decades historical approaches have become more and more topical in the field of disaster studies. This emerged from the growing recognition that history provides a useful tool to better understand how societies deal with shocks and hazards, and which factors play a role in fostering their resilience. As historian Bas van Bavel and others have shown, the past may be used as a 'laboratory' to understand which societies are better at coping with crises than others.[6]

A heightened awareness of the relevance of historical disciplines can also be witnessed in the research on epidemics. Historians Frank Snowden and Samuel Cohn have convincingly shown that epidemics need to be studied from a broad, multidisciplinary perspective: the way people react to new diseases, the medical and political interventions, and the search for causes are deeply influenced by social and cultural circumstances. In *Epidemics and Society. From the Black Death to the Present* Snowden argues that epidemics have an enormous impact on societies: what starts as a medical problem

5 Caroline S. Clauss-Ehlers, 'Cultural Resilience', in Clauss-Ehlers (ed.), *Encyclopedia of Cross-Cultural School Psychology* (Boston, MA: Springer, 2010), https://doi.org/10.1007/978-0-387-71799-9_115.

6 Bas van Bavel, Daniel R. Curtis, Jessica Dijkman, Matthew Hannaford, Maïka de Keyzer, Eline van Onacker, and Tim Soens, *Disasters and History. The Vulnerability and Resilience of Past Societies* (Cambridge: Cambridge University Press, 2020).

may turn out to have huge social, political, and cultural consequences in the longer run. Snowden pays much attention to the representation of diseases in the arts: according to him, works of art reveal how societies reflected upon matters of life and death.[7] Cohn also points out the importance of contextualising contagious diseases in *Epidemics. Hate and Compassion from the Plague of Athens to AIDS*.[8] His case studies stem from different geographical settings and time spans but show certain patterns: outbreaks of new infectious diseases always went hand in hand with searching for scapegoats, political turmoil, and the rise of social reform movements. At the same time, there were tendencies to strengthen bonds within communities through religious practices, rituals, and charity. Cohn calls them 'mechanisms for unity'.[9]

Responses like these – from scapegoating to fostering solidarity – belong to the wide range of coping mechanisms that human beings develop when confronted with catastrophes such as floods, earthquakes, fires, and epidemics. In his study *Cultures of Disaster* (2003), Greg Bankoff differentiates three coping strategies: preventive strategies, strategies which reduce the material impact of disasters, and strategies which reduce psychological stress. His research focuses on the Philippines, where people have to live with the risks of floods and consequential outbreaks of diseases. As a consequence, they have developed a 'disaster culture', in which the reduction of risks prevails.[10] While the first two strategies entail practical measures (e.g. building with new materials, moving to other living areas), strategies to reduce psychological stress are to be sought at the emotional and cultural level. Here, the entire variety of cultural forms – ranging from religious practices and collections to musical events and poetry – played an important role: they not only offered an outlet for emotions such as fear and uncertainty but also provided comfort and meaning because they connected people. In addition, they helped to collect money to alleviate the direct needs.[11]

7 Frank M. Snowden, *Epidemics and Society. From the Black Death to the Present* (New Haven, CT and London: Yale University Press, 2020).

8 Samuel K. Cohn, Jr., *Epidemics. Hate and Compassion from the Plague of Athens to AIDS* (Oxford: Oxford University Press, 2018).

9 Ibid.

10 Greg Bankoff, *Cultures of Disasters. Society and Natural Hazards in the Philippines* (London: Routledge, 2003).

11 See also Beatrice de Graaf, Lotte Jensen, Rina Knoeff, and Catrien Santing, 'Dancing with Death. A Historical Perspective on Coping with Covid-19', *Risk, Hazards & Crisis in Public Policy* 12: 3 (2021), 346–67.

All three coping strategies can be witnessed during the cholera pandemics in the Netherlands. The country was hit by three waves: 1832–33, 1848–49, and 1866–67. At the end of June 1832, the first death cases were reported in the Netherlands. Over the next months, the number rose explosively: in the years 1832–33, a total of 10,000 people died of cholera in the province of Holland, of which 1,273 were in Amsterdam.[12] The outbreak in 1848–49 resulted in 22,708 dead (2,256 in Amsterdam). Then the illness remained latently present, until it erupted again in 1866–67, when approximately 20,000 people died of the disease.[13] The number of casualties in Amsterdam dropped significantly (to 1,104) thanks to the installation of water pipes. In 1854 the British scientist John Snow had discovered that drinking contaminated water played a large role in the spread of the disease After the 1867 outbreak, cholera disappeared in the Netherlands but it still circulated in other European countries. The discovery in 1883 of the cholera bacteria by the German physician Robert Koch, which proved the importance of good hygiene in preventing the disease, led to a further decrease in the number of victims.[14]

During the three waves which hit the Netherlands, the local governments took many measures: large gatherings and events such as fun-fairs and food markets were cancelled and health regulations were issued. During the first outbreak of 1832, the local authorities of Rotterdam published a list of forty-one actions to be taken.[15] They demanded that inhabitants keep their houses clean, ventilate regularly, and reduce the number of people in one room. They also advised against eating fish, fruit, and watery vegetables, in particular melons, cucumbers, and cabbage. People should avoid eating meat, because the animals could suffer from the same symptoms as human beings. Coffee, tea, and red wine, on the contrary, could be healing. Some pharmacists took advantage of the situation and offered 'cholera powder', which would strengthen resilience and keep the consumer in a good humour. The miraculous medicine was quite expensive: one bottle cost ƒ 1.25 (€ 13.41), which was far too much for the average citizen, let alone the poor.[16]

12 Janwillem Koten, 'Verloop en gevolgen van de cholera, 1817-1923', *NGV*, published online 3 January 2021, https://www.ngvnieuws.nl/4-verloop-en-gevolgen-van-de-cholera-1817-1923 (accessed 3 April 2023).

13 Ibid.

14 Today cholera outbreaks still occur, mostly after catastrophes such as earthquakes and floods, for example in Haiti in 2010.

15 *Rotterdamsche courant* (16 August 1832).

16 *Oprechte Haerlemsche courant* (27 March 1832).

Such practical measures were meant to prevent the further spread of the disease. It was uncertain to what extent they actually helped, but from the experience with other contagious diseases such as the plague, people knew that preventive measures with regard to hygiene, food, and social gatherings could slow the spread of the disease. At the same time, people had to find ways to cope with feelings of uncertainty, fear, and grief. Historical research shows that cultural media played an important role in steering emotional and social reactions to catastrophes.[17] They not only offered an outlet for emotions such as fear and uncertainty but also provided comfort and meaning because they connected people. Authors and artists also encouraged a sense of local and national community: they aimed at increasing solidarity among citizens by heightening empathy with victims and by fundraising. With regard to the cholera outbreaks in the Netherlands, the cultural media fulfilled four functions, which I will discuss below: they regulated emotions, offered moral and religious lessons, propagated solidarity, and gave shape to a memorial culture.

Expressing Emotions

While the newspapers published new reports on the number of casualties, authors and artists expressed their feelings of fear and uncertainty by means of poems, stories, and images. They made use of a specific stylistic device that was often applied to diseases: personification. In medieval and early modern times, for example, the plague was represented as an old, bent woman, sometimes wearing a black cape and carrying a broom and a rake. Cholera was visualised as a figure with frightening characteristics, who travelled around and brought death. It took different shapes: the disease could be portrayed as a masked ghost, a devil, an apocalyptic horseman, or a violin-playing skeleton. Sometimes, cholera and death were represented as one and the same figure and visualised as a skeleton, carrying a scythe with which it brutally ended people's lives.

Another way of depicting the disease was by way of comparison. Adriaan van der Hoop, a young and ambitious poet from the city of Rotterdam, compared cholera with other disasters that were familiar to his audience: the plague and floods: cholera was like 'the cruel plague, whose power cannot be described with words'. This plague now threatened the Dutch

17 See, for example, Gerrit Jasper Schenk, 'Images of Disaster. Art and the Medialization of Disaster Experiences', in Kondo Kenichi (ed.), *Catastrophe and the Power of Art* (Tokyo: Mori Art Museum, 2018), 145–49 and de Graaf et al., 'Dancing with Death'.

coasts, like a flood posing a danger to a dyke: 'A cruel plague, o Netherlands, / threatens your coasts, / Like the flood the constructed dam'.[18] By comparing the new disease with other disruptive events, he tried to get a grip on the new situation.

The Dutch minister and poet J.J.L. ten Kate also wondered who this unknown and unwanted guest was. He described cholera as a female traveller, who departed from the world's end and managed to enter Europe without a passport:

Cholera! 't Is the Unknown,
Who triumphs over lock and key:
The wanderer of world's end,
Who finds access without a passport.
She silently sneaks in.[19]

Similar characteristics were used by another minister and poet: J.P. Hasebroek. He also described cholera as an unwanted perpetrator, who brutally manifested herself by wandering around, entering ships, and pushing poor infants into the grave. She wore a black mask over a pale face. The twinkling eyes betrayed her eastern origin (cholera had its origins in Asia).[20]

Printmakers and draughtsmen also made use of personification. Sometimes the disease was represented as an apocalyptic horseman who was trampling people to death; other times the disease took the shape of a devil with a tale and wings. Cholera could also be one and the same person as Death, who was visualised as a skeleton with different attributes. A widespread print showed Death handing out contaminated water to people. Artists also portrayed cholera sufferers to show the disease's terrible consequences: patients turned pale, screamed from pain, threw up white substance, and displayed blue and red spots on their skins. Their situation was so unbearable that they longed for death in the end.

Images like these circulated across nations.[21] This was also the case with the iconic figure of the 'danse macabre', which already circulated during

18 'Een wreede pest, ô Neêrland! Dreigt uw kusten, / Gelijk de vloed de opgeworpen dam'. Adriaan van der Hoop jr., *De cholera; graf- en boeteklanken* (Amsterdam: Brest van Kempen, 1832), 24.
19 'De Cholera! ... 't is de Onbekende, / Die slot en grendel overwint: / De wandlares van 's waerelds ende, / Die zonder paspoort toegang vindt. / Zij laat zich zwijgend binnensluiken'. J.J.L. ten Kate, *De cholera in Nederland. Zang des tijds* (Amsterdam: W.C.H. Willems, 1849), 6.
20 J.P. Hasebroek, 'De onbekende', in J.P. Hasebroek, *Winde-kelken. Gedichten* (Amsterdam: H. Höveker, 1856), 36.
21 See, for instance, Olaf Briese, *Angst in der Zeiten der Cholera* (Berlin: De Gruyter, 2003).

Figure 5.2 *Cholera as Apocalyptic Horseman*, illustration in *L'espiègle. Journal satirique, politique, artistique et littéraire* (Brussels, 1866). Source: Online Museum De Bilt, https://onlinemuseumdebilt.nl/de-cholera-in-de-regio-de-bilt/ (accessed 26 April 2022)

the medieval plagues in churches, cemeteries, and other public spaces. It symbolised the idea that Death did not distinguish between the poor and rich or the young and old and was re-used during the cholera outbreaks. The German artist Alfred Rethel, for example, made a drawing in which Cholera and Death make their entrance at a masked ball in Paris in 1832. Death plays the violin made of human bones, whereas Cholera, dressed as an Egyptian mummy, sits on a throne carrying the scourge of disease as a sceptre.[22] Such references to dances of death reminded the audience of the transience of life and the inevitability of death. They also entailed a social and moral message in showing the moral obligation of charity and solidarity.[23]

Several Dutch authors implicitly referred to the motif of the 'danse macabre' by elaborating on the dangers of musical gatherings and festivities: the sounds of music and laughter attracted many people, but they could lead to death. The poet J.J. A Gouverneur painted a macabre picture of cholera and the horrible pains which the sufferers experienced. These screaming

22 For an analysis and contextualisation of this print, see Christiane Hertel, 'Dis/Continuities in Dresden's Dances of Death', *The Art Bulletin* 82 (2000), 83–116.

23 Cf. de Graaf et al., 'Dancing with Death', 347–48.

Figure 5.3 M. Bos, after Alexander Ver Huell, *The Guard at Our Borders*, 1854–85, lithography, printed by Gouda Quint, 27 × 37 cm. Gelders Archief, Arnhem, 2039 – 4206-0042, A.W.M.C. Ver Huell / M. Bos

people sounded like 'festive music' in the ears of Cholera, who cheerfully waved his scythe.[24] This type of horrific music was also mentioned in an 1849 pamphlet, in which a pharmacist and a barkeeper discussed whether the fair had to be cancelled that year. The sympathy of the (anonymous) author clearly lies with the pharmacist, who dramatically sketches the death of a young mother with her two children; the partying crowd lets the burial procession pass, while the sound of violin music triumphantly fills the streets.[25] In another leaflet the people were also warned against visiting the fair with a little rhyme: 'Here entertainment is calling you, there the grave is waiting. Thus, refrain from the fair's joy'.[26] In such writings a sharp contrast was drawn between the temporary joys of entertainment and the long-lasting consequences they could have.

That fear could be instrumentalised in a political way becomes clear from an 1866 print by the draughtsman Alexander Ver Huell, entitled 'The guard at our borders'. It showed a ghostly skeleton in a large white cape, who was waving a flag with the text 'cholera'. The disease functioned as a guard against foreign enemies. At that time the Prussian ruler Otto von Bismarck was at war with Austria and strived for the creation of a German Confederation. Ver Huell feared that Bismarck would declare war on the Netherlands as well and portrayed cholera as a frightening enemy to the Germans.

By personifying the new disease and comparing it with other catastrophes, poets and artists tried to get grip on the new and unknown killer. By depicting cholera as a monster and equating it with Death, they may have increased fears amongst their audiences. At the same time, they offered all sorts of moral and religious lessons to steer these emotions in the right direction.

Moral and Religious Lessons

Most authors urged people to repent their sins and pray more. Van der Hoop stated that penance was the best remedy against the disease: people had to pray as much as they could. The title of his work, *De cholera; graf- en boeteklanken* ('Cholera. Sounds of the Grave and of Penance', 1832) underlined his religious message.

24 J.J.A. Goeverneur, *De dichtwerken* (Amsterdam: Amsterdamsche Courant, 1889).

25 *Moet het kermis of geen kermis zijn te Amsterdam?* (Amsterdam: Wed. C. Kok, geb. Van Kolm, k1849).

26 'Hier wenkt het vermaak – gindsch wacht U het graf. Zie dus volgaarn van de Kermisvreugd af'. *Kermis of geen kermis?* (1849, Collection Atlas van Stolk, Rotterdam).

Many considered cholera as a punishment from God, who warned the people to behave more virtuously. The theologian and poet Bernard ter Haar represented cholera as a Luciferian figure, the personification of evil, who raged through the universe on his wagon. This monster held a long and furious monologue in which he blamed human beings for their sins. When people asked him how long he would stay, he did not have a clear answer: 'You ask, how long I will stay? O mortal being, no human being nor angel knows, only God knows the number of hours'.[27] The minister and poet J.J.L. ten Kate went even one step further and considered cholera as a sign that the end times had arrived. He stated that God sent these punishments as a means of purification and urged the Dutch to repent their sins.

The religious message was also spread in songs and sermons. The central message of the song *De cholera regeert niet* ('Cholera Does Not Rule', 1848) was that the Dutch nation was not ruled by cholera but, ultimately, by God alone. In his sermon *De cholera: eene roede in de hand des heeren* ('Cholera. A Rod in the Hands of God', 1853) the minister Jan de Liefde repeatedly stated that the people were to blame for the new cholera outbreaks. If they did not improve their behaviour, God would keep on sending his thorn. Some authors adopted a milder tone. The poet C.G. Withuys, for example, did not speak of an angry but a loving God. Death should be welcomed and not feared: 'The love of God is just as endless as great; / The grave brings his Essence closer'.[28]

One sin was singled out in particular: alcohol abuse. The well-known preacher and philanthropist O.G. Heldring gave the debates a new twist by arguing that drinking gin was more harmful than cholera.[29] He argued that the overconsumption of alcohol was a structural societal problem whereas the cholera epidemics lasted only short periods of time.[30] When the fair was forbidden in 1849 in Amsterdam, Heldring became a point of reference in the heated discussions. The local shopkeepers and pub owners strongly disagreed with his plea against alcohol; they considered him to be the enemy of all kinds of pleasure in life.[31]

27 Bernard ter Haar, 'De cholera. Bij hare wederverschijning in 1849', in Bernard ter Haar, *Dichtwerken*, vol. 1 (Leiden: Sijthoff, 1880), 225 and J.J.L. ten Kate, *De cholera in Nederland*, 225.

28 C.G. Withuys, 'Bij het woeden der cholera', in *Aurora. Jaarboekje voor 1850* (Amsterdam: J.H. Laarman, 1850), 258.

29 O.W. Dubois, 'Ottho Gerhardt Heldring, 1804-1876, predikant en filantroop', in *Biografisch woordenboek van Gelderland*, ed. C.A.M. Gietman et al., vol. 3 of 7 vols (Hilversum: Verloren, 2002), 59–61, online: http://www.biografischwoordenboekgelderland.nl/bio/3_Ottho_Gerhard_Heldring (accessed 3 April 2023).

30 O.G. Heldring, *De jenever erger dan de cholera. Een volksboek, in voorbeelden en cijfers, voor arm en rijk, oud en jong* (Arnhem: Is. An. Nijhoff, 1838).

31 *Moet het kermis*, 2.

Only very few authors claimed that the cholera outbreaks had nothing to do with moral sins. In 1832 an anonymous critic stated that it was ridiculous to assume that God used disasters to punish people for their ways: 'Are the people of Groningen guilty of more violations than the inhabitants of Amsterdam? Are the people from Zealand better than those of Holland? … It's not righteous to consider disasters as a real punishment or revenge of the evil'.[32] The physician Hendrik Jan Broers, who at the time of the cholera outbreaks worked in Utrecht, went even one step further by pointing to medical and socio-economic causes: the reason some neighbourhoods were hit harder than others was poverty. He spoke of the 'poor people's disease'. The only remedy was to improve the living conditions of the poor people in society. His socio-politically engaged call for a better distribution of welfare, however, was an exceptional voice.[33]

Solidarity and Charity

Besides expressing emotions and offering moral and religious lessons, cultural responses also aimed at fostering a sense of community by raising money for the sufferers. Citizens were called to action in the local and national newspapers, poems, songs, and sermons. This led to impressive amounts being collected in a short time. This can be explained by the fact that charity was at the heart of Dutch society in the nineteenth century: it was considered every citizen's duty to help the needy.[34] Charity peaked in times of disaster: whenever a flood occurred, citizens' committees went to great lengths to collect money and goods. The infrastructure to help the cholera sufferers was already there.

Poets considered it part of their profession to contribute to the general welfare. Hendrik Tollens, for instance, published many poems on behalf of

32 Review 'A. van der Hoop jr., *De cholera; graf- en boeteklanken* (1832)', *Vaderlandsche letteroefeningen* (1832), 728: 'Zijn de Groningers aan meer overtredingen schuldig dan Amsterdammers? Zijn de Zeeuwen beter Christenen dan de Hollanders? Wie zou dergelijke stellingen durven beweren? … maar men is daarom niet geregtigd, die rampen als eene eigenlijke *straf*, eene *vergelding* van het kwade te beschouwen'.

33 Hendrik Jan Broers, 'De cholera', in J.P. de Keyser, *Bloemlezing, ten gebruike bij de beoefening onzer letterkunde* (The Hague: Henri J. Stemberg, 1877), vol. 1, 688–94. On cholera and resilience in Utrecht, see Beatrice de Graaf, '"Dat een ieder zich beijvert zijn zwakke krachten in te spannen". Veerkracht en cholera in de negentiende eeuw', *De moderne tijd* 6: 4 (2022), 272–98.

34 See Lotte Jensen, 'Floods as Shapers of Dutch Cultural Identity. Media, Theories and Practices', *Water History* 13: 2 (2021), 217–33, at 227–28; Fons Meijer, *Verbonden door rampspoed. Rampen en natievorming in negentiende-eeuws Nederland* (Hilversum: Verloren, 2022).

the distressed. Whether it was on behalf of the wounded, the poor, or the victims of an illness or flood, he was the first to express his sympathy by means of his work. Several ministers also published sermons in order to collect money for the cholera sufferers. In Utrecht J.D.A. Molster published a collection of songs and prayers, *Liederen en gebeden* ('Songs and Prayers', 1849), in support of the distressed, and in Sluipwijk (in the province of South Holland) Abraham ten Bosch published *Twee preken, naar aanleiding der heerschende cholera* ('Two Sermons on the Occasion of the Current Cholera Outbreak', 1849). He paid for the printing costs himself and donated all the income to local committees.

Money was also raised by organising benefit concerts. On 5 October 1832 the members of the rifle company played a tribute in the theatre of Amsterdam to the soldiers who were fighting against the Belgians. Although the performance had nothing to do with the cholera epidemic, the revenues nevertheless went to the victims. Another benefit concert was held on 30 September in Zandvoort, a village located on the west coast of the Netherlands. The entrance price was ƒ 1.50 (€ 18.09), and extra trains were available for visitors outside the village.[35] In 1854 a big concert hall in Rotterdam was the venue of a large event: an orchestra of 100 musicians and a choir of 125 singers went on stage, at the initiative of the Jewish choir Jubal. Amongst other pieces, they performed a cantata by Willem Hutschenruyter, with lyrics by Hendrik Tollens. The audience was cheering and applauding when the old poet (he was seventy-four years old) came on the stage, hand in hand with the composer. The cantata was a call for charity and solidarity and ended with prayers. A total of ƒ 800 (€ 8,176.51) was collected on behalf of the needy.[36]

Memory Culture

A fourth function of cultural responses was the shaping of a memory culture. Although the number of publications and material tokens is much smaller than with other disasters that occurred in the Netherlands, such as floods, the remembrance of severe epidemics was passed on to next generation by means of cultural media.

In several cities memorial books were published with the aim to document what exactly happened. The minister G.H. van Senden from Zwolle,

35 *Algemeen handelsblad* (26 September 1849).

36 *Nieuwe Rotterdamsche courant* (28 January 1954).

for example, published an overview of the way cholera had spread in the province of Overijssel during the first outbreak. His intention was to preserve the events in the regional collective memory: 'This booklet has the purpose, to keep the memory of the ordeal, and the practical use which it can provide, for a long time'.[37] He published his book at the request of others, who had been very much impressed by the sermon he gave on 14 November 1832. The regional, communal, and church authorities provided him with detailed information, names of the victims, and all sorts of statistics. He also provided a historical overview of earlier epidemics in the Netherlands, the measures local authorities took, and the relief actions in Overijssel. Van Senden's memorial book furthermore contained a list of subscribers; most of them came from his hometown.

Such compendia were also published in other cities of the country, such as Leiden and Scheveningen (where the disease occurred for the first time on Dutch soil in 1832).[38] The course of events was accurately documented by tables, statistics, and official regulations. They also contained stories of people who survived, such as the cure of a certain lady in Leiden, who managed to stop vomiting by drinking *bouillon de veau*.[39] In this way these books also provided practical advice in case the disease would return.

The cholera epidemics were not only remembered in memorial books but also by means of medals. On 23 November 1867 King William III awarded bronze medals to persons who had excelled in helping people during the outbreaks, for example physicians. One side showed a Greek male figure holding a staff of Aesculapius with the text 'Ob cives servatos' ('For the saving of citizens'). On the other side, a laurel was engraved with the text: 'For good care and help during the cholera asiatica in 1866'. However, the number of memorial books and medals is very low in comparison to those issued to commemorate floods. And unlike in other European cities such as Sheffield, Hamburg, or Dresden, there are no monuments in the Netherlands which commemorate the cholera deaths in the Netherlands. This probably has to do with the fact that epidemics are long-onset disasters, which have

37 'Het werkje strekke, om, met geheugenis der bezoeking, ook het nut, dat zij te weeg zou brengen, nog lang te bewaren'. G.H. van Senden, *Uren van godsdienst ter gelegenheid van de cholera te Zwolle; met bijvoegselen* (Zwolle: J. de Vri en M. Brinkman de Vri, 1833), 260.

38 J.G. Waardenburg, *Praktische aanteekeningen betreffende de cholera te Scheveningen* (Leiden: J.C. Cyfveer, 1832); J.F. d'Aumerie, *Herinneringen uit de cholera-epidemie te Scheveningen en proeve eener oplossing der raadsels van de Aziatische cholera* (The Hague and Amsterdam: gebroeders Van Cleef, 1833); C. Pruys van der Hoeven, C.W.H. Kaathoven, and G. Salomon, *Geschiedverhaal van de cholera-epidemie te Leiden, in 1832* (Leiden: C.C. van der Hoek, 1833).

39 Ibid.

Figure 5.4 Jacob Samuel Cohen Elion, medal to commemorate the 1866 cholera pandemic, signed on the obverse: J ELION F, 1866, bronze, diam. 5.7 cm, 91.2 g. Museum Lakenhal, Leiden, 3244

no clear beginning and ending. Unlike floods, where there is a sudden disturbance of life, clear heroic actions, and instant recovery, epidemics lack this 'narrative' structure. And the geographical location is more difficult to determine: Whereas flood monuments are erected at the exact spots of the disastrous event, epidemics are everywhere and thus nowhere in particular.

Parallels Between the Past and Present

Cultural responses to the Dutch cholera outbreaks show that coping with the disease entailed more than practical strategies, such as hygienic and social measures. People sought ways to express their emotions, make sense of the distressing events, and help the victims. Cultural representations appealed to solidarity and charity and can be considered as one of the 'mechanisms for unity', as Cohn calls them.[40]

Inspired by the historical, contextualised approaches of Snowden and Cohn, historians Beatrice de Graaf, Rina Knoeff, Catrien Santing, and I have argued that history may provide valuable lessons for present day crises, such as the COVID-19 pandemic.[41] Societal responses to epidemics, we argue, reveal striking similarities across countries and time and entail knowledge about successful and unsuccessful ('failure paths') efforts at coping with diseases. One of the historical insights that is still applicable today is that

40 Cohn, *Epidemics*, 68.
41 de Graaf et al., 'Dancing with Death', 346–67.

legitimacy is crucial in explaining the success or failure of governmental measures: tensions, calls for reform, and civil protest are an integral part of the societal repertoires that occur in times of long-lasting epidemics. Therefore, medical interventions alone will never suffice: a pandemic is never about viruses and bacteria alone, and it poses a so-called wicked problem for society: some solutions (e.g. closing schools, shops, and museums) will inevitably lead to new problems. A second insight which may be derived from history is that the developing cultural and social rituals, symbols, and repertoires can contribute to the resilience of societies. History offers many examples of how local and national communities tried to adapt to crises through cultural practices. The variety of cultural representations helped people to find a shared narrative and language for coping with fear and grief, and for fostering solidarity.[42]

The cultural responses to the Dutch cholera outbreaks provide a telling example: authors and artists gave direction to suffering people, tried to make sense of the events by pointing to higher meaning, offered different explanatory models, and called for action to help those in need. Stories, poems, music, theatre, and other art forms offer a medium in which people can express emotions, escape, reflect, and mobilise help. For that very reason, culture remains indispensable in today's society when confronted with new diseases.

About the Author

Lotte Jensen is Professor of Dutch Cultural and Literary History at Radboud University, Nijmegen and Principal Investigator of the NWO-funded Vici project *Dealing with Disasters. The Shaping of Local and National Identities in the Netherlands* (*1421–1890*), which investigates the impact of natural disasters on Dutch local and national identities from a cultural-historical perspective. She has published books on Napoleon's legacy in the Netherlands, Dutch literary history and the emergence of Dutch identity. Her most recent book is a cultural history of Dutch floods: *Wij en het water. Een Nederlandse geschiedenis* (De Bezige Bij 2022).

42 de Graaf et al., 'Dancing with Death', 360.

Part 2

Disaster and Blame

6 Dealing in Disasters

Selling Apocalyptic Interpretations of Disasters in the Sixteenth-Century Low Countries

Marieke van Egeraat

Abstract

Sixteenth-century pamphlet sellers made a profit by marketing disasters. This contribution takes a closer look at this activity by focusing on apocalyptic motives and the commercial consequences of putting these themes in the pamphlet. By comparing the reactions to the All Saints Flood (1570) in Dutch and German pamphlets, the article highlights the religious and political contexts of the two regions to explain the different uses of the Apocalypse. In Lutheran regions in Germany, printers chose the Apocalypse as an interpretation, while in Catholic regions disasters were seen as God's punishment. Because of the clear religious choice in Germany (at least after 1555), writers knew exactly what interpretation to put in their disaster pamphlets. In the Low Countries, however, authors focused on God's wrath but, considering the diverse religious landscape, still included apocalyptic notions. In this way, they expanded the possible readership of the pamphlet, resulting in potentially bigger profits.

Keywords

Apocalypse – Low Countries – All Saints Flood – providence – news – pamphlets

When we speak of dealing *with* disasters, we should sometimes also speak of dealing *in* disasters. Sixteenth-century pamphlet sellers certainly made a profit by marketing disasters.[1] Almost immediately after catastrophe

1 Sara K. Barker, '"Newes lately come". European News Books in English Translation', in Sara K. Barker and Brenda M. Hosington (eds), *Renaissance Cultural Crossroads. Translation, Print and Culture in Britain, 1473–1640* (Leiden: Brill, 2013), 227–44, at 235.

Asperen, van, Hanneke and Lotte Jensen (eds): *Dealing with Disasters from Early Modern to Modern Times. Cultural Responses to Catastrophes.* Taylor & Francis Group, 2023
DOI: 10.5117/9789463725798_CH06

had struck, broadsheets, pamphlets, and prints could be bought at the local book shop. Although sixteenth-century pamphleteers might have had high-minded motives for publishing their text (warning people about the wrath of God or maybe to arouse compassion in readers), the commercial motivations behind publishing disaster pamphlets should not be forgotten.

This contribution takes a closer look at this activity by focusing on apocalyptic motives and the commercial consequences of putting these themes in the pamphlet's interpretation of the disaster. I do this by contrasting the apocalyptic reactions to disasters in the sixteenth-century Low Countries with the reactions to the same event found in multiple places in the German-speaking parts of the Holy Roman Empire (for the most part, present-day Germany). The latter context has already been heavily studied by multiple scholars. Just as in the Low Countries, these German-speaking regions had a lively pamphlet production. The apocalyptic reaction in these pamphlets was, however, completely different. A possible explanation for this difference lies in the dissimilar commercial possibilities that the apocalyptic interpretation could offer.

In this contribution I will first offer a short historiographic overview of apocalyptic ideas in sixteenth-century Europe, and the Low Countries and the German-speaking parts of the Holy Roman Empire more specifically. Secondly, I will show how Dutch pamphlet writers used the Apocalypse as an explanation for disasters in their pamphlets, contrasting this with how they used the idea of God's providence. Then I will go on to a case study of the All Saints Flood, a flood that hit the North Sea coast in 1570. This will make clear the differences in apocalyptic reactions between German and Dutch pamphlets writers. Lastly, I will try to explain the discrepancies by focusing on the religious and political situations in the two regions. I will argue that Dutch pamphlet sellers had to downplay apocalyptic interpretations for better sales figures, while their German counterparts highlighted the Apocalypse to sell theirs.

State of Research

Research into apocalyptic ideas and reactions in the sixteenth-century Low Countries lags behind the same research done for other parts of Europe. Some smaller case studies have been carried out, but a general overview is missing.[2] An interesting case study for this article is the contribution by Bart Ramakers, professor of historical Dutch literature, on apocalyptic ideas in Dutch rhetoricians' plays. Looking at dramatic performances in Catholic Antwerp and Reformed Gouda, he concludes that the Book of Revelation could have been a source of inspiration for both. The sharp divisions that other scholars detect – which I will come back to shortly – are, according to Ramakers's research, not as relevant for the sixteenth-century Low Countries.[3]

The lack of a good overview might surprise us, since the sixteenth-century Low Countries have many extraordinary features which could advance the scholarship on apocalyptic thinking. The purpose of this article is not to give such a broad overview. It does, however, seek to highlight some of the promising aspects the Low Countries can offer to this field of study. The diversity of the religious landscape of this region and – in the last decades of the century – the freedom of conscience that allowed people to believe what they wanted to believe (with restrictions, because practising was allowed only to Calvinists) make for an entirely different context that resulted in a different way of dealing with apocalyptic thoughts.

The international research into apocalyptic thinking took off with the publication of Norman Cohn's *The Pursuit of the Millennium. Revolutionary Millenarians and Mystical Anarchists of the Middle Ages* (1957). In this work Cohn presents the idea that apocalyptic thinking was mostly used by rebellious groups to justify their actions. One example he gives is the Münster rebellion by the Anabaptists.[4] Since the publication of his book, others have

2 For some smaller case studies, see C.P.M. Burger, 'Eindtijdverwachting aan het einde van de vijftiende eeuw en bij Maarten Luther' in Willemien Otten and Gerard Rouwhorst (eds), *Het einde nabij? Toekomstverwachting en de angst voor het oordeel in de geschiedenis van het christendom* (Nijmegen: Valkhof Pers, 1999), 181–98; Willem Frijhoff, 'Katholieke toekomstverwachting ten tijde van de Republiek. Structuur, en grondlijnen tot een interpretatie', *BMGN – Low Countries Historical Review* 98 (1983), 430–60; Willem Frijhoff, 'Het Gelders Antichrist-tractaat (1524) en zijn auteur', *Archief voor de geschiedenis van de katholieke kerk in Nederland* 28 (1986), 192–217; Peter Nissen, '"Onse salicheit is seer nabi getreden". Toekomst- en eindtijdverwachtingen in het Nederlandse doperdom', in Otten and Rouwhorst, *Het einde nabij?*, 215–24. Huib Zegwaart noted this lack of a general study of the subject by Dutch scholars in Bible studies: 'Apocalyptiek. Overzicht van een verwaarloosd gebied in het Nederlandse bijbelonderzoek', in Cornelis Houtman and L.J. Lietaert Peerbolte (eds), *Joden, christenen en hun Schrift. Een bundel opstellen aangeboden bij het afscheid van C.J. den Heyer* (Baarn: VBK Media, 2001), 130–46.

3 Bart A.M. Ramakers, 'Apocalyptiek op de planken. Twee rederijkersspelen over het Boek Openbaring', *Ons Geestelijk Erf* 66 (1992), 187–223.

4 Norman Cohn, *The Pursuit of the Millennium. Revolutionary Millenarians and Mystical Anarchists of the Middle Ages* (1957; 2nd ed., Oxford: Oxford University Press, 1970), 281–86. Michael Driedger criticised Cohn's study of the Anabaptists because Cohn used sources written

shown that this idea no longer holds. Bernard McGinn, for example, shows that the social history of apocalyptic thinking cannot be reduced to one group. He identifies apocalyptic ideas in all parts of society: from rich to poor and from the powerful to the powerless.[5]

With the assertion about specific groups using the Apocalypse superseded, researchers have tried to find other explanations for sudden bursts of apocalyptic thinking. One suggestion was that periods of crises and disaster led to apocalyptic fervour. This idea was nuanced almost immediately by researchers pointing out that the same crisis was dealt with differently by different people. A crisis alone does not sufficiently explain the rise of apocalyptic thinking.[6] Still, even recent studies sometimes equate crisis situations with apocalyptic feelings.[7]

International research into apocalyptic thinking, including the apocalyptic interpretations of disasters (the focus of the present article), has concentrated particularly on the German-speaking lands in the sixteenth century. Robin Barnes, Volker Leppin, and Philip Soergel all argue that apocalyptic thinking and the interpretation of disasters as signs of the End Times took hold within Lutheranism in particular.[8] Barnes took this idea one step further by proposing that the Apocalypse could be used by confessional groups to strengthen the differences between the confessions in an almost polemical way. According to Barnes, Lutherans were the most fervent advocates of an approaching Apocalypse, while Catholics did not truly believe in this idea. Barnes positioned the Calvinist religion somewhere in between.[9]

In this article I will come back to these last two trends: apocalyptic thinking (not) resulting from crisis and using apocalyptic ideas polemically. By

by enemies of the Anabaptists: Michael Driedger, 'Thinking Inside the Cages. Norman Cohn, Anabaptist Münster, and Polemically Inspired Assumptions about Apocalyptic Violence', *Nova Religio* 21 (2018), 38–62.

5 Bernard McGinn, *Visions of the End. Apocalyptic Traditions in the Middle Ages* (New York: Columbia University Press, 1979), 3–4.

6 Volker Leppin, *Antichrist und Jüngster Tag. Das Profil apokalyptischer Flugschriftenpublizistik im deutschen Luthertum 1548–1618* (Heidelberg: Gütersloher Verlaghaus, 1999), 161; McGinn, *Visions of the End*, 31.

7 Andrew Cunningham and Ole Peter Grell, *The Four Horsemen of the Apocalypse. Religion, War, Famine and Death in Reformation Europe* (Cambridge: Cambridge University Press, 2000).

8 Philip Soergel, 'Die Wahrnehmung der Endzeit in monströsen Anfängen', in Hartmut Lehmann and Anne-Charlott Trepp (eds), *Im Zeichen der Krise. Religiosität im Europa des 17. Jahrhunderts* (Göttingen: Vandenhoeck & Ruprecht, 1999), 33–52, at 33; Robin B. Barnes, *Prophecy and Gnosis. Apocalypticism in the Wake of the Lutheran Reformation* (Stanford, CA: Stanford University Press, 1988), 2; Volker Leppin, *Antichrist und Jüngster Tag.*

9 Robin B. Barnes, 'Varieties of Apocalyptic Experience in Reformation Europe', *Journal of Interdisciplinary History* 33: 2 (2002), 261–74.

focusing on a different religious and political context, I will try to add more body to existing scholarship. My sources, pamphlets written on disasters and portents, do not allow me to go into depth on what people thought the Apocalypse would be. They do offer a view on how apocalyptic words and concepts (such as the Day of Judgement and the Lord's Day) could be used to help present the author's or publisher's point.[10]

Apocalyptic Reactions in Dutch Disaster Pamphlets

The number of apocalyptic reactions to disasters in Dutch pamphlets in the sixteenth century is limited. Out of the eighty-three disaster pamphlets I collected, thirty-one mention something of an apocalyptic nature. These mentions range from biblical references to quick citations of apocalyptic words. Most of the time, the passages are very short and consist of merely a few words. Contemporaries reading them could easily ignore or overlook them. It is only because of the many associations and connotations that these words could truly stand out. They could call to mind the sermons of the priest or pastor about the Day of Judgement[11], or the depictions of the Apocalypse.[12] In these cases, a few words on the Apocalypse could make people aware of the signs that would inevitably show up when the Last Days would be near. However, these words did nothing for others who had not listened to a Doomsday priest or pastor and had not looked at depictions of the End Times. The beauty of these short mentions is that they targeted those readers who already carried the mental baggage of apocalyptic thinking, while others could easily filter them out.

In general, there were three different places in the pamphlets where people could encounter these words: the title page, the preface, or other accompanying text and in the running text. Combinations were possible as well. All three locations bring with them different effects on the reading experience. A reference on the title page guides the reader into the work and therefore stands out, while a mention in the running text could more easily be overlooked.

10 Thomas Long, 'Revising the Revelation. Early Modern Appropriations of Medieval Apocalypticism', in Michael A. Ryan (ed.), *A Companion to Premodern Apocalypse* (Leiden: Brill, 2016), 378–425.

11 e.g. *Dagboek van Cornelis en Philip van Campene*, ed. Frans de Potter (Ghent: Annoot-Braeckman, 1870), 370.

12 e.g. the famous woodcut series by the artist Albrecht Dürer from 1498. See *Die Apokalyps. Faksimile der Deutschen Urausgabe von 1498. 'Die Heimlich Offenbarung Johannis'*, ed. Ludwig Grote (Munich: Prestel, 1970).

The title page was one place where writers or publishers could easily add an apocalyptic note. Many title pages, whether from a disaster pamphlet or not, included some citations of biblical origin or important lessons from church fathers, saints, or other eminent (religious) figures. Apocalyptic title pages mostly fell back on the common apocalyptic biblical texts, especially on the prophetic elements in the gospels of Matthew and Luke.[13] Jan de Coster, a Louvain member of a chamber of rhetoric, published a poem on the floods that had hit his city in 1573. On the title page, he included a poetic rendering of the gospel of Luke 21:

Everyone fears the Lord as he begins to strike
All that are on earth, because all kinds of
Signs will be seen, in stars, sun and moon,
And in the flood of waters on the land.[14]

According to de Coster, the floods were forebodings of the Day of the Lord. Yet, in the running text, the Apocalypse is never mentioned again. The poem on the title page seems to serve merely as an advertisement to promote the sale of the pamphlet.

In addition to the title page, the preface or other accompanying text (or paratext[15]) could also provide an apocalyptic framing of the event. However, I did not find any cases of an author dealing with the Apocalypse at length in their prefaces. At most, they limited themselves to a couple of words or a single sentence to get the warning message across. Sentences like 'God, make sure that we will be prepared for the coming of the Lord' or 'our poor people, for whom the end of the world has arrived' are scattered throughout the prefaces and included without any context or explanation.[16]

13 For an overview of biblical books with apocalyptic features, see Stephen L. Cook, *The Apocalyptic Literature* (Nashville, TN: Abingdon, 2003).

14 Jan de Coster (also known as Jan Stroosnijder), *Warachtighe ende wonderlijcke beschrijvinghe die gheschiet sijn tot Lueven*, ed. Gilbert Huybens, *Memorabilia Lovaniensia* (Leuven: Peeters, 2013). The text in Dutch reads:

Elck vreese den Heere als hy beghint te slane,
Al die in aertrijck sijt, want menigherande
Teeckenen salmen sien, in sterren sonne en mane,
En in den vloet der wateren binnen den lande.

15 Gerard Genette, *Paratexts. Thresholds of Interpretation* (Cambridge: Cambridge University Press, 1997), 1–2.

16 *Wonderlijcke nieumaren uut Calabren geprophiteert* (Gouda: Jan Zas Hoenz, voor P.S., 1587). The text in Dutch reads: 'God maeck dat wy moghen bereyt zijn tegen de toecomste des Heeren'. *Wonderlicke nieuwe tijdinghe ... vande ... teeckenen ... in ... Gendt* (Dordrecht: Peeter Verhaghen,

Other accompanying text also framed the readers' experience. Take, for example, the following poem that accompanied a pamphlet about a wondrous birth in Purmerend in 1599:

> Give up your sins, move away from evil
> Always remember the Day of the Lord
> As Matthew xxiii does show
> No matter how noble, how rich, or how highborn, no one will escape
> God's judgement will come, swiftly and unforeseen
> Like a thief in the night
> So watch and pray, all you people
> That you may suffice before your heavenly father
> Oh, there will be seen such despair
> Like it was ever thought or seen on earth
> Those who did well and virtuous
> Will go with Him in the eternal bliss[17]

After this poem, the main text of the news pamphlet began. The message (move away from sins, because the Day of Judgement is approaching) would still be ringing in the readers' ears while reading about a wondrous birth in Purmerend. Clearly, the author saw this event as a sign of the coming Apocalypse, underlining the point of the poem.

1587). The text in Dutch reads: 'ons arme menschen, op welcke het eynde der werelt ghecomen is'.

17 *Een mirakel ende geschiedenisse hoe dat binnen de stadt Purmurent ... geboren is een kint* (Antwerp: s.pub., 1599).

The text in Dutch reads:

Staet van sonden af, van t'quaet wilt keeren
Sijt altijt denckende op den dach des Heeren
Soo ons Mattheus xxiii doet verstaen
Hoe edel, hoe rijc, hoe stam niemant falt ontgaen
Gods oordeel sal comen so haest en onverhoet
Ghelijck een dief inder nacht doet
Dus waect en bidt ghy menschen allegaer
Dat ghy mocht volstaen voor u hemelschen vaer
Och daer sal alsdan sulcken druc geschien
Als het op aerden oyt was gedacht ende ghesien
Die dan goet ghedaen hebben ende deucht
Sullen met hem gaen in de eeuwighe vreucht

Interestingly, this is the only example in my corpus where the approaching Apocalypse also has a positive ring (those who did well will go to paradise). Compare with Charles Zika, 'Disaster, Apocalypse, Emotions and Time in Sixteenth-Century Pamphlets', in Jennifer Spinks and Charles Zika (eds), *Disaster, Death and the Emotions in the Shadow of the Apocalypse, 1400–1700* (London: Palgrave Macmillan, 2016), 69–90, at 76.

Title pages and prefaces serve to guide the reader into the text, but what if the apocalyptic notion was put right inside the text? In the case of a pamphlet on earthquakes in Italy, the writer put the following words into the mouth of the narrator: 'I believe for sure that I have seen a likeness of the final day of judgement'.[18] The function of this kind of apocalyptic text is unclear. It could be to focus the attention of the reader or to make them aware that they are still dealing with divine happenings and not with mere natural events. Yet, these instances are of such a short nature and left so completely unexplained that they could be easily overlooked or ignored unless the reader was already attuned to them.

All in all, the apocalyptic reactions to disasters in Dutch pamphlets are limited. What stands out the most is that the cases in which there are some forms of apocalyptic expectancy are very concise. The readers had to make up their own minds on what to think about these threatening words and sentences. As we will see in the example below, Dutch pamphlets deal with God's providence and the disaster as a punishment for sins differently. There the interpretation is explained further using examples from history and the Bible. In German disaster pamphlets, the roles are reversed: the Apocalypse is given much room, while God's punishment is dealt with only briefly.

Case Study: All Saints Flood 1570

Let us explore this difference by looking in detail at the All Saints Flood of 1570 and its pamphlets. On the first of November, this flood hit the Low Countries but also the coast of northwestern Germany and Denmark. Storm surges caused dykes to break and water to flow over the land. It resulted in damage, dead cattle, and, above all, many human lives lost. The disaster also struck at a critical time: the Iconoclastic Fury had just swept through the Low Countries; the Duke of Alba had arrived in the Low Countries and tried to issue a new tax ('de tiende penning'); Alba used his Council of Troubles (also known as Council of Blood) to prosecute and execute heretics. All in all, the Netherlands were in a state of war, chaos, and insecurity. The flood only made things worse.

Despite the disastrous timing of the flood and the wave of destruction it caused, only one printed Dutch pamphlet remains that goes into detail

18 *Vander verschrickelijcker aertbevinghen in de stadt van Scharbarien* (Antwerp: Mattheus Crom, 1542). The text in Dutch reads: 'Ick gheloove sekerlijcken dat ick wel een gelijckenisse vanden uutersten dach des oordeels ghesien hebbe'.

about what occurred on that horrible day. In a poem the Calvinist writer Johan Fruytiers told his readers all about the damage that happened in every corner of the Low Countries and even further away (he also discussed Hamburg and Denmark).[19] Fruytiers had experienced the flood, while staying in exile in Emden, in the very northwest of present-day Germany. Afterwards, he collected information from all over the Low Countries and put it together in a lengthy poem. The poem therefore mostly deals with numbers of casualties, dead cattle, damaged buildings, and other rather factual information. In the preface, on the other hand, Fruytiers offered his readers an interpretation of why this flood had struck the Low Countries.

Without any doubt, this flood was a punishment from God. That was the sole reason for Fruytiers to write his pamphlet: 'that is how this little book of mine began, to show us the wondrous punishment from God'.[20] By telling people about the disaster and how it had been a punishment, he wanted to leave a message that served as an 'admonition, lesson, and warning'.[21] The preface is used to further explain this warning by adding examples from history, comparisons with biblical stories, and other current events that showed God's anger. In this introduction, Fruytiers, for the most part, dealt with the disaster as a punishment from God.

There are, however, some minor elements in the pamphlet that pointed towards an apocalyptic explanation. On the title page, Fruytiers included two biblical citations:

> Habakkuk 3.8–9
> The Lord has proven his anger in the flood; his grimness in the water, his wrath in the sea and he divided the currents in the land.
>
> Luke 21.26
> The sea and the water flood will roar, and the people will vanish because of fear and expectation of things that will happen to the world.[22]

19 Johan Fruytiers, *Corte Beschrijvinghe vanden ellendighen ende seer beclaghelicken watervloet* ([Emden]: s.pub., 1571).

20 Ibid. The text in Dutch reads: 'Soo is dan dit cleyne boecxken van my begonst, op dat het ons de wonderbaerlijcke straffe Godes voorstelle'.

21 Ibid. The text in Dutch reads: 'Tot vermaninghe, leere ende waerschouwinghe'.

22 Ibid. The text in Dutch with references to the Bible reads:

Habbacuc 3.8-9: De Heere heft syn gramschap in den vloet, syn grimmicheyt in het water, synen toorn einder Zee bewesen, ende de stroomen in het lant verdeylt.

Luce 21.26: De zee ende watervloet sal brieschen, ende de menschen sullen verdwynen door de vreese ende verwachtinghe der dinghen die der werelt overcomen sullen'.

Not only are both verses well chosen for dealing with an overflowing of water; they are also apocalyptic in nature. Habakkuk 3 tells of the prophetic dream Habakkuk had in which he saw God coming down to earth, causing all kinds of disaster and wiping away entire nations but saving an elected few. The verses of the gospel of Luke are concerned with the return of Jesus to earth in the final days. Again, the world will experience disasters and people will be full of fear. The message on the front page of Fruytiers's work was clear: the flood should be seen as a sign of the coming Day of Judgement.

In the preface this apocalyptic interpretation pops up two more times. At the very beginning, Fruytiers called the flood one of many 'harbingers of the coming of His son Jesus Christ'.[23] Later on, the flood is the 'biggest warning for the Last Day'.[24] Even though these instances imply that Fruytiers thought this interpretation to be important, he gave no further explanation on how this flood was a clear sign of the Apocalypse. In contrast with his other interpretation of the disaster as a punishment from God, he gave no biblical examples, no historical comparisons, and no other current signs to strengthen his argument about the disaster as a harbinger of the Apocalypse.

It might be that these two interpretations (punishment or Apocalypse) followed naturally from each other. Church historian and theologian Volker Leppin, in his work *Antichrist und Jüngster Tag* (1999), explains the connection between Apocalypse and providence in two ways.[25] On the one hand, the two interpretations should be seen as different when looking at them from a theological perspective. Church officials definitely made a distinction between a 'normal' punishment and the all-encompassing Apocalypse. On the other hand, for most lay people the division was not that strict. And according to Leppin, for the average person, the difference between dying because of the disasters of the Apocalypse or because of disasters as punishment was not that important. The more important aspect was death itself.

Although Leppin makes a valid point, I believe it is striking that the balance between the two interpretations shifts when looking at different regions. For German pamphlets, this balance was completely reversed: the Apocalypse got most of the attention, while the disaster as punishment was dealt with only briefly. A couple of German pamphlets, also dealing with the All Saints Flood, will clarify this.

23 Ibid. The text in Dutch reads: 'voorboden der toecoemsten synes Soons Jesu Christi'.

24 Ibid. The text in Dutch reads: 'aldergrootste waerschouwinghe des joncxsten daghes'.

25 This paragraph is based on Leppin, *Antichrist und Jüngster Tag*.

At least four German-language disaster pamphlets were published after this flood. Each of them mentions the punishing God. The anonymous pamphlet *Warhafftige doch erschröckenliche beschreybung der grewlichen Wassernot* ('True but Terrifying Description of the Gruesome Flood', 1570) first calls the flood a 'punishment from God'.[26] Immediately after, the writer compared the flood to the Deluge and a major part of the text deals with an explanation of the disaster as a 'sign of the Day of Judgement'.[27] Like Fruytiers, this author used both the Apocalypse and the idea of a punishing God in interpreting the flood. However, the balance is reversed: this time, the Apocalypse demands most of the reader's attention.

Most tellingly, where Fruytiers used contemporary signs and disasters to show that God was angry, the German authors used the same signs to point to the nearness of the Last Days. The German broadsheet *Warhaffige doch Grewliche und Erschröckenliche Geschicht* ('True but Horrible and Terrifying History', 1570) offers a nice example:

> When had the world suffered more from water, since the Deluge? The same applies to earthquakes, which ruin entire cities and villages, the same applies to the horrible signs in the sun and the moon, comets, and other horrible signs, the same applies to rains of grain and blood that we saw this year. Also war, the shedding of blood, also severe sickness, pestilence, and other corrupt air, together with severe inflation and bad foods; from all these signs the world now sees the Last Days must be near.[28]

Earthquakes, pestilence, signs in the sun and moon: all were omens of the Last Days to this author.

Also, where Fruytiers put apocalyptic, biblical texts on his title page without any explanation, the German authors gave ample room to the

26 *Warhafftige doch erschröckenliche beschreybung der grewlichen Wassernot* (Augsburg: Michael Manger, 1570). The text in German reads: 'ein Straff Gottes'.

27 *Warhafftige doch erschröckenliche beschreybung.* The text in German reads: 'zeychen vor dem Jüngsten Tag'.

28 Daniel Holtzmann, *Warhafftige, doch grewliche und erschröckenliche Geschicht, so geschehen ist zu Antdorff* (Augsburg: Hans Moser, [1570]). The text in German reads: 'Wann hat die Welt vor mehr solche Wasser erlitten, seyt der Sündfluß? Deßgleichen auch Erdbidem, darinne gantze Stett und Dörffer verfallen, deßgleichen erschröckliche gesicht am Sonn unnd Mon, Cometen, auch andere schröckliche Zaichen, deßgleichen auch Koren und Blütregnen, wie diß Jar gesehen ist worden. Darzü auch Krieg, Blüt vergiessen, auch schwere Kranckheiten, Pestilentz, unnd andere verderbende Suchten, sampt schwerer Thewrung und spröder Narung, an denen Zaichen allen, der Welt abnemung unnd letstes End, wel zu spüren ist'. I have interpreted 'Thewrung' as *Teuerung*, meaning inflation, and 'Narung' as *Nahrung*, meaning food.

different prophetic Bible texts and especially to the apocalyptic passages in the gospels. The anonymous pamphlet *Jammerliche und erschröckliche Zeittung auß Niderland* ('Wretched and Terrifying Tidings from the Low Countries', 1570) concludes with an elaborate account in which the author summarised the most important points regarding the Final Days from the gospels of Matthew, Luke, and Mark.[29] Reading this summary at the end of the news about a terrible flood, the reader could not deny the similarities between their own times and the prophecies of the Evangelists.

In short, the difference between the Dutch printed reaction and its German counterparts is striking. The German pamphlets devote much more room to the Apocalypse, while they deal with the disaster as punishment only briefly. In the Low Countries, this balance is completely reversed, with ample room for the idea of punishment while the Apocalypse comes up rarely and, even then, only briefly. To find an explanation, we need to look at the political and religious context of the regions in which these texts were produced.

Explaining the Difference

An obvious reason could be the difference in religious backgrounds of the authors of these pamphlets. After all, Barnes and others already pointed out that the three main Christian denominations in the sixteenth century (Catholicism, Lutheranism, and Calvinism) all dealt with the Apocalypse differently. However, for the Dutch pamphlets, not much difference can be found. Calvinists wrote most Dutch pamphlets with apocalyptic notions, but authors of other denominations are certainly also present.[30] There is no clear difference between the Dutch disaster pamphlets written from different religious backgrounds: all of them contain short apocalyptic notions but do not explain those notions in any detail. In short, for the Dutch disaster pamphlets, the religious affiliation of the author is no indication of whether a pamphlet uses the Apocalypse, just as Ramakers showed for the plays of the chambers of rhetoric.

We should also question the view that the denomination of the writer mattered to the reader of the pamphlet. Not only are most pamphlets

29 *Jammerliche und erschröckliche Zeittung, auß Niderland* (s.l.: s.pub., [1570]).

30 In my corpus, thirty-one pamphlets contain apocalyptic notions. Of those thirty-one, ten have a Calvinist background, four Catholic, four Lutheran, one Mennonite, five more generally Protestant, and ten remain unclear. These religious backgrounds are determined by looking at the religious background of the author and/or the printer.

anonymous and the identity of the author unknown to the reader; it is also hard to recognise their confessional affiliation in most pamphlets. Notwithstanding that some disaster pamphlets were indeed written with confessional polemic in mind, many others were much more neutral in their description of the event.[31] In those cases, readers from many confessional backgrounds could read the pamphlet without regard for the confessional background of the author. Take, for example, the reader Godevaert van Haecht, a Lutheran inhabitant of Antwerp. He noted down in his chronicle that he read Fruytiers's pamphlet of the All Saints Flood. He even copies large parts of it into his text. Clearly, he had no trouble with the disaster pamphlet, even though it was written by someone of an opposing denomination. He might not even have known that Fruytiers had Calvinistic sympathies. Authors and publishers may have even downplayed the confessional polemic for their own benefit. By keeping the purport of a pamphlet more neutral and choosing an interpretation that fitted all confessional backgrounds, they maximised their potential readership and therefore their potential profit.

For example, printer Cornelis Claesz. from Amsterdam was a fervent supporter of the Calvinist religion. His pamphlets, on the other hand, do not reveal this denomination. A disaster pamphlet he produced on floods in England in 1607 included some apocalyptic notions, even though Calvinism in its early stages was not known for its Apocalypticism.[32] By including these notions, Claesz. could sell his pamphlet to people outside of his own confessional circle.[33] Why, then, did Dutch authors use the Apocalypse only as a side note, while their German counterparts devoted entire paragraphs to it?

Another possible explanation for this difference lies in the religious and political constellations in these regions from the second half of the sixteenth century onwards. In the Holy Roman Empire, it was established during the Peace of Augsburg (1555) that the ruler of every territory decided their own religion. The choice was, however, limited: they could choose between Catholicism and Lutheranism. Only from 1648 onwards was Calvinism one

31 On polemical use of disaster pamphlets in the Dutch context, see Raingard Esser, 'Fear of Water and Floods in the Low Countries', in William G. Naphy and Penny Roberts (eds), *Fear in Early Modern Society* (New York: Manchester University Press: 1997), 62–77; Marieke van Egeraat and Hanneke van Asperen, '"Wat dat sy ons bedien". Walvisstrandingen in pamflet en prent rond 1600', *Jaarboek de zeventiende eeuw* (2020), 119–37.

32 Barnes, 'Varieties', 269; *Een warachtich verhael van de schrickelicke springh-vloedt in het Landtschap van Summerset* (Amsterdam: Cornelis Claesz, 1607).

33 Bert Hofman, *Liedekens vol gheestich confoort. Een bijdrage tot de kennis van de zestiende-eeuwse Schriftuurlijke lyriek* (Hilversum: Verloren, 1993), 154.

of the options.[34] This meant that, at least on paper, the religious situation in each territory was very clear. Inhabitants knew what was expected of them: Catholic ritual or Lutheran prayer.

A consequence for the authors and publishers of disaster pamphlets was that they could single out one specific disaster interpretation, dictated by the established confession. In Lutheran regions the Apocalypse was a beloved theme, whereas in Catholic districts the wrath of God was much more popular. In Strasbourg, for example, printer Thiebold Berger published *Warhaffte und erbermliche newe Zeitung von dem erschrecklichen Feurzeichen und Erdbidem* ('True and Pitiful New Tidings about the Terrifying Fire Sign and Earthquake') in 1584. It dealt with an earthquake that surprised the city of Bern. The title page is adorned with a warning message from the gospel of Matthew. The actual text begins with factual news but is immediately followed by an interpretation: the natural signs that had happened in the last couple of months, including this earthquake, should be seen as 'special signs and harbingers of great misfortune ... and all kinds of sadness, especially also the nearing Last Day'.[35] This apocalyptic message fitted well in the Lutheran context of Strasbourg.

In Catholic areas of the Holy Roman Empire, the Apocalypse got little or no mention in the disaster pamphlets. In Munich, for example, Catholic printer Adam Berg published two such pamphlets: one broadsheet about a famine in 'Reussen und Littau', places in present-day Lithuania, in 1573 and another on a flood near Munich in 1582.[36] In both publications the idea of the disaster as punishment prevails. The events should be seen as signs of an angry God. The Last Days are mentioned once in the flood pamphlet, but only to compare the contemporary events to the events that would take place during the Apocalypse. Nowhere is it indicated that the contemporary events should also be seen as precursors of the Apocalypse. In this Catholic city, it was more advantageous to point towards sins than to the Day of Judgement.

34 Violet Soen, 'From the Interim of Augsburg until the Treaty of Augsburg (1548–1555)', in A. Melloni (ed.), *Martin Luther. A Christian Between Reforms and Modernity (1517–2017)* (Berlin: De Gruyter, 2017), 548–64, at 548.

35 *Warhaffte und erbermliche newe Zeitung von dem erschrecklichen Feurzeichen und Erdbidem* (Strasbourg: Thiebold Berger, 1584). The text in German reads: 'sondere Zeichen und vorbotten künsstigen grossen onglücks, ..., und allerley jamers, sonderlich auch des nahenden Jüngsten tags'.

36 *Ein erschröckenliche doch warhafftige graüsame Hüngers nott, und Pestilentzliche blag, so im Landt Reissen unnd Luttau fürgangen im 1573 Jar* (Munich: Adam Berg, 1573); *Wahrhafftige und erschroeckliche Zeitung von dem grausamlichen waetter und schauwr* (Munich: Adam Berg, 1582). For more information on the Catholic printer of this pamphlet, see Josef Benzing, 'Berg (Montanus), Adam', in *Neue Deutsche Biographie* 2 (1955), 72–73, https://www.deutsche-biographie.de/pnd103140522.html (accessed 3 April 2023).

In the Low Countries, the religious and political situation was completely different. In these areas, no definite decision was made as to the accepted confession. Only with the signing of the Union of Utrecht (1579) did the northern provinces agree that there would be freedom of conscience. In reality this meant that everyone was allowed to believe what they wanted, but not everyone was allowed to practise their confession openly. The Calvinists were the only ones who could have public ceremonies and church services. Their religion, Calvinism, became the state-sponsored confession. Other confessions were, however, not forbidden, resulting in a very diverse confessional landscape. In the southern parts of the Netherlands, the situation remained unclear for quite some time. The cities, and especially Antwerp, had been strongholds of Protestantism, but the fall of Antwerp in 1585 and Alexander Farnese's demand that all Protestants leave the city within four years' time or convert to Catholicism caused an exodus of adherents of the new confessions. From that moment on, the Southern Netherlands were predominantly Catholic. Furthermore, the Counter-Reformation ensured that the Catholic faith regained its former glory.[37]

As a result, for authors and publishers the news market in the Low Countries was, at least until 1585, unclear and unstable. After 1585 the confessional landscape in the northern provinces remained diverse, while in the Southern Netherlands printers had to cater to a Catholic market. To enlarge the potential readership, it could be profitable to keep the interpretation of a disaster widely accessible. Since all religions supported the idea of God's wrath, but only Lutherans were fervent adherents to the Apocalypse, it was common business logic to mostly maintain the idea of punishment. A handful of references to apocalypticism were enough to interest a Lutheran reader while not scaring away Calvinists or Catholics.

The aforementioned poem by Johan Fruytiers offers an excellent example. For the most part, this Calvinist writer interpreted the disaster as a punishment from God. He does, however, add some minor apocalyptic remarks, which do not line up with his Calvinist faith, since in this confession – at least in the sixteenth-century Low Countries – the Apocalypse was not a popular topic.[38] The interpretation of events as direct signs of a nearing Apocalypse could not count on much support. The inclusion of minor apocalyptic remarks, however, meant that readers like the Lutheran Godevaert van Haecht could also read

37 Geert H. Janssen, *The Dutch Revolt and Catholic Exile in Reformation Europe* (Cambridge: Cambridge University Press, 2014), 131–32; Judith Pollmann, *Catholic Identity and the Revolt of the Netherlands* (Oxford: Oxford University Press, 2011), 143–53.

38 Barnes, 'Varieties', 269.

this interpretation with agreement. By incorporating two different kinds of explanations for the disaster (punishment and Apocalypse), the pamphlet could attract different confessional groups and, therefore, a larger public.

Another example is offered by a pamphlet printed in Antwerp concerning the aforementioned earthquake in Switzerland in 1584. Whereas the author of the German pamphlet printed in Lutheran Strasbourg chose an explicitly apocalyptic interpretation, the Dutch pamphlet printed in Antwerp had to cater to a much more diverse religious landscape. *Waerachtich verhael vande eertbevinghe gheschiet in ... Berne* ('Truthful Story of the Earthquake that Happened in Bern', 1584) reports on the earthquake. The anonymous author had to keep in mind a Protestant as well as a Catholic readership. A citation of Augustine of Hippo, also known as St Augustine, on the front page is therefore meant to lure in Catholics as well as Protestants, since this theologian and philosopher was important to both denominations. The text itself mostly focuses on the wrath of God, with minor remarks on the Last Judgement. In this way the author kept readers from all backgrounds satisfied.

Conclusion

Although the Low Countries in the sixteenth century experienced crisis after crisis (ranging from rebellion, iconoclasm, and war to floods and epidemics), the apocalyptic reaction in Dutch pamphlets remained limited. In many texts the Apocalypse showed up only very briefly or not at all. A much more popular explanation for all these disasters was the wrath of God: He punished the people of the Low Countries for their sinful behaviour. This interpretation was not only more frequently used; it was also used more elaborately. By giving examples from the Bible and from current times, authors gave more weight to this interpretation. This stands in stark contrast with the use of the apocalyptic interpretation. When used, writers mentioned the End of Times only in passing with minor remarks, consisting of only a few words or a single sentence. Ways to do this included biblical citations on the title page, warning remarks in the preface, or apocalyptic utterances in the text itself. The apocalyptic interpretation does not consist of much more than this. This is all the more remarkable when compared to other countries, such as the Holy Roman Empire. In these regions the apocalyptic interpretation was used regularly. Writers elaborated on this interpretation by adding and explaining biblical prophecy and detailing the kinds of signs that were seen every day.

Even when we look at the Holy Roman Empire, the favoured interpretation differed by territory. After 1555 the ruler determined the religion of the subjects. This meant that the dominant confession was clear, and this determined which interpretation was more acceptable. In Lutheran regions, the Apocalypse became most important, while in Catholic areas the idea of punishment was popular. Authors and printers of pamphlets could easily tailor the interpretation of disasters to the confessional convictions of their public.

In the Low Countries, the confessional and political context was completely different. In these territories no clear choice was made for one confession. Although in the northern provinces Calvinism was the state-sponsored confession, freedom of conscience resulted in a diverse landscape. In the Southern Netherlands, it was only after 1585 that Catholicism regained its hold on the people. For authors and publishers, this meant that they had to cater to a much more diverse confessional reading public if they wanted to sell their pamphlets. Choosing an interpretation that was acceptable to most Christian confessions increased sales figures and could therefore make good business sense. This resulted in pamphlets that elaborate on the idea of punishment rather than the apocalyptic interpretation. Still, the Apocalypse was not ignored completely either. With minor remarks, authors and publishers tried to compromise in order not to scare away Catholics or Calvinists and, at the same time, to keep Lutherans interested as well.

Coming back to Barnes's concept of the Apocalypse as a polemical explanation of disasters, it turns out that this explanation does not fit the Low Countries in the sixteenth century. The diversity of the confessional landscape meant that authors and publishers probably did not choose to use the Apocalypse polemically, because it would scare away potential buyers. By focusing on punishment with only minor apocalyptic notions, they maximised their potential market and therefore their profits. It seems that choosing a specific interpretation of a disaster was not always ideologically bound. It could just as well be about profit, business, and economics. Dealing *in* disasters was as important as dealing *with* disasters.

About the Author

Marieke van Egeraat is a historian interested in news, disasters, and identity formation. From 2018 to 2022, she was a PhD candidate in the NWO-funded Vici project *Dealing with Disasters. The Shaping of Local and National Identities in the Netherlands (1421–1890)*. Within this project, she focused on the sixteenth-century Low Countries and the way providential readings of disasters could be used to strengthen group identities.

7 The Ten Plagues of the New World

The Sensemaking of Epidemic Depopulation in Sixteenth-Century Mesoamerica

Florian Wieser

Abstract

Through a study of the epidemics that devastated Mesoamerica's indigenous population in the sixteenth century, this chapter develops 'sensemaking' as a useful analytical category for the history of disasters. To do so, it revisits established sources like chronicles, medical handbooks, and indigenous painted annals to investigate the discourses produced by different social groups in Spanish colonial society in their attempts to explain why and how the epidemics had occurred. Comparing these groups (medical professionals, clerics, conquest elites, indigenous intellectuals) under the premise of competition for dominance in a 'fractal society', the chapter draws conclusions regarding the social dimension of disaster. It suggests that while different social groups shared certain basic ideas about epidemic disease, they deployed these in different ways to support their respective positions. These disparate discourses are connected by their status as sensemaking, a category that serves an established function within the cycle of disaster and response.

Keywords

New Spain – Spanish empire – indigenous Mexicans – epidemic disease – disaster cycle

In 1519 Spanish *conquistador* Hernán Cortés entered Mesoamerica, a cultural sphere corresponding to today's central and southern Mexico, Guatemala, Nicaragua, Honduras, and Belize. Within the span of two years, he and his indigenous allies overthrew the 'Aztec' Triple Alliance that had been the dominant power in the region until then and claimed its territory for

---

Asperen, van, Hanneke and Lotte Jensen (eds): *Dealing with Disasters from Early Modern to Modern Times. Cultural Responses to Catastrophes.* Taylor & Francis Group, 2023
DOI: 10.5117/9789463725798_CH07

the Spanish crown. Mesoamerica became the centrepiece of the colony of New Spain, yet as it did, the number of its indigenous inhabitants dropped rapidly. Exact figures continue to be debated, but even lower estimates put the rate of decline over the course of the sixteenth century at 50 per cent, with the highest at 96 per cent.[1]

Multiple factors played into this depopulation – the war of conquest itself, the colonial labour regime, etc. An undeniably major one, however, was disease. Throughout the sixteenth century, Mesoamerica was struck by numerous epidemics. Scholars have devoted copious efforts over the past decades to reconstructing the chronology of these epidemics and calculating their demographic effect.[2] This work has been trailblazing, firmly establishing the significance of disease in indigenous depopulation and making immense contributions to the history of medicine. I summarise some of its results in Table 1, outlining the major epidemics identified.

However, the focus on nosology and demography has caused scholars to neglect other aspects of the epidemic experience. In particular, the social meaning-making surrounding sixteenth-century Mesoamerican epidemics remains barely explored.[3] In the following study, I offer a preliminary analysis of that aspect. I investigate the discourses of rationalisation and explanation that contemporaries created in the face of devastating disease. To do so, I tap into many of the same sources that have been used for quantitative analyses of the epidemics: the chronicles of Spanish and indigenous Mesoamerican authors as well as medical handbooks and reports to the crown, from sixteenth-century eyewitness accounts to seventeenth-century retrospectives.

1 See the overview in Robert McCaa, 'The Peopling of Mexico from Origins to Revolution', in Michael R. Haines and Richard H. Steckel (eds), *A Population History of North America* (Cambridge: Cambridge University Press, 2000), 241–304, at 252–58.

2 See the overviews in Noble David Cook, *Born to Die. Disease and New World Conquest, 1492–1650* (Cambridge: Cambridge University Press, 1998); shorter but more recent: W. George Lovell, 'Columbus and Covid-19. Amerindian Antecedents to the Global Pandemic', *Americanía. Revista de Estudios Latinoamericanos de la Universidad Pablo de Olavide de Sevilla* 11 (2020), 4–31.

3 The studies that come closest to this topic are Martha Few, 'Indian Autopsy and Epidemic Disease in Early Colonial Mexico', in Rebecca P. Brienen and Margaret A. Jackson (eds), *Invasion and Transformation. Interdisciplinary Perspectives on the Conquest of Mexico* (Boulder: University of Colorado, 2008), 153–65, which offers thoughts on the connection of epidemic discourse to processes of colonial othering, and the two pieces by Sandra Elena Guevara Flores, 'La construcción sociocultural del cocoliztli en la epidemia de 1545 a 1548 en la Nueva España', PhD dissertation, Universitat Autònoma de Barcelona, 2017 and 'A través de sus ojos. Médicos indígenas y el cocoliztli de 1545 en la Nueva España', *eHumanista* 39 (2018), 36–52, which discuss the diagnostic perspectives indigenous Mesoamericans themselves developed.

I thus engage in the research approach historian Matthew Restall calls 'revisitation', a re-reading of established, mostly Spanish sources that breaks through their seemingly monolithic agenda.[4] Indeed, the society of sixteenth-century New Spain was not unified in its purpose but, as historian Serge Gruzinski phrases it, 'fractal', meaning unstable, heterogeneous, and hybrid. Not just colonisers and colonised clashed; the Spanish themselves were competing against each other.[5] Below I will show how different subsets of that society had different bases for and different aims in explaining the epidemics.

To do so, I rely on some systematisations I made in a previous study, where I outlined the major dynamism of discourse as that between the *conquistadores* and their settler successors, the missionaries, and indigenous intellectuals.[6] In the case of epidemic rationalisation, other currents must also be considered, most notably that of medical specialists. Even though competing, these different groups existed within a common social framework and conditioned each other. To explore their dynamic, I introduce thoughts and methodology from the field of history of disasters, most importantly the category of *sensemaking*. This concept allows me to trace larger patterns manifest within the Mesoamerican epidemic experience.

My study thus has two primary aims: First, to offer an outline and analysis of the sensemaking of the sixteenth-century Mesoamerican epidemics as a complex social contest. Second, to use these insights into the idea of sensemaking to explore its value as a general analytical tool in the history of disasters. I begin my study by defining what I mean by sensemaking and how it fits into broader explanations of disaster response. I then examine the practical responses of New Spain's authorities to the sixteenth-century epidemics as a manifestation of their ideas about disease. This serves as a basis to discuss how four different social groups – medical professionals, members of the clergy, conquest elites, and indigenous intellectuals – applied and evaluated these ideas. Throughout, I show that sensemaking was always not just an attempt to understand but also to impact society and one's status within it.

4 Matthew Restall, 'The New Conquest History', *History Compass* 10: 2 (2012), 151–60, at 155.

5 Serge Gruzinski, 'Las repercusiones de la conquista. La experiencia novohispana', in Carmen Bernard (ed.), *Descubrimiento, conquista y colonización de América a quinientos años* (Mexico City: Fondo de Cultura Economica, 1998), 148–71, at 151, 153, 167.

6 Florian Wieser, '"... und ich weiß, dass sie alle Sodomiten sind". Diskurse von Macht, Männlichkeit und Homosexualitäten in Darstellungen des frühkolonialen Neuspanien (Mexiko)', *Invertito. Jahrbuch für die Geschichte der Homosexualitäten* 19 (2017), 10–39.

Theoretical Considerations

I derive the concept of sensemaking from economic historian John Singleton. In a 2015 study, he formulated his Disaster Management Cycle (DMC), a multi-stage model illustrating how societies react to disasters and how these reactions shape the likelihood of future catastrophe. The DMC articulates crucial ways of thinking about disasters, including epidemics, highlighting the human contribution to their unfolding. It builds upon various earlier disaster-cycle models, including ones used by governments to help understand and combat disasters in the present. However, Singleton argues that these have neglected an important element of disaster response: that of 'apportionment of blame', in which certain individuals or groups within society are made into scapegoats for what has happened. He makes this stage the sixth in his cycle, occurring in the aftermath of relief measures.[7]

As important as Singleton finds the 'apportionment of blame', it is precisely this aspect of his DMC that has been most sharply criticised. The authors of *Disasters and History* argue that scapegoating is not a necessary part of disaster response and challenge the notion of a uniform cycle through which all disasters develop.[8] It is in reaction to these criticisms that I propose Singleton's notion of sensemaking can be made useful. Singleton writes that sensemaking is necessary for 'decision making', the third stage of his cycle, which directly follows the disastrous event itself. To initiate any measures that combat the unfolding of disaster, society must first develop an understanding of what is happening and why.

However, I suggest that Singleton misses an important implication here. What he calls 'apportionment of blame' is itself a form of sensemaking, a search for disaster causation. This search does not necessarily end with certain individuals serving as scapegoats. Rather, it is a quest for understanding that manifests throughout the disaster experience, not just at a single stage of it, and that informs the measures taken from first relief efforts to the long-term rebuilding of society. As the authors of *Disasters and History* show, various social groups engage in self-serving negotiation and competition to formulate such measures.[9] Sensemaking, in the broadened

7 John Singleton, 'Using the Disaster Cycle in Economic and Social History', in A.T. Brown, Andy Burn, and Rob Doherty (eds), *Crises in Economic and Social History. A Comparative Perspective* (Woodbridge: The Boydell Press, 2015), 53–78.

8 Bas van Bavel, Daniel R. Curtis, Jessica Dijkman, Matthew Hannaford, Maïka de Keyzer, Eline van Onacker, and Tim Soens, *Disasters and History. The Vulnerability and Resilience of Past Societies* (Cambridge: Cambridge University Press, 2020), 33.

9 van Bavel et al., *Disasters and History*, 3, 6–7, 26, 36–38, 86, 99, 145, 152, 178–79.

definition I develop here, is thus any discourse that investigates the causes of a specific disaster. As such, it serves a double purpose. It is certainly an attempt at explanation and thus enables informed action. At the same time, however, it is also an attempt to use that action to change society to one's own benefit.

Thus, while I borrow the concept of sensemaking from Singleton, I redefine it and use it to move away from the formulaic rigidity of his DMC. In doing so, I create an analytical tool that helps me frame the sources I examine and relate them to each other as expressions of a certain historical position: that of members of a society affected by disaster who tackle the questions that their situation entails. The actions they take are based on the answers they find; as disaster historian Virginia García Acosta observes, discursive and practical disaster responses condition each other.[10] Thus, it is appropriate for me to begin my exploration of the sensemaking of sixteenth-century Mesoamerican epidemics not with theories of causation but with the relief measures taken.

Combatting Epidemic Disease, Physically and Spiritually

In the context of sixteenth-century New Spain, relief measures are best attested for the 1576 epidemic. During that period, Viceroy Martín Enríquez de Almanza provided flour and bread to the sick, had Mexico City cleaned of refuse, and convened 'all doctors' to investigate the origins of the disease.[11] This included ordering autopsies.[12] Monastic orders also provided medical and spiritual assistance to the sick.[13] Their application of the sacraments

10 Virginia García Acosta, 'Divinidad y desastres. Interpretaciones, manifestaciones y respuestas', *Revista de Historia Moderna. Anales de la Universidad de Alicante* 35 (2017), 46–82, at 53.

11 'todos los medicos', Archivo General de Indias, Seville (hereafter AGI), México, 19, N. 177, no. 22 (31 October 1576). See also AGI, México, 20, N. 43 (30 October 1580); *Geschichte der Azteken. Codex Aubin und verwandte Dokumente*, ed. and trans. Walter Lehmann and Gerdt Kutscher (Berlin: Gebrüder-Mann-Verlag, 1981), 54 (= *Codex Aubin*, fol. 65v); Fr. Jerónimo de Mendieta, *Historia eclesiástica indiana*, ed. Joaquín García Icazbalceta (Mexico City: Antigua Librería, 1870), 517.

12 Few, 'Indian Autopsy', 155. Compare Alonso López de Hinojoso, *Svmma y recopilacion de cirvgia, con vn arte para sangrar, y examen de barberos* (Mexico City: Pedro Balli, 1595), fol. 150r.

13 Agustín Dávila Padilla, *Historia de la fvndacion y discvrso de la provincia, de Santiago de Mexico de la Orden de los Predicadores por las vidas de sus varones insignes casos notables de Nueva España* (2nd ed., Brussels: Jan van Meerbeeck, 1625), 157; Fr. Juan de Grijalva, *Crónica de la orden de N.P.S. Augustin en las prouincias de la nueua españa* (n.p., n.d.), fol. 68v; Mendieta, *Historia eclesiástica*, 515–17; Juan Sánchez Baquero, S.I., 'Relación breve del principio y progreso de la provincia de Nueva España de la Compañía de Jesús', in Francisco González de Cossío (ed.),

was seen as at least as important as physical healing and was recommended by doctors just as much as by priests.[14]

Such sincere belief in religion's healing power explains why processions were one of the most prominent steps taken against epidemics. Historian Martha Few considers them merely 'designed to calm the rattled capital',[15] but that overlooks the significance assigned to them by contemporaries. Historian Abigail Agresta notes that, for the plague epidemics in late medieval València, processions were organised even before the seemingly more practical measure of street cleaning.[16] Indeed, among contemporaries, Augustinian friar Juan de Grijalva called the 1576 procession of the image of the Virgin of Remedios to Mexico City the 'only remedy' that showed effect, and Tlaxcaltec historiographer Diego Muñoz Camargo credited it with lowering the death toll in that year compared to other epidemics.[17]

Such measures reveal a worldview that combined the Galenic-Hippocratic framework of medicine with encompassing Christian piety. The medical paradigm at work here postulated that the human body functioned because of the fluctuations of its humours. Sickness was caused by humours being either imbalanced or outright corrupted. Viewed through the lens of Catholic Christianity, this corruption could have both physical and spiritual causes. One material cause was the inhalation of miasma, malignant air generated through astronomical and seismological phenomena.[18] Yet miasma could also exude from an immoral person.[19]

Colonial officials fielding disaster response thus operated within a framework in which medicine and religion were both considered authoritative on the nature of epidemic disease. The claims they made were sometimes competing and sometimes complementary, but at no point without ulterior

Crónicas de la Compañía de Jesús en la Nueva España (1957; first reprint of 3rd ed., Mexico City: Universidad Nacional Autónoma de México, 2007), 83.

14 See López de Hinojoso, *Svmma*, fol. 152r.

15 Few, 'Indian Autopsy', 155.

16 Abigail Agresta, 'From Purification to Protection. Plague Response in Late Medieval Valencia', *Speculum* 95: 2 (2020), 371–95, at 371.

17 'vnico remedio', Grijalva, *Crónica de la orden*, fol. 85v; 'Diego Muñoz Camargos Chronik "Descripción de la Provincia y Ciudad de Tlaxcala" (ca. 1562–1592)', vol. 2, ed. and trans. Christine von Wantoch Rekowski-Santos, PhD dissertation, University of the Saarland, 2002, 36.

18 Agresta, 'From Purification to Protection', 378–79; Jorge Cañizares-Esguerra, 'New World, New Stars. Patriotic Astrology and the Invention of Indian and Creole Bodies in Colonial Spanish America, 1600–1650', *American Historical Review* 104 (1999), 33–68, at 36–37; Elsa Malvido and Carlos Viesca Treviño, 'La epidemia de cocoliztli de 1576', *Historias México* 11 (1985), 27–33, at 31.

19 Annemarie Kinzelbach, 'Infection, Contagion, and Public Health in Late Medieval and Early Modern German Imperial Towns', *Journal of the History of Medicine and Allied Sciences* 16: 3 (2006), 360–89, at 377.

motives. When doctors and clergy as well as other social groups engaged in sensemaking by developing theories of epidemic causation, they did so not just to aid in combatting this and future disasters but also to protect their own long-term interests in a constantly struggling 'fractal society'.

The Medical Professionals

Medical professionals of different backgrounds were among those who received most official legitimacy in explaining and combatting the epidemics. Viceroys and even the Spanish king trusted their expertise and employed them to apply it. This endowed their statements with a privileged status; the authority they carried made their academic sensemaking the basis for that of others.

Once again, the sources are clearest for the 1576 epidemic. Two primary medical eyewitnesses report on it, eminent experts hired to analyse the epidemic through autopsies of indigenous victims: the physician Francisco Hernández de Toledo and the surgeon Alonso López de Hinojoso. Their attempts at sensemaking are overall similar but differ on some details. Hernández speaks of the 'dry and calm' weather in 1576 having created 'a real breeding ground for rotting and corruption'.[20] This explanation found its way into the official account given by Viceroy Enríquez to the king.[21] Similarly analysing environmental causes, López specifies that the corruption had come about because of 'conjunctions of stars or ... comets'.[22] This already implies a certain value judgement, as historian Jorge Cañizares-Esguerra has pointed out that the early modern discourse of American astrology drew a connection between the more 'malignant' stars above that part of the world and the weak, sinful, inferior nature of its native inhabitants.[23]

Indeed, both Hernández and López go on to blame the epidemic on indigenous victims themselves. Hernández considers the 'most sordid food' the natives ingested, including alcohol, chili, and corn, a principal cause of their humoral imbalance.[24] López meanwhile sees indigenous tobacco consumption as the culprit. He explains that smoking not only upsets

20 Francisco Hernández de Toledo, *The Mexican Treasury. The Writings of Dr. Francisco Hernández*, ed. Simon Varey, trans. Rafael Chabrán, Cynthia L. Chamberlin, and Simon Varey (Stanford, CA: Stanford University Press, 2000), 84.

21 AGI, México, 19, N. 177, no. 22 (31 October 1576).

22 'conjunciones de estrellas o ... cometas', López de Hinojoso, *Svmma*, fol. 151v.

23 Cañizares-Esguerra, 'New World, New Stars', 40–45.

24 Hernández de Toledo, *The Mexican Treasury*, 84. This can be compared to a 1571 example from the German city of Nördlingen, where the dietary habits of the city's poor were blamed

the body's functions but also, due to a trick of the Devil, masks the pain that would normally warn one of disease.[25] Thus, physiological and moral explanations are again blended here. That both Hernández and López assign concrete blame for the epidemic reveals the power play at work in their sensemaking.

Both men were highly respected, Hernández as former royal physician and López as a veteran of hospital practice in Mexico.[26] Yet both nonetheless had to defend the validity of their craft and the social position derived from it, not least against each other. By the late sixteenth century, an intense conflict for authority was raging in the Spanish world between physicians trained in the classical theories of Hippocrates and Galen and surgeons who assimilated new knowledge from empirical observation.[27] As Few has argued, by physiologically and morally othering indigenous Mesoamericans and showing the superiority of the European scholar to the Native American disease victim and thus the former's ability to comprehend the latter's alien afflictions, both Hernández and López insisted on the continued need for their respective fields of medical practice.[28] They made claims to authority and thus created a discourse of sensemaking that others drew upon.

The Clergy

Besides medical professionals, the clerics were most directly on the frontlines of the battle against epidemic disease. In investigating the causes of that disease, they were far from ignoring the Hippocratic-Galenic model. Yet they usually emphasised explanations based on Christian moralism over physiological ones. Only the Jesuit Juan Sánchez Baquero, writing at the turn of the seventeenth century, presents a clear synthesis of Hippocratic-Galenic and Christian religious approaches. He mentions that a conjunction of Mars and Saturn which specifically affected the melancholic complexions of the natives was blamed for the 1576 epidemic.

for the epidemic they were suffering from; see Kinzelbach, 'Infection, Contagion, and Public Health', 383.

25 López de Hinojoso, *Svmma*, fol. 151r–v.

26 Few, 'Indian Autopsy', 156–57.

27 See Michele L. Clouse, *Medicine, Government and Public Health in Philip II's Spain. Shared Interests, Competing Authorities* (Oxon and New York: Routledge, 2011), 75; José Luis Fresquet Febrer, 'La práctica médica en los textos quirúrgicos españoles en el siglo XVI', *Dynamis* 22 (2002), 251–77.

28 Few, 'Indian Autopsy', 159–62.

However, there were also, he says, those who 'attributed it to the scourge and punishment of God, who wanted to take vengeance for the idolatry and cruelty of these people'.[29]

This argument, viewing the near extinction of indigenous Mesoamericans as just punishment for their pagan past, may seem the most familiar to the modern observer, but in truth the situation is remarkably complex. Its most prominent rendition was delivered by Franciscan missionary Motolinía shortly after the Conquest. He considered the epidemics occurring in the 1520s and 1530s the first of the Ten Plagues of the New World.[30] With these plagues, Motolinía argues, God was punishing indigenous Mesoamericans for the human sacrifices they had practised before Christianisation. Yet Motolinía also notes that the Ten Plagues of the New World were harsher than those of Egypt, with all ten of them being deadly instead of one, implying some doubt about how deserved they were.[31] Thus, he suggests that God was punishing the whole world for 'our sins', not those of a specific group.[32]

This opened the way for a different kind of moral argument, one followed by most religious authors after Motolinía. Sánchez himself states that 'there are very strong reasons to think that it is not just the faults of the Indians that Our Lord is punishing with this scourge'.[33] The natives were going to Heaven, he argues, whereas the Spanish languished in the absence of their labourers.[34] The Augustinian Grijalva made this same point even more strongly. Referencing the plague that struck Israel under King David (1 Chronicles 21), he argues that the Spanish were the true subject of God's wrath, their vassals taken away as punishment for their vanity and greed. At the same time, however, God 'always mixes justice with His mercies'. In death, the natives 'were freed from a subject life' as reward for their speedy conversion to Christianity.[35]

The Franciscan Jerónimo de Mendieta wholeheartedly agreed, stating that 'God sent them the pestilence not for their ill but for their good'. This way, 'our

29 'lo atribuyeron a azote y castigo de Dios, que quiso tomas venganza de la idolatría y crueldad de esta gente', Sánchez Baquero, 'Relación breve', 80.

30 Motolinía (= Fr. Toribio de Benavente), *Memoriales* (*Libro de oro, MS JGI 31*), ed. Nancy Joe Dyer (Mexico City: El Colegio de México, 1996), 137–38.

31 Motolinía, *Memoriales*, 146.

32 Motolinía, *Memoriales*, 405, 408.

33 'hay muy fuertes razones para pensar no fueran solas las culpas de los indios las que castigó Nuestro Señor con este azote', Sánchez Baquero, 'Relación breve', 81.

34 Sánchez Baquero, 'Relación breve', 86.

35 'siempre mezcla la justicia con sus misericordias', 'se librauà de una vida sujeta', Grijalva, *Crónica de la orden*, fol. 68v.

God is filling the seats of Heaven in order to end the world'.[36] Far from being punished sinners, to these missionaries, the natives were beloved by God and taken up to Him in preparation for the coming apocalypse. This point had already been stated by the Dominican Domingo de Betanzos in 1545; when considering the prophecy in Matthew 24, he links the disappearance of Native Americans to the coming of the end times in seventy years or less.[37]

The clergy of colonial New Spain thus cited the Bible as authority in much the same way that medical men did the classics of medicine. Their respective worldviews were not exclusive but rather rested on common assumptions.[38] Yet they drew different conclusions from this shared basis, ones that mapped onto their own interests as parties to early colonial power struggles. The missionary fathers of New Spain were aware of how incomplete their evangelisation of indigenous Mesoamericans was and at times complained bitterly about this situation. However, they had no interest in conceiving of the natives as unredeemable pagans or naturally disposed towards wickedness.[39] That God was just and disaster a manifestation of His justice was unquestioned.[40] Yet by shifting the cause of divine wrath from indigenous victims of disease to Spanish victims of labour shortage, the clergymen could acknowledge that wrath without delegitimising their missionary efforts as wasted upon the sinful and unworthy.

Nonetheless, this was not a unified line of argument. Motolinía, one of the first to attempt a sensemaking of the epidemics, remained ambiguous about whose sins had angered God. Meanwhile, the Dominican Bernardo de Santa María seems to have fully embraced the view of indigenous bodies as anatomically inferior. He co-authored the *Relación Geográfica* ('Geographic Account') for Nexapa in Oaxaca, one of a series of descriptions of colonial towns for administrative use, which were created in the 1570s at the order of King Philip II. This *Relación de Nexapa* ('Account of Nexapa') suggests that epidemics occurred 'because by their nature [the natives] always carry a terrible stench with them, enough to engender by itself whatever pestilence'.[41]

36 'la pestilencia se la envia Dios, no por su mal sino por su bien', 'va hinchiendo muestro Dios de ellos las sillas del cielo para concluir con el mundo', Mendieta, *Historia eclesiástica*, 518.

37 Fr. Domingo de Betanzos, 'Carta de Fray Domingo de Betanzos', in Joaquín García Icazbalceta (ed.), *Colección de documentos para la historia de México*, vol. 2 of 2 vols (Mexico City: Antigua Librería, 1866), 198–201. See also Dávila Padilla, *Historia de la fvndacion*, 156.

38 e.g. Sánchez Baquero's appraisal of astrology as an instrument of God at 'Relación breve', 80.

39 Wieser, '"... und ich weiß"', 29–30.

40 Compare García Acosta, 'Divinidad y desastres', 47–50.

41 'porq[ue de] su natural traen siempre consigo una hedentina pésima, bastante para engendrar en sí cualquiera pestilencia', René Acuña (ed.), *Relaciones geográficas del siglo XVI*, vol. 2 (Mexico City: Universidad Nacional Autónoma de México, 1982–88), 352.

This is a miasmic explanation and, unlike most authored by members of the clergy, one reflecting negatively upon indigenous Mesoamericans. Thus, the sensemaking discourses produced by different groups at all times overlapped and fed into each other.

The Conquest Elite

Although Cortés rarely passed up an opportunity to justify his conquest of Mesoamerica with the alleged sinfulness of the natives, he offered no explanation for the 1520 smallpox epidemic, even though it aided him greatly in achieving his victory.[42] Possibly, he remained silent because his own indigenous allies were affected by the disease before his enemies were. Some of those around Cortés, however, were more candid. The *conquistador* Francisco de Aguilar plainly stated that 'our God was served, as the Christians were very much tired by the war, to send them smallpox, and among the Indians came a great pestilence'.[43] Here, epidemic disease is God's way of helping Christianity triumph over paganism. Somewhat pettier yet still invoking higher justice, Cortés' biographer Francisco López de Gómara wrote that 'here they paid for the syphilis that they had given to ours'.[44]

Yet López also delivered a complementary explanation for how smallpox came to Mesoamerica. Cortés had begun his conquest of Mexico without official license (explaining his heightened need to justify himself), and his superior, Governor of Cuba Diego Velázquez de Cuéllar, had sent Pánfilo de Narváez to capture him. Cortés swiftly defeated Narváez, but by López's account, there was still some damage done. Among Narváez's troops, López relates, had been 'a black man with smallpox' who subsequently passed the illness on to the natives.[45] This account is echoed by the *conquistador*

42 Compare Hernán Cortés, *Cartas de relación*, ed. Mario Hernández (Madrid: Historia 16, 1985), 188. For an example of Cortés' rhetoric of native sin, see Wieser, '"... und ich weiß"', 19. For the military importance of the epidemic, see Cook, *Born to Die*, 67–69; Robert McCaa, 'Spanish and Nahuatl Views on Smallpox and Demographic Catastrophe in Mexico', *The Journal of Interdisciplinary History* 25: 3 (1995), 397–431, at 411.

43 'con esto fue nuestro Dios servido, estando los cristianos harto fatigados de la guerra, de enviarles viruelas, y entre los indios vino una grande pestilencia', Fr. Francisco de Aguilar, *Relación breve de la conquista de la Nueva España*, ed. Jorge Gurría Lacroix (Mexico City: Universidad Nacional Autónoma de México, 1977), 96.

44 'pagaron aquí las bubas que pegaron a los nuestros', Francisco López de Gómara, *La conquista de México*, ed. José Luis de Rojas (Madrid: Historia 16, 1987), 227. This reason is also cited, though in a more doubting tone, in Mendieta, *Historia eclesiástica*, 514.

45 'un negro con viruelas', López de Gómara, *La conquista de México*, 226.

Bernal Díaz del Castillo and by Motolinía, who as one of the first Franciscan missionaries was personally favoured by Cortés.[46]

It is tempting to assume greater validity for this account, which seems to conform to modern ideas of contagion. Yet it coexisted with theories of divine punishment, sometimes in the same text, and served the same purpose as these. Those who had led the conquest of Mexico needed to justify it to continue enjoying their stewardship of indigenous labour. This justification had to be both general, depicting indigenous Mesoamericans as sinners worthy of subjugation, and specific, presenting Cortés and his men as in the right and those who had opposed them in the wrong. The 1520 epidemic among the natives was well deserved, these authors affirmed, yet it had been caused not by the conqueror Cortés but his rival Narváez. Thus, the sensemaking of the conquest elite explained epidemic causation from social relations and treated disaster as a confirmation of existing hierarchies. That the concrete agent of contagion is here a black man seems particularly appropriate.[47] Sickness was passed from one racial inferior to another and thus served as proof of that very inferiority, in turn justifying the exploitation of indigenous (and black) labour on which the conquest elite depended.

The Indigenous Intellectuals

The fact that a large portion of sensemaking for the Mesoamerican epidemics revolved around the relative culpability of indigenous people makes the question of their own perspective all the more pressing. However, while indigenous Mesoamerican societies had their own intricate hierarchies and factionalisms, which were not so much superseded by as incorporated into the colonial state, the available sources represent only a fraction of that complexity. Whether first-hand accounts in indigenous languages and Spanish or second-hand reports communicated by Spanish observers, they come overwhelmingly from the indigenous nobility.

This indigenous elite was divided among countless ethnic groups, each seeking to advance the agenda of their respective family and city-state, yet

46 Bernal Díaz del Castillo, *Historia verdadera de la conquista de la Nueva España*, vol. B, ed. Miguel León-Portilla (Madrid: Historia 16, 1984), 479; Motolinía, *Memoriales*, 137.

47 For more thoughts on this, see Thomas Franke, 'The Other Early Modern Pandemic. Slavery, Colonialism, and Novel Coronavirus', *The Electronic Sixteenth Century Journal*, published online 25 August 2020, https://www.escj.org/blog/other-early-modern-pandemic-slavery-colonialism-and-novel-coronavirus.html (accessed 3 April 2023).

it was also united by its relationship to the colonisers. Indigenous nobles were cultural mediators. They collected taxes for their Spanish superiors and received a share. They thus also closely associated with missionaries and adopted elements of Spanish culture, becoming multiculturally and multilingually adept. They used these cultural resources to defend their own position. Their aim was to integrate with the colonial state in a way advantageous to themselves, and thus they acted both as enforcers of oppression and as champions of traditional rights as it suited them.[48]

Perhaps the most direct statements of indigenous mentality are the central Mexican painted chronicles, as they were a part of a pre-Conquest tradition that continued to be produced throughout the sixteenth century. Some were combined with written text in the dominant Nahuatl language, while others remained purely pictographic. Their primary function was to serve as the memory of their communities, recording noteworthy events, including disasters.[49] However, by their very nature, they do not offer theoretical explanations of the epidemics like Spanish accounts. As art historian Elizabeth Hill Boone has explained, the painted chronicles circumvent language and instead create meaning through the relationships of symbols to each other.[50]

Numerous painted chronicles contain pictographs depicting the epidemics. These have been studied by historian of science Sandra Elena Guevara Flores for information on disease symptoms.[51] Two, however, can also be used to approach indigenous sensemaking, as both single out epidemic disease as special and different from other causes of death. The usual pictograph marking someone's death is a body wrapped up in cloth for burial, and both chronicles record most deaths this way. Yet for certain records of epidemic disease, they vary the symbol. Thus, the *Codex Telleriano-Remensis* visualises the 1545 epidemic with a large pile of such bodies, one stacked upon the other (Fig. 7.1), expressing a nameless

48 Gabriela Ramos and Yanna Yannakakis, 'Introduction', in Ramos and Yannakakis (eds), *Indigenous Intellectuals. Knowledge, Power, and Colonial Culture in Mexico and the Andes* (Durham, NC: Duke University Press, 2014), 1–17; Peter B. Villella, *Indigenous Elites and Creole Identity in Colonial Mexico, 1500–1800* (Cambridge: Cambridge University Press, 2016), 4–5, 98–99, 113.

49 James Lockhart, *The Nahuas after the Conquest. A Social and Cultural History of the Indians of Central Mexico, Sixteenth through Eighteenth Centuries* (Stanford, CA: Stanford University Press, 1992), 345–51, 376–78.

50 Elizabeth Hill Boone, 'Introduction. Writing and Recording Knowledge', in Elizabeth Hill Boone and Walter D. Mignolo (eds), *Writing without Words. Alternative Literacies in Mesoamerica and the Andes* (London: Duke University Press, 1994), 3–26, at 8–9, 15. See also the discussion of Boone's method in Guevara Flores, 'La construcción sociocultural', 127.

51 Guevara Flores, 'La construcción sociocultural' and Guevara Flores, 'A través de sus ojos'.

Figure 7.1 *Pile of Epidemic Victims and an Individual Death*, pictographs in *Codex Telleriano-Remensis*, c. 1562, fol. 46v. Bibliothèque nationale de France, Paris, Département des Manuscrits, Mexicain 385. Source: Gallica BnF, https://gallica.bnf.fr/ark:/12148/btv1b8458267s (accessed 26 April 2022)

multitude. This is contrasted by a normal, single death on the same page. Meanwhile, the *Codex Aubin*, which records the history of Tenochtitlan (today's Mexico City), marks rulers who died of epidemic disease with reddish bulbs that imitate the pustules of smallpox or measles (compare Figs 7.2 and 7.3).

This attitude, which considers the epidemics as a shocking and unusual form of death, is continued by written accounts in indigenous languages, which began to replace painted chronicles by the late sixteenth century. The *Annals of the Kaqchikel*, a Guatemalan Maya account, announce for the 1520 epidemic: 'It was truly terrible, the number of dead there were in

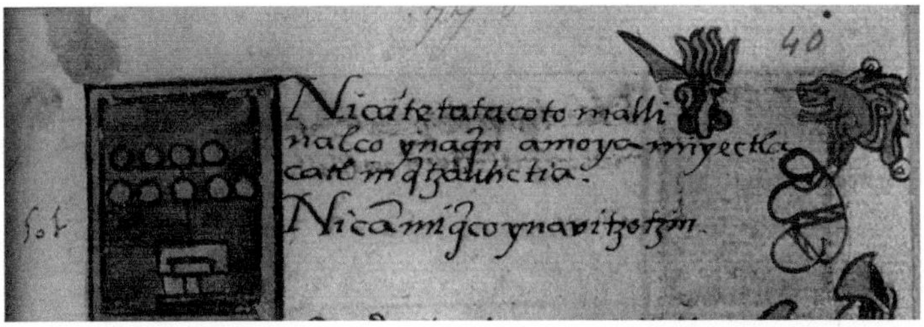

Figure 7.2 *Death of King Ahuizotl* (who died of a head injury), pictograph in *Codex Aubin*, c. 1576–1607, fol. 40r. British Museum, London, Am2006,Drg.31219. © The Trustees of the British Museum

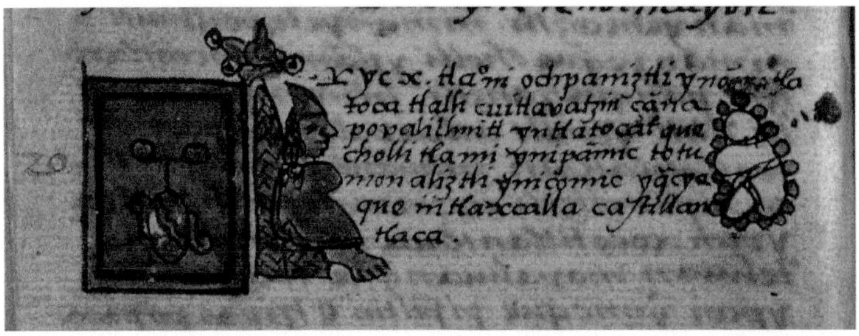

Figure 7.3 *Death of King Cuitlahuac* (who died of smallpox), pictograph in *Codex Aubin*, c. 1576–1607, fol. 44v. British Museum, London, Am2006,Drg.31219. © The Trustees of the British Museum

that period. ... Little by little heavy shadows and black night enveloped our fathers and grandfathers and us also, oh, my sons! when the plague raged. ... We were born to die!'[52] The Nahuatl-language *Annals of Tecamachalco* even record smallpox as the only event for 1520 – although the conquest itself was ongoing in that year![53] Part of what made the epidemics so horrifying to indigenous authors was their entirely unprecedented nature. 'Never once had this been seen; never had it been suffered in Mexico', say the Nahuatl materials collected from Tlatelolca nobles by Franciscan friar Bernardino de Sahagún.[54]

Given the novelty of what they were experiencing, many indigenous Mesoamericans professed ignorance of its causes as well, as several *Relaciones*

52 *The Annals of the Cakchiquels / Title of the Lords of Totonicapán*, trans. Adrián Recinos, Delia Goetz, and Dionisio Jóse Chonay (Norman: University of Oklahoma Press, 1953), 115–16.

53 *Anales de Tecamachalco, 1398–1590*, ed. and trans. Eustaquio Celestino and Luis Reyes García (Mexico City: Fondo de Cultura Económica, and Puebla: Gobierno del Estado de Puebla, 1992), 24. Compare Cook, *Born to Die*, 69; McCaa, 'Spanish and Nahuatl Views', 410.

54 Fr. Bernardino de Sahagún, *Florentine Codex. General History of the Things of New Spain*, vol. 9 of 13 vols, ed. and trans. Charles E. Dibble and Arthur J.O. Anderson (Santa Fe, NM: School of American Research, 1954), 4.

Geográficas report.[55] Although usually penned by Spanish officials, these reports drew upon interviews with local indigenous elites and can thus be used to complement their own writings.[56] Based on this material, anthropologist Barry Isaac suggests that '[n]ot daring to attribute mortality and morbidity to indigenous gods ... but not yet committed to assigning these powers to the Christian god, many of the native informants avoided assigning ultimate causality'.[57]

However, at least some of the indigenous chronicles do exactly that, speaking of the will of God. Divine causation is implied by the *Annals of Tecamachalco* calling smallpox *teozahuatl* ('divine rash') and outright stated in the *Annals of the Kaqchikel*, which assert for the 1560 epidemic that 'a fearful death fell on our heads by the will of our powerful God'.[58] Similarly, Chimalpahin, chronicler of the Nahua polity of Chalco, states for 1576 that it 'was thus the will of God Our Lord that we suffered'.[59] Notably, most sources that invoke the wrath of God remain silent on what had roused it in the first place. Only Muñoz, a descendant of Tlaxcaltec nobility, ambiguously speaks of '[our] sins'.[60] Meanwhile, the *Relación Geográfica* for Tiripetío in Michoacán even makes the opposite claim, taking mass depopulation as a sign of God's favour. It relates that the natives of Tiripetío believed they had lived longer before the conquest because God had wanted to give them more time to reject the Devil. Now that they were good Christians, He was fetching them more quickly.[61]

By adopting such Christian language, indigenous intellectuals could present themselves as deserving and pious subjects of the Catholic monarchy and their cause as allied to the Spanish one. Their claim to status as good Christians undercut the rhetoric with which the conquest elite were justifying their usurpation of native territories and labour.[62] When

55 Acuña, *Relaciones geográficas*, vol. 8, 102 (*Relación de Texcoco*), 162 (*Relación de Totolapan y su partido*); Acuña, *Relaciones geográficas*, vol. 10, 238 (*Relación de Tenamaxtlan*).

56 Villella, *Indigenous Elites and Creole Identity*, 74, 104.

57 Barry L. Isaac, 'Witnesses to Demographic Catastrophe. Indigenous Testimony in the Relaciones Geográficas of 1577–86 for Central Mexico', *Ethnohistory* 62: 2 (2015), 309–31, at 326.

58 Recinos et al., *The Annals of the Cakchiquels*, 143.

59 'Fue pues la voluntad de Dios Nuestro Señor que la sufriéramos', Francisco de San Antón Muñón Chimalpahin Cuauhtlehuanitzin, *Relaciones originales de Chalco Amaquemecan*, ed. and trans. Silvia Rendón (Mexico City and Buenos Aires: Fondo de Cultura Económica, 1965), 282.

60 'n[uest]ras culpas', Muñoz Camargo, 'Diego Muñoz Camargos Chronik', 36. For Muñoz's status within the indigenous nobility, see Villella, *Indigenous Elites*, 134.

61 Acuña, *Relaciones geográficas*, vol. 9, 346.

62 Villella, *Indigenous Elites*, 4–5, 9–10.

the indigenous nobility engaged in sensemaking regarding the epidemics bereaving their people, they were thus simultaneously fighting a battle over the titles and privileges inherited from their ancestors. This heritage also had an intellectual component, which they combined with European thought to create hybrid discourses that resonated across different cultures.

This can be seen in the Nahuatl concept of *tlazolli*, which describes both physical refuse and moral failing. Similar to how sin could create miasma in the European framework, a person accumulating *tlazolli* could affect others, drawing negative influences, including disease, towards them.[63] One aspect of *tlazolli* was bad smell, and so Santa María's account of native odour is paralleled by the indigenous town council of Tlaxcala, which worried about the sweaty stench of textile workers contributing to disease.[64] Meanwhile, when Chimalpahin echoed the surgeon López in linking the worsening spread of an epidemic in 1577 to the appearance of a comet, he was also in line with the traditional association of *tlazolli* with inauspicious celestial phenomena.[65]

Overall, indigenous intellectuals largely performed their sensemaking within the epistemic confines of the colonial state. Outright defiance of the dominant Spanish ideologies is attested only from indirect accounts. In these, Spanish authors refer to unnamed and unnumbered natives, whose status and background remain unclear. In two *Relaciones Geográficas*, one from Tuxcacuesco in Jalisco and the other from Huatulco in Oaxaca, disease is blamed on the cessation of pre-Christian worship, challenging the project of evangelisation. In the old times, the people of Huatulco supposedly said that their gods 'told them what they had to do to cure those who fell ill'.[66] Equally unspecific is the account of Dominican Agustín Dávila Padilla, who reports that, jealous of the health among the Spanish, some natives in Mexico City threw the dead bodies of epidemic victims into the water supply to poison it.[67]

These examples suggest that outside the intellectual elite, other discourses of sensemaking were developed among indigenous Mesoamericans, ones that rejected the colonial regime and blamed it for the epidemics. The

63 Louise M. Burkhart, *The Slippery Earth. Nahua-Christian Moral Dialogue in Sixteenth-Century Mexico* (Tucson: University of Arizona Press, 1989), 87, 91, 95–97, 171.

64 AGI, México, 1091, L. 11 (3 March 1585).

65 Chimalpahin Cuauhtlehuanitzin, *Relaciones originales de Chalco Amaquemecan*, 283.

66 'les decían lo que habían de hacer p[ar]a sanar cuando caían enfermos'. *Chimalpahin Cuauhtlehuanitzin, Relaciones originales de Chalco Amaquemecan*, vol. 2, 190. For the Jalisco case, see the same, vol. 10, 54.

67 Dávila Padilla, *Historia de la fvndacion*, 157–58.

prevalence and inner logic of these discourses, however, remain difficult to assess and thus reveal one of many directions for further research on sixteenth-century Mesoamerica's epidemic sensemaking.

Conclusions

This study has been, as cautioned before, no more than surface level. In its investigation of the social dimension of Mesoamerica's sixteenth-century epidemics, it has been a 'revisitation', limiting itself to well-known and oft-used sources examined through a new lens, that of sensemaking. Through this approach, the study has revealed how different theories of epidemic causation mapped onto different social groups, each struggling for self-assertion and dominance within the 'fractal society' of early colonial New Spain.

A shared centre of their sensemaking was Christian religion. Members of the clergy, conquest elites, and indigenous intellectuals all referred to God in their theories, and medical professionals recommended the sacraments as means of healing. However, the precise role assigned to divine power varies. Medical professionals foregrounded climatic and dietetic causation that emphasised the need for their expertise. Clerics, conquest elites, and many indigenous intellectuals agreed that the epidemics were divine punishment provoked by sin, but whose sin it was depended on the speaker's social position. The clergy blamed Spanish greed and spoke prophecies that required their spiritual work to be fulfilled. The conquerors blamed indigenous debauchery and each other, justifying their own privileges. Indigenous intellectuals produced the most varied and contradictory responses, bearing witness both to the shock of their losses and the strenuous work of cultural hybridisation. Together, these four groups give an idea of the many meanings assigned to the epidemics across colonial Mesoamerican society. Their discourses hint at the existence of many others, especially those of common people, that remain to be discovered through archival and indigenous-language sources.

With my study I have attempted to lay a groundwork upon which such further investigation can build. Part of this has been my refashioning of the idea of sensemaking as a discourse concerned with explaining disaster. This has allowed me to group together statements from across the epidemic experience and connect them to the larger processes of disaster response. Sensemaking stresses the importance of understanding for the formulation of relief measures while simultaneously showing how these explanations and the measures derived from them are products of social conflict and negotiation. It helps describe and analyse a style of discourse recurrent in various kinds of disaster situations. Thus, sensemaking serves as a tool to connect and compare social responses not just in the history of colonial Spanish American epidemics, but in the history of disasters as a whole.

Table 1 Overview of epidemics in sixteenth-century Mesoamerica, based on data from Cook, Born to Die; Lovell, 'Disease and Depopulation'; and Prem, 'Disease Outbreaks in Central Mexico'; compiled by the author

Time frame	**Affected regions**	**Spanish identifications**	**Indigenous names**
1520–21	all of Mesoamerica	*viruelas* (smallpox)	Nahuatl: *huey zahuatl* (great rash), *totomonaliztli* (pustules and blisters); K'iche': *chaac*
1530–31	Central Mexico	*sarampión* (measles)	Nahuatl: *zahuatl* (rash)
1532	Central Mexico	-	Nahuatl: *zahuatl* (rash), *totomonaliztli* (pustules and blisters)
1533	Guatemala, Nicaragua	*sarampión* (measles), *viruelas* (smallpox)	-
1545–48	all of Mesoamerica	*tabardillo* (typhus)	Nahuatl: *huey cocoliztli* (great sickness); K'iche': *[gu]cumatz* (snake); P'urépecha: *tepari* (swelling)
1550	Central Mexico	*paperas* (mumps)	Nahuatl: *quechpozahualiztli* (throat sickness)
1558–62	Guatemala	*sarampión* (measles), *tabardillo* (typhus), *viruelas* (smallpox)	-
1560	Veracruz, Guatemala	-	-
1563–64	Central Mexico, Guatemala	*sarampión* (measles), *viruelas* (smallpox)	Nahuatl: *matlatotonqui* (green fever), *zahuatl* (rash)
1576–80	all of Mesoamerica	*sarampión* (measles), *tabardillo* (typhus), *viruelas* (smallpox)	Nahuatl: *matlatotonqui* (green fever)
1587	Central Mexico	-	Nahuatl: *cocoliztli* (sickness)
1595	Central Mexico	*paperas* (mumps), *sarampión* (measles), *tabardillo* (typhus)	-

About the Author

Florian Wieser is a PhD student in History at the University of Edinburgh. He previously studied at Ludwig-Maximilians-University Munich and Universidad Autónoma de Madrid and worked as a research assistant in projects of Ludwig-Maximilians-University and of the Deutsches Medizinhistorisches Museum Ingolstadt. His research interests include early modern colonialism, history of marginalised peoples, history of the body, gender, and sexuality, and discursive approaches to these topics. His research is currently being funded by the Scottish Graduate School for the Arts and Humanities.

8 'Hungry Balliz Wants Weel Fillin'

The Visualisation of the Great Irish Famine (1845–1851) and the Lancashire Cotton Famine (1861–1865) in the Victorian Illustrated Press

Sophie van Os

Abstract

Between 1862 and 1864, the *Illustrated London News* (1842) published several wood engravings which directly or indirectly referenced the Lancashire Cotton Famine (1861–65). Though poetic responses to this disaster have recently received scholarly attention and their significance has been recognised, visual representations of the Cotton Famine have remained largely unexplored. To uncover how early Victorian illustrated periodicals attempted to visualise famine, this article will examine press renderings of three nineteenth-century instances of hunger – the Great Irish Famine (1845–51), rural poverty in Dorset (1846), and the Lancashire Cotton Famine (1861–65) – drawing parallels between these three case studies in terms of subject, form, and stylistic registers. It argues that illustrations of the Lancashire Cotton Famine belonged to a larger nineteenth-century cultural repertoire of depicting hunger, poverty, and famine in the Victorian illustrated press that dated back to the 1840s, a decade that has become synonymous with hunger in historical discourse.

Keywords

Lancashire Cotton Famine – Great Irish Famine – wood engraving – stylistic registers – nineteenth century – illustrated press

Tears wer rowlin daun ther faces,
Fallin ontuth' graund like rain,
Un ther limbs fergeet ther places
Uz they trembl'd fra ther pain.

Asperen, van, Hanneke and Lotte Jensen (eds): *Dealing with Disasters from Early Modern to Modern Times. Cultural Responses to Catastrophes.* Taylor & Francis Group, 2023
DOI: 10.5117/9789463725798_CH08

Cowd un hungry een kept starin
Savagely, like owt ut's mad;
Even th' mother seem'd dispairin
Uz hoo seet there, cowd un sad.[1]

On 29 August 1863 the *Burnley Advertiser*, a local newspaper in East Lancashire, published an anonymous poem titled 'Th' Poor Mon un Hiz Childer', written in Lanky dialect.[2] Victorian dialect poetry has often been characterised as 'conciliatory and conservative ... homely rhyming' turned inward 'by embracing the ideology of domesticity, and participating in an apotheosis of the home and family'.[3] This poem, however, subverts traditional domestic scenes of marriage and love to underline how the warmth and security of the hearth and homestead have been stripped away by a debilitating hunger that leaves this family 'Cowd un hungry'.[4] 'Th' Poor Mon un Hiz Childer', like hundreds of recently recovered regional poems, was inspired by the Lancashire Cotton Famine (1861–65), a depression in the textile industry of North West England due to an overproduction of cotton, and 'the blockade of the southern ports of the United States' caused by the American Civil War (1861–65).[5] The challenging trading conditions and the inaccessibility of raw cotton brought about a change in the social circumstances of the regional cotton mill workforce. Factory owners no longer bought large quantities of raw cotton to process and, as a result, workers from considerable parts of Lancashire and its surrounding regions became unemployed.[6] The consequent suffering of these workers was widely discussed in English national and regional newspapers of the time.

Though poetic responses to the distress in Lancashire have recently received more scholarly attention and their significance has been recognised, visual representations of the Lancashire Cotton Famine have

1 'Th' Poor Mon un Hiz Childer', *Burnley Advertiser* (29 August 1863).

2 'Lanky' refers to the Northern vernacular speech spoken in the county of Lancashire.

3 See Brian Maidment, *The Poorhouse Fugitives. Self-Taught Poets and Poetry in Victorian England* (Manchester: Carcanet, 1987), 16 and Susan Zlotnick, 'Dialect, Domesticity, and Working-Class Women's Poetry', *Victorian Studies* 35 (1991), 7–27, at 9.

4 'Th' Poor Mon un Hiz Childer', 17.

5 'Country News', *Illustrated London News* (1 November 1862), 455. For more information about the Lancashire Cotton Famine, see Eugene A. Brady, 'A Reconsideration of the Lancashire "Cotton Famine"', *Agricultural History* 37: 3 (1963), 156–62, at 156, and Douglas Farnie, *The English Cotton Industry and the World Market 1815–1896* (Oxford: Clarendon Press, 1979), 135–70.

6 Farnie, *The English Cotton Industry*, 135–70.

remained largely unexplored.[7] Between 1862 and 1864, the *Illustrated London News* (1842–), the first newspaper to make illustrations its most prominent marketing feature, published several wood engravings which directly or indirectly represented or referenced the distress in Lancashire. Closer analysis of these illustrations reveals that they belong to a larger nineteenth-century cultural repertoire of depicting hunger, poverty, and famine in the Victorian illustrated press, linking back to the 1840s, a decade that has become synonymous with hunger in historical discourse.[8] During this period, a wealth of British newspaper reports were dedicated to the theme of starvation and politicised 'questions of access and entitlement to food'.[9] These columns, often accompanied by black-and-white sketches and wood engravings which functioned as modes of visual reportage and communicated a rhetoric of immediacy and authenticity, were inspired by several factors, including bad harvests, crop diseases, protectionist economic policies, and, from 1845 onwards, a catastrophic famine in Ireland.[10] While some newspapers associated with a variety of competing parties and interest groups, such as the Chartists and the Anti-Corn Law League in the case of Ireland, strategically deployed representations of famine to instigate political reforms, nineteenth-century concerns about hunger were not solely limited to the radical or working-class press.[11]Instead, this study considers famine and, more specifically, the problem of its representation in a selection of early Victorian periodicals that targeted a prosperous middle-class readership.

To uncover how early Victorian illustrated periodicals attempted to visualise famine, this article will examine press renderings of three nineteenth-century instances of hunger in the United Kingdom of Great Britain and Ireland, namely that of the Great Irish Famine (1845–51), rural poverty in Dorset (1846), and the Lancashire Cotton Famine (1861–65). The argument has been

7 For a discussion of Lancashire Cotton Famine poetry and its significance, see F. Elizabeth Grey, 'Journalism and Poetry in the Nineteenth Century. Calls to Action', *Journalism Studies* 18: 7 (2017), 807–25; Simon Rennie, 'This "Merikay War". Poetic Responses in Lancashire to the American Civil War', *Journal of Victorian Culture* 25: 1 (2020), 126–43; and *The Poetry of the Lancashire Cotton Famine*, database published by University of Exeter AHRC-funded project, http://cottonfaminepoetry.exeter.ac.uk (accessed 3 April 2023).

8 Peter J. Gurney, '"Rejoicing in Potatoes". The Politics of Consumption in England During the "Hungry Forties"', *Past & Present* 203: 1 (2009), 99–136.

9 Charlotte Boyce, 'Representing the "Hungry Forties" in Image and Verse. The Politics of Hunger in Early-Victorian Illustrated Periodicals', *Victorian Literature and Culture* 40: 2 (2012), 421–49, at 421 and Gurney, 'Rejoicing in Potatoes', 110.

10 Cormac Ó Gráda, *Black '47 and Beyond. The Great Irish Famine in History, Economy, and Memory* (Princeton, NJ: Princeton University Press, 2000).

11 Boyce, 'Representing the "Hungry Forties"', 421.

Figure 8.1 'Sketches from Life in the Agricultural Districts – The Labourer's Home-Evening', *Graphic* (13 July 1872), 3. Author's collection

structured thematically instead of chronologically to underline the parallels between these three case studies in terms of subject, form, and stylistic registers. The analysis of these case studies will cover several themes, including representations of the interior/exterior, the technical characteristics of wood engraving as a distinct medium, and the layout of the printed page. In exploring why these traditions were adopted, this study argues that several aesthetic genres, such as the picturesque and local colour, were invoked, and various representational and distancing strategies were employed to introduce the experience of famine and hunger to a middle-class readership that had often not experienced its miseries. In doing so, these illustrated periodicals affirmed their loyalty to a middle-class audience – 'the real, faithful, and influential patrons, – the Respectable Families of England' – while contending 'to keep continually before the eye of the world a living and moving panorama of all its actions and influences', as the *Illustrated London News* stated in its first issue.[12]

The focus of this study is primarily on wood engravings from the *Illustrated London News*, which gave by far the most extensive visual coverage of these three historical events, but I will also make comparisons to responses in other competing English newspapers, illustrated and otherwise, such as the *Times*

12 'Our Address', *Illustrated London News* (14 May 1842), 1.

(1788–), *Pictorial Times* (1843–63), and the *Graphic* (1869–1932). The selection of illustrations has been based on art historian Emily Mark-FitzGerald's distinction of famine imagery, which points to 'a larger genealogy' of 'an array of categories' that can be classified as 'famine images' including, but not limited to, depictions of starvation, charity and relief efforts, scenes of evictions, and emigration.[13] Employing a multi-focal approach, in which both the iconography and the physicality of illustrations are taken into consideration, the aim of this study is not to question the (historical) accuracy of famine pictures but rather to point to the functions, effects, and implications of such press representations.

Domestic Interior

Between 1862 and 1864, during the height of the Cotton Famine, the *Illustrated London News* published at least twenty-eight illustrations which directly or indirectly represented or referenced the distress in Lancashire. The popularity of illustrative media was driven by successive advances in the use of new technologies of reproduction, including lithography, wood engraving, and the use of stereotyping (which facilitated reprinting).[14] Many nineteenth-century periodicals not only adopted opportunities for incorporating illustration and visual embellishments but were often branded through and became popular because of their visual content. Nineteenth-century audiences often assumed that wood engravings were 'unproblematically naturalistic' and expected wood engravings to authentically depict the world.[15] This was, however, 'a tall order given that wood engraving was essentially a monochrome and linear medium poorly suited to tonal complexity and was long associated with vernacular and popular culture, where simplified images were the rule'.[16] Wood engravings were thus limited in how they could realistically depict the world, 'even for generations of Victorian readers well acquainted and at ease with its formal qualities'.[17]

13 Emily Mark-FitzGerald, 'Visualizing the Famine. Nineteenth-Century Image, Reception and Legacy', in Emily Mark-FitzGerald (ed.), *Commemorating the Irish Famine. Memory and the Monument* (Liverpool: Liverpool University Press, 2013), 11–56, at 14–15.

14 Lorraine Janzen Kooistra, 'Illustration', in Joanne Shattock (ed.), *Journalism and the Periodical Press in Nineteenth-Century Britain* (Cambridge: Cambridge University Press, 2017), 104–25, at 102.

15 Brian Maidment, 'Illustration', in Andrew King, Alexis Easley, and John Morton (eds), *The Routledge Handbook to Nineteenth-Century British Periodicals and Newspapers* (London: Routledge, 2016), 102–24, at 104.

16 Maidment, 'Illustration', 108.

17 Ibid.

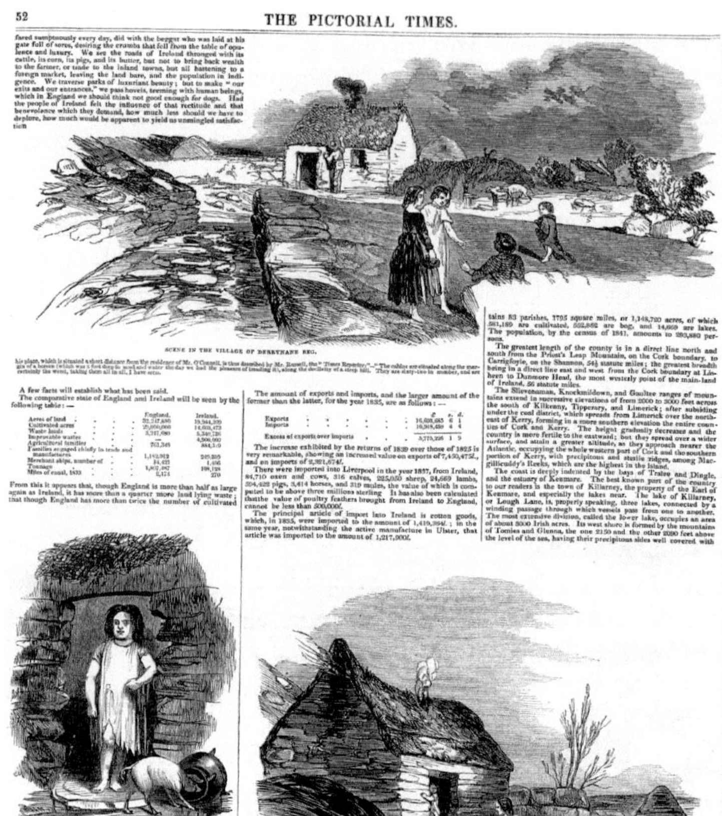

52 THE PICTORIAL TIMES.

fared sumptuously every day, did with the beggar who was laid at his gate full of sores, desiring the crumbs that fell from the table of opulence and luxury. We see the roads of Ireland thronged with its cattle, its corn, its pigs, and its butter, but not to bring back wealth to the farmer, or trade to the inland towns, but all hastening to a foreign market, leaving the land bare, and the population in indigence. We traverse parks of luxuriant beauty; but to make "our exits and our entrances," we pass hovels, teeming with human beings, which in England we should think not good enough for dogs. Had the people of Ireland felt the influence of that rectitude and that benevolence which they demand, how much less should we have to deplore, how much would be apparent to yield us unmingled satisfaction

SCENE IN THE VILLAGE OF DERRYNANE BEG.

his place, which is situated a short distance from the residence of Mr. O'Connell, is thus described by Mr. Russell, the "Times Reporter,"—"The cabins are situated along the margin of a boreen (which was [illegible] feet deep in mud and water the day we had the pleasure of treading it), along the declivity of a steep hill. They are sixty-two in number, and are certainly the worst, taking them all in all, I have seen.

A few facts will establish what has been said.

The comparative state of England and Ireland will be seen by the following table:—

	England.	Ireland.
Acres of land	32,247,880	19,344,309
Cultivated acres	25,000,000	14,603,473
Waste lands	3,317,980	5,340,736
Improvable wastes	—	4,900,000
Agricultural families	761,348	884,339
Families engaged chiefly in trade and manufactures	1,192,912	249,359
Merchant ships, number of	14,421	1,456
Tonnage	1,807,487	108,198
Miles of canal, 1833	2,473	270

From this it appears that, though England is more than half as large again as Ireland, it has more than a quarter more land lying waste; that though England has more than twice the number of cultivated acres, the number of agricultural families is 100,000 less; that though the population of England is nearly twice as much as that of Ireland, it has five times the number of persons employed in trade, manufactures, and handicraft; and that there are in Ireland a greater number of labourers absolutely than in the whole of Great Britain, more than double the number relatively to cultivated land, and more than *four times* the number relatively to produce.

The number of families as employed in different ways is computed at the following amount:—those engaged in agriculture, 884,339; those employed in trade, manufactures, and handicraft, 249,359; those not comprised in the preceding classes, 251,368. Of these, 3,794,880 were males, 3,972,521 were females; making a total of 7,767,401 persons, forming 1,385,066 families, inhabiting 1,240,818 houses.

The agricultural produce of Ireland, in 1832, was estimated at 36,000,000*l.* per annum, raised from 14,603,473 acres. This falls short, by nearly one half, of the amount yielded by an equal area in Great Britain; and yet, in the latter country, there are only *two* for every *five* for the same quantity of land in Ireland; so that the productive powers of the soil of the Emerald Isle, as compared with those of Great Britain, are little more than half developed.

The proportion of the tonnage of the shipping is about 1 to 14; in the merchant service of England, it is as 1 to nearly 18, the difference being accounted for by the superior size and management of the English ships, requiring fewer men.

The amount of exports and imports, and the larger amount of the former than the latter, for the year 1835, are as follows:—

	£	s.	d.
Exports	16,608,685	6	1
Imports	10,918,459	4	4
Excess of exports over imports	5,770,226	1	9

The increase exhibited by the returns of 1839 over those of 1825 is very remarkable, showing an increased value on exports of 7,430,475*l.*, and on imports of 2,321,674*l.*

There were imported into Liverpool in the year 1837, from Ireland, 84,710 oxen and cows, 316 calves, 225,050 sheep, 24,669 lambs, 394,422 pigs, 3,414 horses, and 319 mules, the value of which is computed to be above three millions sterling. It has also been calculated that the value of poultry feathers brought from Ireland to England, cannot be less than 500,000*l.*

The principal article of import into Ireland is cotton goods, which, in 1835, were imported to the amount of 1,419,394*l.*; in the same year, notwithstanding the active manufacture in Ulster, that article was imported to the amount of 1,217,900*l.*

tains 83 parishes, 1795 square miles, or 1,148,720 acres, of which 581,189 are cultivated, 552,862 are bog, and 14,669 are lakes. The population, by the census of 1841, amounts to 293,880 persons.

The greatest length of the county is in a direct line north and south from the Priest's Leap Mountain, on the Cork boundary, to Carrigfoyle, on the Shannon, 54½ statute miles; the greatest breadth being in a direct line east and west from the Cork boundary at Lisheen to Dunmore Head, the most westerly point of the main-land of Ireland, 56 statute miles.

The Slievenaman, Knockmildown, and Gaultee ranges of mountains extend in successive elevations of from 2000 to 3000 feet across the south of Kilkenny, Tipperary, and Limerick; after subsiding under the coal district, which spreads from Limerick over the north-east of Kerry, forming in a more southern elevation the entire counties of Cork and Kerry. The height gradually decreases and the country is more fertile to the eastward; but they spread over a wider surface, and attain a greater altitude, as they approach nearer the Atlantic, occupying the whole western part of Cork and the southern portion of Kerry, with precipitous and sterile ridges, among Macgillicuddy's Reeks, which are the highest in the island.

The coast is deeply indented by the bays of Tralee and Dingle, and the estuary of Kenmare. The best known part of the country to our readers is the town of Killarney, the property of the Earl of Kenmare, and especially the lakes near. The lake of Killarney, or Lough Lane, is, properly speaking, three lakes, connected by a winding passage through which vessels pass from one to another. The most extensive division, called the lower lake, occupies an area of about 5000 Irish acres. Its west shore is formed by the mountains of Tomies and Glenna, the one 2150 and the other 2090 feet above the level of the sea, having their precipitous sides well covered with

A "STREET DOOR" IN TARMONS.

Speaking of the hovel to which it belongs, Mr. Russell says, "It was built of loose stones, without mortar or cement, and thatched with half-decayed potato-stalks and rushes. In the front was an opening, about 3½ feet high by 2½ broad, intended as a doorway. Inside, the darkness was so great that it required a considerable time for the eye to discern any thing distinctly."

CABIN IN DERRYNANE BEG.

Little Derrynane, of which the above cut presents an average picture, is thus described by Mr. Foster, the "Times Commissioner,"—"I entered several of the cottages; the distress of the people was horrible. There is not a pane of glass in the parish, nor a window of any kind in half the cottages. Some have got a hole in the wall for light, with a board to stop it up. In not one in a dozen is there a chair to sit upon, or anything whatever in the cottages beyond an iron pot, and a rude bedstead with some straw on it; and not always that. In many of them the smoke is coming out of the doorway, for they have no chimneys. In one that I entered, the door was taken off the hinges and made a table of, by placing it on two turf-baskets. [illegible] the poor creatures are in the lowest degree of squalid poverty I have yet seen, and this within sight of Derrynane House. As one of the tenants told me, 'they were eating one another's heads off; and if they did not get some assistance they would starve, and the jails would be full.'"

The proportion of offenders, in the year 1836, to the entire population, was 1 in 325, and the male offenders were to the female 0·82 to 0·18. Of the total number of culprits, 6744 males and 490 females could neither read nor write; and of 1542 males and 275 females, the instruction could not be ascertained. The total number of convictions in that year was 18,110.

THE COUNTY OF KERRY.

In the survey on which we intend to enter, we proceed first to the county of Kerry, in the south-west of Ireland. This county con-

forest trees. On the opposite shore is the striking contrast of a plain of flat land belonging to Lord Kenmare. Glenna is a beautiful little place, situated in a small cove, where Lady Kenmare has exercised her taste in the construction of a house, and the planning of grounds in the midst of a number of fine forest trees.

The venerable ruins of Mucruss Abbey are well worthy of notice. From Denis Island the long winding passage is exceedingly beautiful. The most striking spot is at the Eagle's Crag, a stupendous and rugged cliff, which bursts suddenly on the view, rising in a pyramidal form from the water. Throughout all the rocks of Killarney, but

Public domain

Figure 8.2 'A 'Street Door' in Tarmons' and 'Cabin in Derrynane Beg', *Pictorial Times* (25 January 1846), 52. Source: British Newspaper Archive (accessed 13 December 2022)

To depict an increasingly complex society, engravers and artists were encouraged to use simplified iconography to explain and categorise the world. To balance the incongruity and instabilities of a form 'which by necessity made so many borders permeable', illustrated periodicals tended to stress formal repetition and visual branding, which created new forms of stability and boundaries.[18] Visual emblems, such as the repeated use of a particular font, page layout, and/or masthead, were marked as signals of self-identification which helped audiences to interpret and place the unfamiliar: 'in their telling of the new, periodicals accounted for new things, events, or phenomena by accommodating them within a world that had already been negotiated with their readers through repeated acts of telling, reading and buying'.[19] For this reason, the *Illustrated London News* repeated particular icons and symbols as well as stylistic registers and aesthetic categories to depict the distress in Lancashire and the hunger in Dorset that had likewise been utilised by the illustrated press during the Great Irish Famine. Within these categories, a distinction between representations of the interior and exterior can be noted: whereas 'Sketches in the West of Ireland' (1847) (Fig. 8.6) and 'Preston, Lancashire' (1862) (Fig. 8.7), for example, present picturesque depictions of the region, as will be discussed in the next section, 'Mullins's Hut, at Scull' (1847) (Fig. 8.3) and 'Tickle-street Deansgate'(1862) highlight the deteriorating living conditions of agricultural labourers and factory workers through depictions of the domestic space.

The domestic space, as Victorianist and literary scholar Peter Sinnema observes, 'has become hegemonic as a type of domestic interior, and has been invariably bound to discussions of the family'.[20] There were few aspects of society 'the Victorians regarded with greater reverence than the home and family life within it'.[21] 'Sketches from Life in the Agricultural Districts' (Fig. 8.1), published in the *Graphic* on 13 July 1872, portrayed such an idealised version of the Victorian peasant home. Created by artist Herbert Johnson, a draughtsman and painter for the *Graphic*, the 33 x 27 cm full-page engraving details a harmonious multi-generational and co-residential well-furnished and stocked household: the patriarch is seated at the table, surrounded by his well-fed, neatly dressed, and robust-looking children, who traditionally played a major part as symbols of purity and innocence in pictures of cottage and village life.[22] His wife, positioned to his right and breastfeeding an infant,

18 Barbara Korte and Stefanie Lethbridge, 'Introduction. Borders and Border Crossings in the Victorian Periodical Press', *Victorian Periodicals Review* 51: 3 (2018), 371–79, at 372.

19 James Mussell, 'Cohering Knowledge in the Nineteenth Century. Form, Genre, and Periodical Studies', *Victorian Periodicals Review* 42: 1 (2009), 93–103, at 95.

20 Peter W. Sinnema, *Dynamics of the Pictured Page. Representing the Nation in the Illustrated London News* (Aldershot: Ashgate, 1998), 44.

21 Anthony S. Wohl, *The Victorian Family. Structures and Stresses* (London: Routledge, 2016), 10.

22 Christopher Wood, *Victorian Painters. Dictionary of British Art*, vol. 4 (Woodbridge: Antique Collectors' Club, 1995), 87, 283.

116 THE ILLUSTRATED LONDON NEWS. [FEB. 20, 1847.

SKETCHES IN THE WEST OF IRELAND.—BY MR. JAMES MAHONY.

BOY AND GIRL AT CAHERA.

THE VILLAGE OF MEINIES.

WE resume from our Journal of last week our Artist's Sketches of Scenes and Incidents from the distressed district of Skibbereen, and its neighbourhood; premising that our main object in the publication of this Series of Illustrations is to direct public sympathy to the suffering poor of these localities, a result that must, inevitably, follow the right appreciation of their extent and severity.

We left our Artist, last week, on the road to Dromdaleague, to inquire into the horrible circumstances of Leahey's death.

The first Sketch is taken on the road, at Cabera, of a famished boy and girl turning up the ground to seek for a potato to appease their hunger. "Not far from the spot where I made this sketch," says Mr. M., "and less than fifty perches from the high road, is another of the many sepulchres above ground, where six dead bodies had lain for twelve days, without the least chance of interment, owing to their being so far from the town. After leaving this fearful spot, we soon reached Dromdaleague, where I called upon the Rev. J. Creedon, and inquired of him as to the fate of Leahey. 'Not only do I know the statement to be true,' replied the reverend gentleman, 'but also prepared the man for death, and am ready to accompany you to the spot.' We, accordingly, started; and, within half an hour's drive, reached the village of Meinies, where the house of Leahey is situated, and of which I send you a sketch. Whilst making this, I learned from Mr. Creedon, and one of the villagers, that not only was the account of Leahey's house in the Diary true, but the case was even more disgusting than there stated; and, horrifying as it was, the man's mother, who found the dogs about him, after having first lain him across the few remaining sparks of fire upon the floor, went out to beg as much as would purchase a coffin to bury him in."

"Having heard much of the wants of Dunmanway, I proceeded thither, and am delighted to say that this large and thriving town (of which I send you a sketch, taken from the bridge on the Cork road) seems to be the barrier to the dreadful want further west. Not, at the same time, but that much want does exist here, though nothing beyond what may be expected upon land where nature is not bountiful. The worst feature presenting itself, at this moment, all through the West, is the entire abandonment of agricultural occupation; and, during my entire excursion from Clonakilty round to Dunmanway, not more than ten or a dozen fields seemed to have been prepared for the spring; and the answer of all those to whom I addressed myself on the subject was, that if they put down, they did not know who would reap; and that, in case the crops were sown, the poor famished wretches would be there to eat them up long before they had time to grow.

"Again, all sympathy between the living and the dead seems completely out of the question; and the revolting practice will, doubtless, go on until it works its own remedy. I certainly saw from 150 to 180 funerals of victims to the want of food, the whole number attended by not more than 50 persons; and so hardened are the men regularly employed in the removal of the dead from the workhouse, that I saw one

MULLINS'S HUT, AT SCULL.

ENTRANCE TO DUNMANWAY, FROM THE BRIDGE ON THE CORK ROAD.

Figure 8.3 'Boy and Girl at Cahera' and 'Mullins's Hut, Scull', *Illustrated London News* (20 February 1847), 116. Author's collection

evokes the image of the *Madonna lactans* ('nursing Madonna'), signifying not only Christian and Victorian notions of humility and motherhood but simultaneously symbolising the fertility and nurturing qualities of the farmed land. The former head of the household, the grandfather, the embodiment of tradition and wisdom, has been positioned to the far-right, making place for the next generation: the buoyant agricultural labourers that have returned from the fields after a day of hard but rewarding work. Following nineteenth-century landscape and cottage-life painting traditions, as exemplified by the works of John Linnell, Samuel Luke Fildes, and George Elgar Hicks, these types of depictions did not detail the harsh realities of life on the farm, and failed to acknowledge, for instance, that most of the depicted children 'would be sent out to work on the fields by the age of eight or nine' or were left 'out in the snow all day to scare birds off the newly-sown corn'.[23] These types of images, as Celina Fox points out, demonstrate the *Graphic*'s 'quest for higher things, a tasteful neutrality above the grubby strife-torn world, [which] suited admirably the growing market for respectable family reading matter'.[24] But what happened to these pictorial representations of harmonious domestic life when these 'cornerstones of civilisation' were invaded by famine and poverty?

In 1846 the *Pictorial Times* sent a commissioner, artist Mr Frederick N. Sheppard, to Derrynane in County Kerry to determine the true 'Condition of the People of Ireland … that justice may be done to Ireland – Ireland must be known as she really is'.[25] Sending commissioners to Ireland was part of a larger trend started by the *Times* in 1845, who were the first to publish a series of articles on the condition of Ireland between August 1845 and January 1846. On 24 January the newspaper printed 'A "Street Door" in Tarmons' and 'Cabin in Derrynane Beg' (Fig. 8.2). These illustrations both depict ragged children with unkempt hair and empty eyes standing on the doorsteps of their cabins. Inducing eye-contact with the viewer, these children would have confronted the middle-class readers of the London-centred newspaper with a disturbing, animalistic image of a human that they could have regarded as 'Other', as was conventional at the time.[26] Visualising these children in this fashion

23 Wood, *Victorian Painters*, 87.

24 Celina Fox, 'The Development of Social Reportage in English Periodical Illustration During the 1840s and Early 1850s', *Past & Present* 49 (1977), 90–111, at 93.

25 'Pictorial Times Commissioner', *Pictorial Times*, 51.

26 As illustrated by, for instance, Michael de Nie in *The Eternal Paddy. Irish Identity and the British Press, 1798–1882* (Madison: Wisconsin University Press, 2004) and Charlotte Boyce, 'Food, Famine and the Abjection of Irish Identity in Victorian Representation', in Marlene Tromp, Maria K. Bachman, and Heidi Kaufman (eds), *Fear, Loathing, and Victorian Xenophobia* (Columbus: Ohio State University Press, 2013), 153–80.

was not without risk of confirming ethnic stereotypes.[27] Traditionally, imperial discourses would identify the Irish peasantry with unruliness and a lack of decorum and self-reliance.[28] For example, John Christian's *Observations of the State in Ireland* (1818), as literary scholar Marguérite Corporaal observes, was 'uncomfortably infused with traditional bias toward the Irish when it addressed the question of whether the extreme poverty witnessed among cottagers should be ascribed to "peculiar improvidence, and want of management"'.[29] These sentiments were reflected in many English newspapers of the period, including the influential London-based daily newspaper *Times* (1788–), which used racial contrasts and stereotyping to interpret the Famine and continuously provoked resentment towards the Irish among their readers, claiming that the Irishman was 'accessory to at least one half of his misery'.[30]

This bias of 'Irish indolence' was often bound up with the idea that the Irish were by nature barbaric, untameable, and uncivilised.[31] The fact that these dishevelled and barely dressed children were staged on the threshold, therefore, is significant. The symbol of the threshold highlights that the class and ethnicity to which the girls belong are considered to be at the margins of civilisation. The reader is forced to look into the lamentable circumstances that have shaped their current destitution, as illustrated by the descriptions of the cabin's interiors, yet a physical barrier keeps them safely apart from the deprivations that are concealed inside. Constituting an example of what David Lloyd calls the indigent sublime – that which 'exceeds conceptualisation and overwhelms the boundaries of the subject' – this enables readers to abstain from any identification with the unbearable and inhuman conditions that characterise their domestic life.[32] While the commissioner claims to have 'entered several of the cottages' of which 'the distress was horrible', 'the poor creatures are in the lowest degree of squalid poverty', and tenants 'were eating another's head off', none of the engravings on the page detail the interiors of these dwellings.[33] While the

27 On such stereotypes, see L. Perry Curtis, *Apes and Angels. The Irishman in Victorian Caricature* (Washington, DC: The Smithsonian Institution Press, 1971).

28 Marguérite Corporaal, *Relocated Memories. The Great Famine in Irish and Diaspora Fiction, 1846–1870* (Syracuse, NY: Syracuse University Press, 2017), ch. 3, 17.

29 Ibid.

30 'Distress of the Dorchester Labourer. (From Our Correspondent. Blandford, June 15)', *Times* (18 June 1846), 5.

31 Corporaal, *Relocated Memories*, 17.

32 David Lloyd, 'The Indigent Sublime. Specters of Irish Hunger', *Representations* 92 (2005): 152–85, at 160.

33 'The Condition of the People of Ireland', *Illustrated London News* (24 January 1846), 52.

painful realities of hunger are addressed in the text, there is the tendency to steer away from too harrowing details in the accompanying engravings, as the reader does not get to see the misery experienced inside. This reveals the paradoxical registers that underlie pictorial representations of famine that can also be witnessed in, for instance, early famine fiction and travel accounts, and are thus part of a larger nineteenth-century convention.[34] In *Narrative of a Journey from Oxford to Skibbereen, in the Year of the Famine* (1847), for instance, travellers Lord Dufferin and G.G. Boyle remained 'on the threshold and looked in', and leading nineteenth-century Irish-born artist and engraver James Mahony (1810–1879), expressed a fear of facing 'the almost indescribable in-door horrors', that 'neither pen nor pencil ever could portray': 'I thrust my head through the hole of the entrance, and had immediately to draw back ... though rendered callous by a companionship for many years with disease and death, yet I was completely unnerved at the humble scene of suffering and misery that was presented to my view'.[35] This illustrates how threshold depictions of famine often literally and metaphorically functioned as boundaries that kept disturbing, dehumanising situations at a remove.

Some engravings and reports, however, did detail the miseries that could be found beyond the threshold. These press renderings of want and misery, nevertheless, were still mediated through several distancing techniques and artistic conventions that would mitigate the horrors to be found beyond the threshold.[36] Traditionally, the famished body was often represented in the domestic interior setting of a cottage or cabin. 'Mullins's Hut' (Fig. 8.3) published in Mahony's 'Sketches in the West of Ireland' (Fig. 8.6) in the *Illustrated London News* in 1847 is one such example. The engraving is accompanied by an article, which details 'the in-door horrors of Scull', where 'a poor man named Mullins ... lay dying in a corner upon a heap of straw, supplied by the Relief Committee, whilst his wretched children crouched over a few embers of turf, as if to raise the last remaining spark of life'.[37] Similarly, 'Dwellings of Manchester Operatives' (Fig. 8.4), published in the *Illustrated London News* on Saturday, 29 November 1862, offers a full-page illustration consisting of a collage of engravings that outline the bleak and

34 Corporaal, *Relocated Memories*.

35 Lord Dufferin and G.G. Boyle, *Narrative of a Journey from Oxford to Skibbereen, in the Year of the Famine* (Oxford: John Henry Parker, 1847) and 'Sketches in the West of Ireland', *Illustrated London News* (13 February 1847).

36 These reports were often written by travellers who were affiliated with relief and charity organisations, such as the Central Executive Cotton Famine Relief Committee in Lancashire.

37 'Mullins's Hut, Scull', *Illustrated London News* (20 February 1847), 116.

Figure 8.4 'Dwellings of Manchester Operatives', *Illustrated London News* (29 November 1862), 585. Source: Google-digitized, original from University of Michigan, via HathiTrust Digital Library (accessed 13 December 2022)

wretched interiors of four industrial labourers' homes. These illustrations of 'touching scenes', of which three are discussed in the report, have been based on the 'two or three cases' from the correspondent's 'notebook', realised by an exterior 'artist'.[38] All things that make these dwellings a home have been stripped away and are replaced by 'the intense pleading of silent want': 'the bedding had all disappeared', 'superfluous furniture was sold', and the head of the household 'is sickly'.[39] Often the commissioners or special artists who crossed the thresholds of these wretched homes came from an entirely different, usually middle-class, background and were relatively unfamiliar with the indigence that the depicted workers endured. Not only did these reporters thus transcend literal boundaries, but these thresholds also symbolised the mental boundary that they had to overcome in confronting human conditions that clashed with dominant ideas about home and family put forward by the *Illustrated London News*, as depicted in 'Sketches from Life in the Agricultural Districts' (Fig. 8.1).

Although the reports 'Mullins's Hut' and 'Dwellings of Manchester Operatives' provide a glimpse of the true horrors behind closed doors,

38 'Manchester Distress and Relief Organisation. Manchester', 582.

39 Ibid.

the captions of the engravings, as do all captions discussed in this article, contain factual and neutral language. They point solely to a specific setting (e.g. Mullins's cabin) or geographical location (e.g. 'Tickle-street Deansgate 2') and contain very few adjectives or modifiers to pass judgement on the depicted victims or instruct the reader on how to interpret the scene. In these press representations, it was also common to visually present the cabins of the rural poor as dim and gloomy spaces: 'Mullins's Hut, at Scull' (Fig. 8.3), 'Dwellings of Manchester Operatives' (Fig. 8.4), but also 'Interior of a Dorsetshire Labourer's Cottage' (Fig. 8.8), published in the *Illustrated London News* on 5 September 1846, are but a few examples. This strategy not only made it possible to conceal the horrifying details of the starved body but also underlined the representation of famine as a phenomenon that exceeded the limits of representation.[40] In all of these illustrations, moreover, the starving inhabitants are seated around a fire or fireplace. The nineteenth-century domestic hearth, as Claudia Kinmonth observes, was perceived as 'the focal point around which household activities revolved' including 'the daily collection and arrangement of the bedding'.[41] This symbolised 'a sense of warm security' as well as 'family continuity and ... hospitality towards the stranger'.[42] In the aforementioned interior depictions, the centrality of the fireplace within the engraving signifies a space of significant comfort and warmth amid wretched living conditions. In 'No. 4, Thornton-Court, Tickle-Street' depicted on the top-right of 'Dwellings of Manchester Operatives' (Fig. 8.4), however, the stools in front of the domestic hearth, once occupied, are now empty, while a sick figure lies on a pile of straw in the corner, 'where death was within a few steps waiting to close the hard but unsuccessful struggle for life'.[43] In this particular engraving, the image of the extinguished and deserted fireplace can be read as a symbol of death and disintegration of the family and all that it traditionally stands for. In 'Sketches from Life in the Agricultural Districts' (Fig. 8.1), interiority is developed as a desirable state of being – all of its articulations, as home, family, companionship, warmth, or isolation, 'showing the Victorian interior to be axiomatic in the organization and preservation of a gratified reading audience'.[44] By contrast, in 'A "Street Door" in Tarmons', 'Cabin in Derrynane

40 As has been underlined in, for instance, Christine Kinealy, *A Death Dealing Famine. The Great Hunger in Ireland* (London: Pluto Press, 1997).

41 Claudia Kinmonth, *Irish Rural Interiors in Art* (New Haven, CT: Yale University Press, 2006), 2–7.

42 E. Estyn Evans, *Irish Folk Ways* (London: Routledge and Paul, 1961), 59 and Kinmonth, *Irish Rural Interiors*, 12.

43 'Manchester Distress', 582.

44 Sinnema, *Dynamics of the Pictured Page*, 108.

Beg' (Fig. 8.2), 'Dwellings of Manchester Operatives', and 'Mullins's Hut' (Fig. 8.4), being on the outside emerges as 'the exempting franchise of the spectator'.[45] In these depictions, different distancing techniques, such as the usage of abstract captions and the interplay between light and shade, are adopted that dehumanise the famine victims, allowing the middle-class readers to dismiss the impoverishment of the interior.

Picturesque Exterior

Engravings of the domestic interior were often offset and mediated through picturesque and local-colour representations of the exterior. Local colour, an often 'under-theorized notion', was essentially characterised by a longing for cultural specificity, focusing on the folklore and customs of specific regions.[46] As a literary genre, it gained currency in the late Romantic period in France but continued to attract artists, photographers, social scientists, and the reading public throughout the nineteenth century in Europe and the United States.[47] Rooted in the belief of a *genius loci*, the notion that a 'spirit of place' could be captured empirically, the local customs and activities of the poor and peasant classes, as evident from the *Illustrated London News* opening issue on 14 May 1842, emerged as appealing and marketable subjects for depiction and provided evidence of social relationships: 'the literature – the customs – the dress – say, the institutions and localities of other lands, shall be brought home to you with spirit, with fidelity, and, we hope, with discretion and taste'.[48] Indeed, the nineteenth-century British illustrated press had a fascination with the region at home and elsewhere and often sought to map out various distinct types of regional character, using both text and engravings to outline specific traits of physiognomy, labour, and dress. As wood engravings were limited in how they could realistically depict the world, artists and engravers used eighteenth-century painting traditions, stylistic registers, and a simplified iconography to explain and categorise the world for an ever-growing mass readership.

45 Ibid.

46 Vladimir Kapor, *Local Colour. A Travelling Concept* (Oxford: Peter Lang, 2009), 8.

47 Maria Grazia Lolla, 'Local Colour and the Grey Aura of Modernity. Photography, Literature, and the Social Sciences in Fin-de-Siècle Italy', in Sarah Patricia Hill and Giuliana Minghelli (eds), *Stillness in Motion. Italy, Photography, and the Meanings of Modernity* (Toronto: University of Toronto Press, 2014), 67–96, at 68.

48 Mark-FitzGerald, 'Visualising the Famine', 16 and 'Our Address', *Illustrated London News* (14 May 1842), 1.

Figure 8.5 'Views of the O'Connell Property in Ireland. (By Our Own Artist)', *Illustrated London News* (10 January 1846), 24–25. Source: Internet Archive (accessed 13 December 2022)

Prevailing artistic genres and aesthetic categories, it seems, were important in selecting what was worth seeing and remembering and in codifying archetypical landscapes. In the engravings 'Waterville', 'Derrynane Abbey', and (perhaps most noteworthy) 'Valentia, From the Ferry point', published in 'Views of the O'Connell Property in Ireland' (Fig. 8.5) in the *Illustrated London News* in 1846, the 'Special Artist' employs the aesthetic category of the sublime to depict the regional landscape, drawing attention to the towering mountain ranges and meandering roads into town, while the tiny figures of local residents emphasise the vastness of nature. In each of his panoramic depictions, the artist plays with light, shadows, and form varieties and introduces small figures of shepherds, peasants, countrywomen, and domestic animals, intended to capture a sense of reality. These additions not only lend local colouring to otherwise anonymous situations or landscapes, but they also belong to a strand of idealised poverty, expressed in the artistic works of, for instance, Scottish contemporary painters Thomas Faed (1826–1900) and Erskine Nicol (1825–1904), in which harsh realities of lower-class existence are muted and depictions of the poor are aestheticised so as to appear as an organic extension of a 'natural' social order.[49]

These visual depictions of local colour in the British press were often mediated through the picturesque, an aesthetic category developed in the eighteenth century as a Romantic reaction against the Enlightenment's rationalising tendencies. Inspired by the work of classical painters, such as

49 Mark-FitzGerald, 'Visualising the Famine', 25.

Claude Lorrain (1600–1682) and Salvator Rosa (1615–1673), various aestheticians, in particular William Gilpin (1724–1804), Uvedale Price (1747–1824), and Richard Payne Knight, (1750–1824), developed their own views about the aesthetic category, altogether leading to the development of a specific set of picturesque codes.[50] A picturesque landscape painting typically included three planes, consisting of a foreground, middle, and background often connected by a river, containing a variety of light, vegetation, and elevation to provide depth of perspective.[51] The foreground invited leisurely habitation by providing some animal and/or human presence, and rocks and trees acted as coulisses to focus the attention of the viewer towards the centre of the painting. Viewers typically considered the scene from an elevated perspective to 'look down on nature and command the landscape'.[52] More generally, a picturesque landscape painting involved an element of roughness, 'an essential feature to ensure visual variety', while the play of light and shade contributed to the harmonisation of the overall scene.[53] All of these elements combined, evoked, it was theorised, a sense of comfort and contentment.

The use and popularity of the picturesque thus owed much to the fact that it provided a way of showing landscapes and their contents to viewers or readers in a fashion that seemed organised and sensibly pictorial.[54] The picturesque was, however, essentially a class-based, cultural construction, stemming from elitist perspectives in nature, art, literature, and land ownership. For those of socio-economic privilege, land represented something to be composed, improved, and constructed into a landscape.[55] In this regard, the picturesque also revealed the politics of exclusion, as only members of the middle and upper classes, educated in literature, art, and landscape touring, held the capacity for ordering and controlling the land.[56] Fittingly, illustrated newspapers such as the *Illustrated London News* and the *Pictorial Times* adopted the picturesque as a socially acceptable way of apprehending the world for its middle-class readership.

50 For an overview of various eighteenth- and nineteenth-century theoretical approaches to the concept of the picturesque, see Stephen Copley and Peter Garside (eds), *The Politics of the Picturesque. Literature, Landscape and Aesthetics Since 1770* (Cambridge: Cambridge University Press, 1994).

51 Greg Gillespie, *Hunting for Empire. Narratives of Sport in Rupert's Land, 1840–70* (Vancouver: University of British Columbia Press, 2011), 81.

52 Gillespie, *Hunting for Empire,* 81.

53 Ibid.

54 Alexander Ross. *The Imprint of the Picturesque on Nineteenth-Century British Fiction* (Waterloo: Wilfrid Laurier University Press, 1986), xiii.

55 Gillespie, *Hunting for Empire,* 82.

56 Ibid.

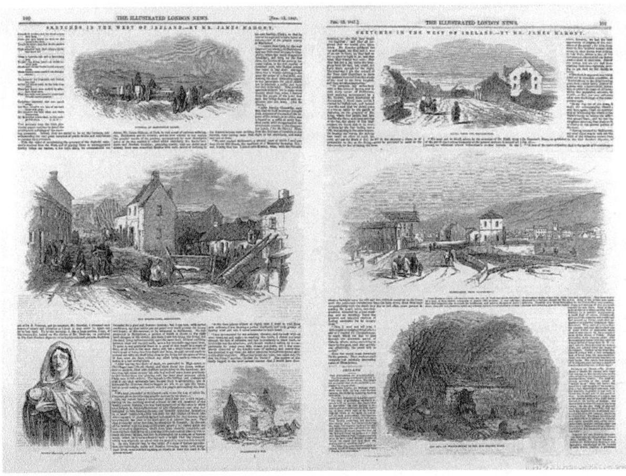

Figure 8.6 'Sketches in the West of Ireland. – By Mr. James Mahony', *Illustrated London News* (13 February 1847), 100–1. Author's collection

Artists and reporters often unambiguously referred to the picturesque in their articles. In 'Views of the O'Connell Property in Ireland' (January 1846) (Fig. 8.5), for instance, the reporter explains that 'the artist sends us a sketch of the town, from the Killarney entrance, which shows it to be rather picturesque'.[57] The aesthetic category, moreover, is mentioned in reports of hunger in England: in 'The Town of Preston, Lancashire' (September 1862), 'the more picturesque sections of the district' are highlighted, and in 'Peasantry in Dorsetshire'(September 1846) (Fig. 8.8), the reporter refers to a village as 'a charmingly picturesque bit for the painter; though its propped up walls and decaying thatched roof, but too closely indicate the privation and suffering of the inmates'.[58]

The excerpt from 'Peasantry in Dorsetshire' (Fig. 8.8) proves especially interesting, as the reference to the duality of both beauty and decay not only adequately underlines the difficulty in defining the genre of the picturesque, but also highlights a friction between the genres of the picturesque and local colour.[59] This friction, which can also be witnessed in many tourist reports written during the Famine, becomes visible when the themes of famine and

57 'Sketches in the West of Ireland. – By Mr. James Mahoney', *Illustrated London News* (13 February 1847), 100–1 and 'Views of the O'Connell Property in Ireland. (By Our Own Artist)', *Illustrated London News* (10 January 1846), 24–25.

58 'Manchester Distress', 582 and 'The Peasantry of Dorsetshire', *Illustrated London News* (5 September 1846), 156–57.

59 Paul Fyfe, 'Illustrating the Accident. Railways and the Catastrophic Picturesque in *The Illustrated London News*', *Victorian Periodicals Review* 46: 1 (2013), 61–91, at 64.

Figure 8.7 'Preston, Lancashire', *Illustrated London News* (6 September 1862), 260. Source: Google-digitized, original from University of Michigan), via HathiTrust Digital Library (accessed 13 December 2022)

indigence are introduced, threatening the idealistic domesticating boundaries meant to present the metropolitan audience with a tranquil, soothing vista.[60] The spectacle of hunger cannot be organised into a 'meaningful series that would deliver the reconciliatory denouement and the structure of identification' that the genre of the picturesque requires.[61] In these press renderings of the picturesque exterior, famine materialises as a reality that cannot be represented in detail but instead twists and shapes the stylistic mode that seeks to contain and express it. Picturesqueness is to the exterior what destitution is to the interior: 'a condition produced (on the face of it, artlessly) by the viewing position, the point of view, assumed by the reader'.[62] While representations of the domestic interior affirm famine and destitution forcing the reader to acknowledge the 'pathos of poverty', picturesque external scenes diminish and divert social responsibility through aesthetic

60 For an analysis of famine tourism and tourist accounts in which this tension also becomes visible, see ch. 4 of Corporaal, *Relocated Memories* and John Urry, *The Tourist Gaze. Leisure and Travel in Contemporary Societies* (Thousand Oaks, CA: Sage, 2002).
61 Lloyd, 'The Indigent Sublime', 162.
62 Sinnema, *Dynamics of the Pictured Page*, 121.

qualities.[63] The genre 'often exposes social penury in such a way that it appeals to the observer's compassion rather than to his social indignation'.[64]

For example, in its representations of the Lancashire Cotton Famine, the *Illustrated London News* utilises a mixture of the picturesque and the industrial to conceal disturbances of industrial modernity with an aesthetic that paradoxically reopened them to view. Beginning in the late eighteenth century, artists and engravers began adapting the picturesque to represent industrial sites, including collieries, factories, and railways – reimagining the political, material, and ideological disruptions of industry in terms of pictorial harmony.[65] Even from its onset, as the excerpt from 'Peasantry in Dorsetshire' (Fig. 8.8) has emphasised, the picturesque was 'already characterized by destructive interests' and 'weathering and deformities'.[66] Industrial sites, as Paul Fyfe observes, progressively offered the potential for producing just such effects: 'they functioned as self-deforming landscapes or as ruins instantly sculpted by catastrophe – odious to some, fascinating to others'.[67]

While more traditional strands of the picturesque evoked 'a satisfying historical awe about degeneration over time, in industrial contexts, such weathering or deforming happened more suddenly'.[68] Victorian engravers and artists therefore repurposed eighteenth-century aesthetics as a strategy for organising modernity and attempting to contain its social consequences. On 6 September 1862, for instance, the *Illustrated London News* published a report titled 'The Town of Preston' about Lancashire's 'chiefly important ... manufacturing and commercial town', containing 'upwards of fifty mills for the conversion of the cotton crop into thread, yarn, and fabric'.[69] Ordinarily, the article finds, 'it happens that the neighbourhood is beheld in its beauty but once a week. The mills are quiet on the Sunday, and not a cloud of smoke interposes its tenebrious shade between the laughing earth and the blue sky'.[70] With the factories now closing due to the Cotton Famine, 'for some months past, however, the valley has been less obscured ... and with a vacant stare questions a distant Providence as to the means of providing

63 Ibid.

64 Hans Lund, *Text as Picture. Studies in the Literary Transformation of Pictures*, trans. Kacke Gotrick (Lewiston, NY: Edwin Mellen Press, 1992), 28.

65 Fyfe, 'Illustrating the Accident', 66.

66 Ibid.

67 Ibid.

68 Ibid.

69 'The Town of Preston', *Illustrated London News* (6 September 1862), 264.

70 Ibid.

against the chill blasts of winter'.[71] The article is accompanied by a full-page wood engraving on page 260, which depicts the industrial town nestled in a valley. The engraving is not a statement of the 'anti-picturesque', nor does it evoke the industrial sublime where human figures seem threatened by man-made constructions. Instead, the illustration suggests stasis between human industry and nature: the smoke rising directly from the factories blends almost seamlessly into the clouds, cattle drink from the same river that powers the factories, and the arch bridge symbolically connects the fields on which the cows graze to the urban city centre of the cotton industry. While the engraving clearly posits an aesthetic harmony through the genre of the picturesque, the accompanying article, by contrast, lays bare the visual repression of the true state of the region: 'the eye of hunger is not cheered by the scene'.[72]

A similar friction can be observed in Mahony's reports of the Famine in Ireland in the *Illustrated London News* in 1847. Mahony was commissioned to produce a series of sketches and written reports on Skibbereen, a village in county Cork, and its surroundings in 'Sketches in the West of Ireland' (Fig. 8.6) 'with the object of ascertaining the accuracy of the frightful statements received from the West, and of placing them in unexaggerated fidelity' before the readers.[73] As with the coverage of famine in Preston, Lancashire, many of Mahony's illustrations do not emphasise the extreme misery experienced in the region, where, according to one witness, people were 'dropping ... daily' from starvation.[74] Instead, the stylistic registers of the picturesque are employed to create a distance between the reader and the famished subject. Like his representations of 'Skull, from the Ballidchob' and 'Skibbereen, from Clover-Hill', Mahony's sketch of Ballydehob is only sparsely populated; the few, small figures face away from the reader, and the distance makes it impossible to detect the visual signifiers of hunger on their bodies. As Charlotte Boyce notes, a large number of the *Illustrated London News*'s illustrations of western Ireland in 1847 and later periods simulated this distanced viewpoint, visually mediating the effects of starvation.[75] This picturesque view of peasant life, like the depictions of the domestic interior, resulted in a non-committal social position, whereby readers of these illustrated newspapers found themselves involved with, at best, a

71 Ibid.

72 Ibid.

73 'Sketches in the West of Ireland', 100–1.

74 Ibid.

75 Boyce, 'Representing the "Hungry Forties"', 434.

tentative progressivism, visually tempered by an overall insistence on the rugged charm of local colour.

Layout of the Printed Page

Famine representations of the domestic interior and picturesque exterior were often brought together on the printed page. This can, for instance, be observed in the double-page spread titled 'The Peasantry of Dorsetshire' (Fig. 8.8), published in the *Illustrated London News* on Saturday, 5 September 1846, which contains four illustrations that portray the deterioration of living conditions for the local agricultural labourers. Inspired by a series of letters to the *Times* regarding the sufferings of Dorset's labouring population, in which comparisons were made to the Famine in Ireland[76], the report is meant to confirm the prevalence of hunger and poverty in the region. During the early decades of the nineteenth century, industrial and agricultural revolutions as well as the effects of the Revolutionary and Napoleonic Wars provoked profound social change.[77] Employment in agriculture, fishing, and forestry dropped from 35.9 per cent of the workforce in 1700 to 21.7 per cent by 1851, while those employed in mining and manufacturing increased from 29.7 per cent to 42.9 per cent.[78] Due to mechanisation and cheaper production in large-scale factories, cottage industries in England declined. This had a significant impact upon many rural communities, and agricultural labourers in the rural south suffered a particular degradation of both their standard of living and quality of life after circa 1780 due to their vulnerability to structural changes and continued population growth.[79]

In 'Peasantry of Dorsetshire' (Fig. 8.8) a general exterior picturesque view of the village of Whitchurch at the top of the printed page is contrasted with and separated from a picture of the 'Interior of a Dorsetshire Labourer's Cottage' by three columns of print. The accompanying text assures the reader that the portrayal of the domestic interior corresponds with the exterior: 'the propped-up walls and decaying thatched roof' of the exterior view 'but too closely indicate the privation and suffering of the inmates',

76 'From Our Correspondent. Blandford, June 15', *Times* (18 June 1846), 5.

77 Bernard Harris, *The Origins of the British Welfare State. Social Welfare in England and Wales,*

78 N.L. Tranter, 'The Labour Supply 1780–1860', in Roderick Floud and Deirdre N. McCloskey (eds), *The Economic History of Britain Since 1700*, 3 vols, vol. 1: *1700–1860* (Cambridge: Cambridge University Press, 1994), 204–26, cited in Harris, *Origins*, 29.

79 Keith David Snell, *Annals of the Labouring Poor. Social Change and Agrarian England 1660–1900* (Cambridge: Cambridge University Press, 1985).

Figure 8.8 'The Peasantry of Dorsetshire', *Illustrated London News* (5 September 1846), 156–57. Source: Internet Archive (accessed 13 December 2022)

emphasising a correlation between external aspects and internal misery.[80] As the breakdown of the domestic interior illustrates, however, the indigence of the peasants residing in Dorset resists being depicted in an aesthetically pleasing fashion. All three illustrations of the exterior, 'Village of Whitchurch, Dorset', 'Dorsetshire Peasantry', and 'Labourer's Cottage, Near Blandford' (Fig. 8.8), confirm, in varying degrees, the aestheticisation of a life that, if approached from the inside, is emphatically 'comfortless and wretched'.[81] While the article, moreover, finds that 'it is, perhaps, worthy of remark that dishes, plates, and other articles of crockery, seem almost unknown' and 'want, famine, and misery' are at the labourers' doorsteps, this is not mirrored in the adjacent engraving of a group of agricultural labourers that, visually, seem to be unaffected physically by the devastation of hunger.[82] Although the report characterises the region's peasants as 'hungry, emaciated and squalid', the engraving that follows is not an unattractive picture; the viewer is not confronted with disturbing or shocking imagery but instead invited to acknowledge and appreciate the carefully constructed, aesthetically framed moment.[83] Similarly, the engraving of 'Whitchurch Lane' does not provide evidence of the 'festering and rotting' stream which is said to 'meander down each street', collecting 'the matter which constantly escapes from

80 'The Peasantry of Dorsetshire', *Illustrated London News* (5 September 1846), 156–57.

81 Ibid.

82 Ibid.

83 Ibid.

pigsties and other receptacles of filth'.[84] These examples underline a clear disharmony, both between interior and exterior representations, as well as between image and text.

As Sinnema observes, the engravings are 'striking for [their] benign composition' and picturesque representation of rural England.[85] While the article clearly underlines 'the frail and miserable' appearance and condition of the labourers and their cottages undermined by partisanship, in the illustration, as Sinnema argues, 'the proud legacy of the yeomanry can be read'.[86] This nostalgia of 'a vital link with a common past' neutralises the descriptions of famine in the accompanying article in line with eighteenth-century painting traditions of the rural poor, in which similar inconsistencies between subject and form could be observed.[87] John Barrell, for instance, emphasises the moral and social tensions disguised as aesthetic choices in the paintings of English portrait and landscape painter Thomas Gainsborough (1727–1788). His study, as observed by Mark-FitzGerald, 'tackles the predicament, unique to "fine art" representations of the poor', and illustrates how they became 'an acceptable part of the décor of the drawing rooms of the polite, when in their own persons they would have been unlikely to gain admission even to the kitchens'.[88]

The organisation of engravings on the printed page proves to be of great importance in mediating depictions of famine. Representations of famine are negotiated, not only in terms of the incongruities between representations of the interior/exterior or imbalance between illustration and text, but also through the layout of the printed page, which directs 'the reader-viewer's attention toward a specific message which may, or may not, be literally depicted'.[89] As cultural historian Patricia Anderson notes, 'the medium for transmitting an image can also be the agency by which that image's meaning is altered. Its position on the page, the caption, text, and name of the publication together provide a context for the picture'.[90] In 'Views of the O'Connell Property in Ireland' (Fig. 8.5), for instance, eight carefully balanced engravings are displayed, and two large illustrations anchor each

84 Ibid.
85 Sinnema, *Dynamics of the Pictured Page*, 101.
86 Ibid.
87 Ibid.
88 See Mark-FitzGerald, 'Visualising the Famine', 17 and John Barrell, *The Dark Side of the Landscape. The Rural Poor in English Painting 1730–1840* (Cambridge: Cambridge University Press, 1980), 5, quoted in Mark-FitzGerald, 'Visualising the Famine', 17.
89 Patricia Anderson, *The Printed Image and the Transformation of Popular Culture, 1790–1860* (Oxford: Clarendon Press, 1994), 57–58.
90 Ibid.

page. The smaller illustrations inserted at the left and right of each page counterbalance each other, thereby creating visual consistency within the article's spread, whereas the two largest illustrations, which immediately capture the reader's eye, offer broad views of the surrounding landscape. Starting at the centre of the issue, on page 8, the engravings colour the overall feel and tone of the issue, while the technique of the double-spread highlights the series' importance within the spread.

The engravings clearly dominate the page, with the text being squeezed into the smaller available spaces. The language of the report, moreover, ranges from the poetic to the explanatory, alternating witness and travel accounts with almost lyrical descriptions of the scenery. This variety of expression reflects the need for the *Illustrated London News* not only to accurately explain the disaster in Ireland to its readers, but also to embellish such explanations with appealing language that had an illustrative function similar to that of the weekly's engravings. Such a layout altered the relationship of text to engraving, reinforcing the dominance of the image and consequently its dramatic possibilities. As a result, an overall 'sense of symmetry, order and control' was created that counterbalanced the harrowing descriptions of hunger through a harmonious layout.[91] The layouts, as much as the engravings themselves, encouraged readers to see illustrations of famine as aesthetic objects and to examine illustrations of disasters from a safe, contemplative remove. This 'compositional equilibrium' can be identified in a number of the illustrated articles examined in this study, including 'Sketches in the West of Ireland' (Fig. 8.6) and 'Peasantry in Dorsetshire' (Fig. 8.8), which position their engravings in similar fashion.[92]

Some engravings, however, did depict the consequences of hunger and poverty at close range, such as Mahony's well-known 'Boy and Girl at Cahera' (Fig. 8.3) and 'A Woman Begging at Clonakilty' (Fig. 8.6) published in the *Illustrated London News* in 1847. More than any other press engravings of hunger in Ireland and England, these portraits induce eye-contact with the readers to evoke a sharing of emotions. In these engravings, a breakdown of the picturesque representation of the ideal rustic peasant, as for instance depicted in 'Sketches from Life in the Agricultural Districts' (Fig. 8.1) and 'Dorsetshire Peasantry' (Fig. 8.8), can be observed. The rural peasantry makes up 'an important feature of many of the narratives and paintings of the genre'

91 Leslie A. Williams, *Daniel O'Connell, the British Press and the Irish Famine. Killing Remarks*, ed. William H.A. Williams (Aldershot: Ashgate, 2003), 339.

92 Sinnema, *Dynamics of the Pictured Page*, 44.

and 'several of Gilpin's iconic sketches display peasants at their margins, providing local colour as well as perspective for the scale of the mountains and distances'.[93] However, Mahony's 'Boy and Girl at Cahera' (Fig. 8.3) and 'A Woman Begging at Clonakilty' (Fig. 8.6) discard aestheticised depictions of poverty and foreground the suffering described in the accompanying article, which, for instance, outlines the horrifying tale of a mother begging alms to buy a coffin for her child. These engravings have received significant critical attention from scholars such as Margaret Kelleher, who has found that these types of famine depictions often concentrated on depictions of starving women and children, who functioned as a ready cultural stimulus to pity.[94] The remarkable illustrations, however, reveal more about the ideology of the periodical in which they were published than about nineteenth-century pictorial traditions. Positioned in the corners, the disproportionate size of 'Boy and Girl at Cahera' (Fig. 8.3) and 'A Woman Begging at Clonakilty' (Fig. 8.6) betrays their rhetorical significance: within the larger layout of the printed page, the engravings appear smaller than depictions of the picturesque local landscapes, which distract the reader and draw attention to the centred illustrations. Consequently, less weight is put on these representations of the ravishing effects of hunger, and considered as a series, the illustrations instead offer an impression of a journey through the region. Finally, these examples also highlight the importance of studying illustrations, not in isolation but through a multi-focal approach in which both the iconography and the physicality of illustrations are taken into consideration. Victorian illustrated periodicals did much more than simply evoke visual pleasure; they also mirrored and critiqued the culture that produced them and must therefore be understood within the aesthetic, economic, and reprographic history of image-making for mass audiences.

Conclusion

In the dialect poem 'Th' Poor Mon un Hiz Childer', an anonymous poet provides an account of a mill operative's family affected by the Lancashire Cotton Famine (1861–65). Even though the speaker assures the reader that their adversity is a story 'weel worth yerin', the exact details of the family's

93 Murray Pittock, *Edinburgh Companion to Scottish Romanticism* (Edinburgh: Edinburgh University Press, 2011), 64.

94 Margaret Kelleher, *The Feminization of Famine. Expressions of the Inexpressible?* (Cork: Cork University Press, 1997), 23.

suffering – 'whot this life tug them hed bin' – remain untold. Instead, the suffering of the family is indirectly narrated, and the speaker decides what is worth revealing to the audience. Similarly, in press coverage of the Great Irish Famine, the hunger in the agricultural districts in Dorset in 1846, and the Lancashire Cotton Famine, the experience of poverty and famine is mitigated to cater to a largely 'respectable' middle-class readership, unaccustomed to the experience of hunger.

Interpretations and descriptions of famine in the Victorian press were shaped in decisive ways by cultural patterns and tied to practices of representation. Several techniques and icons, such as depictions of dark interiors and picturesque exteriors, the avoidance of eye contact, and the symbols of the threshold and hearth, were employed to create a distance between reader and famine victim. Even the more pronounced wood engravings of famine and poverty, such as 'Mullins's Hut, Scull' (Fig. 8.3), 'Boy and Girl at Cahera' (Fig. 8.3), and 'A Woman Begging at Clonakilty' (Fig. 8.6), were tempered on the printed page by representations of local colour and the picturesque, which often underlined a general remoteness of perspective and allowed the readers to shift their gaze.

This study demonstrates that while visual and textual press responses to disaster might follow individual genealogical trajectories and draw on specific cultural and artistic traditions, they nevertheless intersect and interact on various levels. The similarities between 'Th' Poor Mon un Hiz Childer' and press engravings demonstrate how visual depictions of the Lancashire Cotton Famine belong to a larger nineteenth-century cultural repertoire of famine, poverty, and hunger and highlight how representations of disaster combined established visual repertoires, in turn influencing how later disasters were depicted and interpreted. While the famines in Ireland and Lancashire took place in different geographical and political contexts, and the ultimate responses and policy outcomes to these crises were markedly different, the similarities in how these historical disasters were depicted in the illustrated press highlight how disasters can bring forth culturally formed patterns, modes of representation, and ways of comprehension. This makes these disasters, as Gerrit Jasper Schenk and Monica Juneja have illustrated, profound socio-cultural events.[95]

95 Monica Juneja and Gerrit Jasper Schenk (eds), *Disaster as Image. Iconographies and Media Strategies Across Europe and Asia* (Regensburg: Schnell & Steiner, 2014), 14.

About the Author

Sophie van Os is a PhD candidate at Radboud University. Her dissertation, part of the NWO-funded Vici project *Redefining the Region*, examines the representation of regional identity in the nineteenth-century illustrated press. She is also a postgraduate researcher for the Irish Women's Writing Network (1880–1920), the postgraduate representative of the Society for the Study of Nineteenth-Century Ireland, and student representative for the Research Society of Victorian Periodicals.

9 Rinderpest in Dutch Regional Fiction

Community, Precariousness, and Blame

Anneloek Scholten

Abstract

This article investigates the portrayal of the social and cultural consequences of disaster in Dutch rural literature during the long nineteenth century, primarily its final decades, in which fiction often emphasised the precariousness of rural existence. Following a discussion of how several regional texts represent both the emotional and economic impact of failed harvests and loss of livestock, the analysis focuses on the representation of an outbreak of rinderpest in Josef Cohen's *Ver van de menschen* (1910), set in rural Overijssel. The events in the novel have a historical referent: they are based on the 1865 outbreak of rinderpest across the Netherlands. This article thus contributes to a better understanding of the nineteenth- and early twentieth-century literary imagination of rural vulnerability and calamity, which, in turn, provides insights into dynamics of social relations in the face of disaster and into how (past) natural disasters are made meaningful (in the present).

Keywords

cattle plague – regional fiction – nineteenth century – literature – rural economy – community

Throughout the eighteenth and nineteenth centuries, farmers in the Netherlands were plagued by outbreaks of rinderpest. Unlike some other national disasters, such as several nineteenth-century cholera epidemics that often originated in urban centres, rinderpest disproportionately affected the countryside. The 1865 rinderpest outbreak resulted in the loss of about half of the livestock in some areas. Imported from England, the cattle plague quickly

Asperen, van, Hanneke and Lotte Jensen (eds): *Dealing with Disasters from Early Modern to Modern Times. Cultural Responses to Catastrophes.* Taylor & Francis Group, 2023
DOI: 10.5117/9789463725798_CH09

Figure 9.1 *God's Punishing Hands in the Netherlands by the Cattle Plague*, copy after a print by Jan Smit, 1745. Rijksmuseum, Amsterdam, RP-P-1920-2588

spread from Schiedam across the rest of the country. Nationwide, almost 157,000 animals were infected.[1] Although particularly disastrous for many livestock farmers, the economic consequences of rinderpest outbreaks in the eighteenth and nineteenth centuries were felt throughout the country.[2]

In general, rural economies were vulnerable to natural disasters: even smaller adverse events, such as hailstorms or periods of drought, could wreak havoc on crops and impact financial stability. Many nineteenth-century works of regional literature, often set in small villages in the countryside, explore this vulnerability of rural communities and economies. Loss of livestock, disease, fires, and storms are commonplace in the genre, which often explores the effects of such events on individuals as well as communities: it attends, in other words, to ramifications beyond the economic. This makes regional fiction a valuable resource for the study of agricultural disasters, as literature functions partly as a way of theorising, and imaginatively

1 Ronald Rommes, '"Geen vrolyk geloei der melkzwaare koeijen". Runderpest in Utrecht in de achttiende eeuw', *Jaarboek Oud-Utrecht* (2001), 87–136, at 120.

2 Landowners suffered as tenant farmers became unable to pay rent. Cattle plagues also resulted in a scarcity of dairy products, with price increases affecting consumers, and the loss of income for farmers even affected tax revenues so that the consequences were felt throughout society. See Rommes, 'Runderpest in Utrecht'; Adam D. Sundberg, 'Floods, Worms, and Cattle Plague. Nature-Induced Disaster at the Closing of the Dutch Golden Age, 1672–1764', PhD dissertation, University of Kansas, 2015.

experimenting with, social experiences, processes, and responses that result from disaster.[3]

This article investigates the portrayal of the social and cultural consequences of disaster in Dutch rural literature during the long nineteenth century, with a focus on its final decades, in which fiction often emphasises the precariousness of rural existence. Following a discussion of the way several regional texts represent both the emotional and economic impact of failed harvests and loss of livestock, this article zooms in on the representation of an outbreak of rinderpest in Josef Cohen's *Ver van de menschen* ('Far from the People', 1910), a novel set in rural Overijssel.[4] While many regional novels consider the impact of adversity, Cohen's representation of rinderpest has a clear historical referent: although this is not mentioned explicitly, the details surrounding the cause and spread of the outbreak in the novel match up with the 1865 rinderpest that was brought over from England and, from there, spread across the rest of the Netherlands.[5] Reviewers of the book refer to it as a regional or 'peasant novel' (*boerenroman*), perhaps because the historical setting of its events is not made explicit, but it can also be read as historical fiction.[6] At a safe remove from the events they describe, historical novels experiment with the meaning of the past in the present and explore the lessons to be drawn from past events. Novels set in the past are necessarily also a reflection on the present moment: literary historian Marita Mathijsen calls this 'double-timeliness' ('dubbeltijdigheid'), referring to the interplay of past and present in historical novels.[7] She also notes the

3 Bruno Latour, *Reassembling the Social. An Introduction to Actor-Network Theory* (Oxford: Oxford University Press, 2005), 82. Latour argues that literary fiction is an important resource for understanding social reality. Building on this, David Alworth advocates for literature as a resource for thinking through the 'complexity of collective life', in *Site Reading. Fiction, Art, Social Form* (Princeton, NJ: Princeton University Press, 2016), 24.

4 All novel title translations from Dutch are mine, except for *The Flaxfield* (translated in 1988), *Pretty Marie* (translated, without its subtitle, in 1919), and *The Manse of Mastland* (translated in 1860).

5 Other outbreaks were primarily caused by infected cattle imported from Denmark and Northern Germany, as most beef cattle was imported from these countries. See also Peter A. Koolmees, 'Epizootic Diseases in the Netherlands, 1713–2002. Veterinary Science, Agricultural Policy, and Public Response', in Karen Brown and Daniel Gilfoyle (eds), *Healing the Herds. Disease, Livestock Economies and the Globalization of Veterinary Medici* (Athens: Ohio University Press, 2010), 23.

6 See 'Literaire kroniek', *Bredasche courant* (29 April 1911); 'Het boek van de week', *Leeuwarder courant* (29 May 1911); 'Letteren', *Het vaderland* (2 September 1911); 'Een uitstekende roman', *Gracieuse. Magazijn voor Neerland's vrouwen* 49 (1911), 16. *Ver van de menschen* is described as a novel of farm life or the Overijssel heath, with no reference to historical dimensions.

7 Marita Mathijsen, *Historiezucht. De obsessie met het verleden in de negentiende eeuw* (Nijmegen: Vantilt, 2013), 288.

value of (historical) fiction for understanding history beyond mere facts, as literature reveals patterns and mindsets.[8]

In *Ver van de menschen*, Cohen explores how a mass outbreak of rinderpest affects the social stability of a rural community and how this relates to notions of identity and belonging. By considering Cohen's novel alongside other contemporary narratives of rural life, this article contributes to a better understanding of the nineteenth- and early twentieth-century literary imagination of agricultural disaster and rural vulnerability. This, in turn, provides insights into dynamics of social relations in the face of disaster, as well as into how (past) natural disasters are made meaningful (in the present).

Regional Literature and the Precariousness of Rural Life

Regional literature, often in the form of village tales, emerged throughout Europe in the nineteenth century. Its popularity is attributed by some scholars to the rise of nationalism as well as industrialisation and urbanisation, processes which increased cultural interest in the countryside as a repository for the nation's traditions.[9] In the Netherlands Cornelis van Koetsveld, who wrote *Schetsen uit de pastorij te Mastland* ('Sketches from the Manse of Mastland', 1843), is identified by some literary historians as the first author of contemporary village life.[10] From the mid-nineteenth century onwards, many other village tales cropped up, describing regions all over the country.[11] Many early authors in this genre were village pastors or school teachers: they moved to the countryside from the city and wrote short stories about their experiences with the rural population, which idealised rural life.[12] As pastors or teachers, these authors were seen as cultural mediators, conveying the essence of rural life to urban readers and transforming the rural population from 'strangers' into 'fellow citizens'.[13]

8 Mathijsen, *Historiezucht*, 289–90.

9 See e.g. Josephine Donovan, *European Local-Color Literature. National Tales, Dorfgeschichten, Romans Champêtres* (New York: Continuum, 2010).

10 This is initially suggested by Jan ten Brink in *Geschiedenis der Noord-Nederlandse letteren in de XIXe eeuw. Eerste deel* (Amsterdam: Tj. van Holkema, 1888), 258–59, and several later literary histories follow this view.

11 See Jan te Winkel, *De ontwikkelingsgang der Nederlandsche letterkunde. Deel 7* (Haarlem: Erven F. Bohn, 1927), 501–6 for an overview of works per region.

12 Examples include Cornelis van Koetsveld, Cornelis van Schaick, and Pieter Heering.

13 'Van vreemdeling wordt hij dan medeburger'. 'Maclaren-van Koetsveld', *Rotterdamsch nieuwsblad* (18 November 1897). See also Oscar Westers, 'Welsprekende burgers. Rederijkers in

Their role in producing national awareness was thus seen as significant. However, although regional literature was very popular in the mid-to-late nineteenth century – van Koetsveld's *Schetsen uit de pastorij* was reprinted twelve times by 1899, and reviewers repeatedly identify village tales or rural novels as contemporary fashion trends – there is a paucity of scholarship on many of these texts.[14] Most of their authors have been all but forgotten, when their work in fact offers fascinating insights into contemporary perspectives on community and identity, and the way these are impacted by disasters.[15]

The economic vulnerability of rural regional communities becomes more pronounced in regional literature of the later nineteenth century. However, despite their reputation for superficiality and sentimentalism,[16] several early village tales already comment on the risk exposure that rural existence entails. For instance, in the short story 'De oude Wessels en zijn gezin' ('The Old Wessels and His Family'), published in *Betuwsche novellen* ('Stories of the Betuwe', 1854), J.J. Cremer describes how a hailstorm ruins Wessels's tobacco harvest: 'it had been the hail, dreaded so by the tobacco planter' that completely destroyed his plants. '[T]he plants, so lush just moments before, were on the ground in tatters and rags'.[17] The sight is heartbreaking to the farmer: 'There lies the product of so many drops of sweat, and so much care, destroyed in mere moments! There lies this whole year's livelihood, ruined!'[18] A single ill-fated hailstorm affects the family's

de negentiende eeuw', PhD dissertation, Universiteit van Amsterdam, 2003, 275.

14 For regionalism as a fashion, see e.g. J.H.C. Heijse, ''t Fingertje naast den duum en Neeltjes Nôl, Texelsche vertellingen door D. Dekker. Enkhuizen, J. Groot. 1871', review in *De gids* 36 (1872), 366; H. van Loon, 'Het vervloekte geslacht, door Josef Cohen ...', review in *Den gulden winckel* 13 (1914), 122.

15 While some regional authors, including van Koetsveld and J.J. Cremer, are comparatively well known, others – such as Cohen and Idsardi – are extremely marginal figures and rarely feature in scholarship.

16 See, for instance, G.P.M. Knuvelder, *Handboek tot de geschiedenis der Nederlandse letterkunde*, 4 vols ('s-Hertogenbosch: Malmberg, 1976), vol. 4, 431; 'dorpsroman', in *Algemeen letterkundig lexicon* (2012–), via DBNL, digitale bibliotheek voor de Nederlandse letteren, https://www.dbnl.org/tekst/dela012alge01_01/dela012alge01_01_00782.php (accessed 3 April 2023). Village tales (or so-called local-colour fiction more broadly) have a similar reputation abroad; see M.H. Abrams and Geoffrey Galt Harpham, *A Glossary of Literary Terms* (9th ed., Boston, MA: Wadsworth, 2005); Donald Dike, 'Notes on Local Color and Its Relation to Realism', *College English* 14: 2 (1952), 81.

17 '[H]et was de, voor den tabaksplanter, zo gevreesde hagel geweest'; 'aan flarden en lompen lagen de straks nog zoo welige planten ter aarde'. J.J. Cremer, *Betuwsche novellen* (Haarlem: Erven Loosjes, 1860), 61. Translation from Dutch is mine.

18 'Daar ligt dan in weinige oogenblikken vernietigd wat zoovele droppelen zweets gekost, zoovele zorgen vereischt heeft! Daar ligt dan het bestaan van den ganschen jaarkring verdorven!' Cremer, *Betuwsche novellen*, 62.

yearly income, illustrating the dependence of the rural community on natural circumstances beyond their control. In Cornelis van Schaick's *Tafereelen uit het Drentsch dorpsleven* ('Scenes of Village Life in Drente', 1848), the village's vulnerability to natural disaster is similarly illustrated: 'In the previous night, when everything lay quiet, lightning had struck the mill, and in less than an hour everything had fallen prey to the flames ... Even the cattle and the wheat had burned. Therefore the unfortunate souls were dirt poor, and had nothing, not even to provide for their own and their children's most urgent needs'.[19] As in Cremer's story, one unfortunate natural incident forces a community into poverty and leaves them unable to care for their children. Even these early Dutch village tales show an awareness of the rural community's exposure to risk: due to the dependence on – sometimes fragile – crops, a single storm can have lasting consequences on a farm's (financial) wellbeing.

In the nineteenth century, the rural region is commonly associated with tradition, a 'rootedness' in the past, and a resistance to change and industrialisation. For instance, Joep Leerssen explains that national identity is shaped by ideals of progress and modernity as well as by continuity and tradition, and that the latter ideal is largely dependent on the region, which is often imagined as rooted in a premodern past.[20] One reviewer of van Schaick's *Geert* (1847) writes that the rural man has 'preserved' '[t]he peculiarity and separateness' of the rural region 'against all improvement and corruption'.[21] In the popular imagination, the countryside is often connected to simplicity and stability and contrasted with the rapidly changing environments of urban spaces.[22] However, even in early regional stories by authors who generally uphold this distinction, rural life is also portrayed

19 'In den vorigen nacht, terwijl alles in diepe rust lag, was de bliksem in den molen geslagen, en in minder dan een uur was alles een prooi der vlammen geworden... Zelfs 't vee en koren was verbrand. Zoodoende waren de ongelukkige zielen doodarm, en hadden zelfs niets om in de eerste en dringendste behoeften des levens voor zich zelve en hunne kinderen te voorzien'. Cornelis van Schaick, *Tafereelen uit het Drentsch dorpsleven* (1848; 2nd ed., Meppel: H. ten Brink, 1858), 64.

20 Joep Leerssen, 'De canonisering van het typische. Tradities, culturen en de cultivering van tradities', in Carlo van der Borgt, Amanda Hermans, and Hugo Jacobs (eds), *Constructie van het eigene. Culturele vormen van regionale identiteit in Nederland* (Amsterdam: Meertens-Instituut, 1996), 45–57.

21 'Het eigenaardige, het afgescheidene, dat [de landman] tegen alle verbetering en verbastering tevens bewaard heeft'. 'Geert. Een verhaal voor het volk, vooral ten platten lande. 't Hoogduitsch gevolgd door C. van Schaick', review in *De gids* 11 (1847), 795.

22 See e.g. Henk Nijkeuter, *Geschiedenis van de Drentsche literatuur, 1816–1956* (Assen: Van Gorcum, 2003), 82.

as precarious. Inhabitants of rural communities are vulnerable to natural events ranging from bad weather to pests affecting crops and livestock, and as both Cremer and van Schaick illustrate, one storm or lightning strike can effectively destroy the financial stability of a family or community. Rural existence is, therefore, characterised by exposure to risk.

As the century progresses, Dutch regional literature becomes increasingly attentive to this vulnerability of the countryside: many later novels in the genre are socially engaged, and explicitly critical of power dynamics and poverty.[23] The precariousness of rural existence becomes increasingly central to literary representations. This might be partly because its authors are increasingly local and thus more aware of social issues in the countryside than outsiders who idealise rural life. For instance, W. van Palmar's *De golden kette* ('The Golden Chain', 1875) narrates a wealthy farmer's exploitation of a growing working class in Oldambt and shows the precarious situation of underpaid field labourers. In the early twentieth century, H.H.J. Maas's early works – including *Verstooteling* ('Outcast', 1907), *Het Goud van de Peel* ('The Treasure of De Peel', 1909), and *Landelijke Eenvoud* ('Rural Simplicity', 1910) – show rural families affected by poverty and failed harvests and comment on how heavy labour is rewarded with very little pay, leaving farmworkers financially vulnerable. His works are particularly critical of mismanagement and power abuse by local authorities, and they were heavily criticised as a result.[24] However, some reviewers praise him for his honesty about corruption in the countryside, and one reviewer commends him for 'having the courage to blow away the pseudo-poetic haze with which earlier writers shrouded Limburg's rural life'; the reviewer explicitly notes that earlier writers idealised the countryside, while Maas focuses on its harsher reality.[25] Both Maas and van Palmar (presumed to be a pseudonym for Wiepke Heikes Bouman-Geertsema) were from the area about which they wrote, and Maas's later work is especially critical of literary works that idealise rural life.[26]

23 This shift is also described by Johanna Maria van Buuren, 'De taal van het hart. Retorica en receptie van de hedendaagse streekroman', PhD dissertation, Universiteit van Groningen, 2005.

24 See, for instance, a scathing piece in the *Limburger koerier* which chastises Maas for his negative portrayal of Limburg (5 May 1915).

25 '[D]ie den moed gehad heeft, het pseudo-dichterlijke waas, door vroegere schrijvers gelegd om het Limburgsche landleven, weg te blazen'. Review of Maas's *Landelijke eenvoud* in *De telegraaf* (13 June 1911).

26 In his later novel *Het goud van de Peel* (Laren: Schoonderbeek, 1944), Maas criticises authors and readers who idealise rural life and 'admire Conscience's descriptions of loam huts which

Besides increasing attention to poverty and misconduct of those in power, exposure to natural events and disasters continues to feature in regional fiction throughout the long nineteenth century. In later works, the impact of these disasters on social relations becomes more important. In Stijn Streuvels's *De vlaschaard* ('The Flaxfield', 1907), published in Amsterdam and widely read in the Netherlands, the destruction of the flax fields after a devastating storm escalates a conflict between father and son. Streuvels was a Flemish author who lived in and wrote about West Flanders, and *De vlaschaard* is one of his most famous works. Streuvels's work is described as exceeding boundaries of regionalism, precisely because the social and generational struggles in his novels are seen as more universal.[27] However, the conflict in *De vlaschaard* is clearly informed by agricultural practices. Early in the novel, the farmer's inability to control the circumstances that determine the health of his crop becomes clear: the elder Vermeulen waits impatiently for summer weather, powerless to change climate conditions.[28] Meanwhile, he struggles to manage his son and to remain in control of his farm. After several disagreements between father and son about when to sow and reap the crops, a storm suddenly hits before the harvest:

> now the raging turmoil came so unexpectedly that people looked on it, dazed and bewildered. The flax, which had grown and matured for so long in the peaceful caressing breeze without disturbance, was seized on this final day, beaten and swayed, rendered defenceless in the power of that same wind that suddenly appeared mad and bad-tempered and wanted to destroy it all.[29]

Angered by his powerlessness against nature, and wounded in his pride, Vermeulen eventually strikes down his son for defying his wishes in an attempt to regain control.[30] An act of nature and its accompanying loss

breathed bliss-and-happiness' ('men bewonderde hoog de geluk-en-tevredenheid-ademende leemen-hut-beschrijvingen van Conscience'), 95.

27 See Jacqueline Bel, *Bloed en rozen. Geschiedenis van de Nederlandse literatuur 1900–1945* (Amsterdam: Prometheus, 2018), 107. Bel writes that *De vlaschaard* 'avoided the accusation of being regional fiction' ('onttrok zich aan het verwijt streekliteratuur te zijn').

28 Stijn Streuvels, *De vlaschaard* (Amsterdam: L.J. Veen, 1907), 5.

29 [N]u kwam zoo onverwacht die razende woeling dat de menschen er verpaft en radeloos stonden op te kijken. Het vlas dat zoo langen tijd in de vreedzaam aaiende lucht, zonder krook of stoornis was opgegroeid en gerijpt, werd nu dien uitersten dag, gepakt, vermooscht, geslegen en geslingerd en weerloos stond het in de macht van dien zelfden wind die ineens zot en baloorde scheen en 't al vernoozelen wilde'. Streuvels, *De vlaschaard*, 251.

30 Streuvels, *De vlaschaard*, 259, 288–90.

thus intensifies existing conflicts and potentially results in the death of the farmer's son: the violence of man and nature mirror each other.[31]

In Maas's *Landelijke eenvoud,* a storm likewise leads to strife. For years, a small village has been spared any disastrous weather events. Farmers are offered insurance against hailstorms but most refuse, except one: Jorissen, a farmer who has embraced modern techniques and machinery on his farm despite his community's contempt for new technologies.[32] When heavy thunder and hail destroy most of the village's fields, his neighbours turn against him: 'it *had* been strange, that hailstorm ... It must have been a punishment. The "striking hand" had touched so many places in the village these past years. Disease had affected animals here, people there ... If *one* family had brought all these misfortunes to the village!'[33] In the absence of belief in science or technology, and lacking insurance to compensate their losses, the community turns to superstition and blame to explain the destruction of their crops. The 'striking hand of God' is a common explanation for disaster,[34] but in this case, the rural community in Maas's novel attributes this wrath to the actions of one family. Adversity due to natural disasters thus challenges community relationships and social cohesion as people resort to blame.

Amidst the multiple losses and disasters that feature in regional texts, loss of livestock is portrayed by several authors as especially devastating, and several rural narratives emphasise the strong emotional connections between farmers and cattle. After a farmer has to kill his horse due to an injury in van Schaick's *Tafereelen uit het Drentsch dorpsleven*, the pastor immediately comes for a visit because he knows 'what the loss of a horse means to a farmer'; there is an unspoken understanding in the community about the intensity of such a loss.[35] In Pieter Heering's *Overijsselsche vertellingen* ('Tales of Overijssel', 1883) a community also bands together over a farmer's loss. When one farmer's cow refuses to eat, he is concerned, because 'these days, one frequently heard of similar symptoms amongst cattle, and many animals had already succumbed'.[36] As the cow's symptoms worsen, members

31 The novel ends with the son in a comatose state but leaves ambiguous whether he will die.

32 Maas, *Landelijke eenvoud*, in *Peel omnibus* (De Bilt: De Fontein, 1969), 192.

33 "t wás vreemd geweest met die hagelslag ... Dat zou wel een straf zijn. De 'kwade hand' was in de laatste jaren op zóveel plaatsen in het dorp geweest. Hier waren het ziekten onder het vee en daar onder de mensen ... Als één zo'n familie dan toch al die ongelukken over het dorp bracht!' Maas, *Landelijke eenvoud*, 190.

34 See Sundberg, 'Floods, Worms, and Cattle Plague', 206.

35 '[D]e Dominé ... wist wat het verlies van een paar voor een boer in heeft'. Van Schaick, *Tafereelen uit het Drentsch dorpsleven,* 69.

36 Pieter Heering, *Overijsselsche vertellingen* (3rd ed., Leiden: E.J. Brill, 1893), 149.

of the community – previously despised by the farmer – help him care for her, but despite their efforts, the cow dies in the night. The farmer wants to deliver the news to the neighbours, but words fail him: instead, '[t]he old man wept, wiping away his tears with his sleeves'.[37] More than worry about his financial dependence on the cow, he grieves the loss of an animal that he had for eight years and that was beloved by the whole family.[38]

Idsardi's *Mooie Marie: Een verhaal uit het Friese volksleven* ('Pretty Marie. A Tale from Frisian Rural Life', 1916) considers the impact of governmental policy on rural communities in the face of agricultural disaster. Idsardi was a pseudonym for G.I. van der Ploeg, a teacher of religion in Friesland who wrote several popular moralistic novels set in the Frisian countryside. In *Mooie Marie*, farmers argue about the requirement to kill any cattle that have been in contact with a cow infected with foot-and-mouth disease. Farmer Steenma, whose cow might be affected, disagrees vehemently with this policy and claims that 'when you mess with the lives of his cattle, you mess with the farmer himself'.[39] He is unwilling to be persuaded by scientific arguments presented by fellow farmers and people in the community and cannot accept the loss of his healthy cows. Not only does he describe the care for his cattle as his life's work, but he compares them to ill children in an attempt to convince others of the injustice of 'murdering them in cold blood' for contracting a disease.[40] To Steensma, his cattle represent his livelihood, but his repeated comparisons of the cows to humans also suggest a deep attachment and a desire to protect them. All three stories share a focus on the importance of animals to rural inhabitants, and the death of livestock is accompanied by grief that exceeds concern about loss of profit, a grief that is instead based on a genuine attachment of farmers to their animals.

The following section will therefore focus on a literary representation of the 1865 rinderpest, a disaster that was both economically and emotionally devastating to rural communities, in Cohen's *Ver van de menschen*. This novel is particularly interesting because, unlike the examples above, it describes the aftermath of a historic national disaster in the rural community (rather than commonly recurring local disasters such as storms and failed harvests) and therefore invites its readers to engage with the connection between nation and region as well as past and present. It also explicitly engages

37 'De oude man schreide en veegde met zijne mouw de tranen af'. Heering, *Overijsselsche vertellingen*, 151.

38 Heering, *Overijsselsche vertellingen*, 151–52.

39 '[W]anneer je aan het leven van zijn vee komt, [kom je] aan het leven van den boer zélf'. Idsardi, *Mooie Marie* (Kampen: J.H. Kok), 158.

40 '[K]oelbloedig te vermoorden', Idsardi, *Mooie Marie*, 159, 162.

with the relationship between the regional, national, and transnational in its representation of this disaster and its aftermath. Literary scholar and genre studies expert June Howard writes that in regional fiction, 'the local and the global are by no means distant ends of a continuum': global events consistently influence regional communities, and so-called 'global culture' cannot annihilate difference.[41] Regional fiction is therefore always relational and engages with places elsewhere. Not only did regional stories themselves travel across borders – many works were translated – but the genre also functions as a way of thinking through the ways national and global actors and events impact the local. (Natural) disasters are powerful examples of this dynamic: agricultural disasters can have a nationwide impact (consider, for instance, the failed potato harvest and famine in the 1840s in the Netherlands and Flanders), while national or transnational events can have deeply felt repercussions in local communities (outbreaks of rinderpest usually came to the Netherlands from abroad). More than the literary examples explored above, Cohen's novel acknowledges these interrelations between the region and the 'outside' in its representation of the ways disaster impacts the community.

Community, Loss, and Blame in *Ver van de menschen* (1910)

Although the rural community in Cohen's *Ver van de menschen* is plagued with multiple disastrous events throughout the novel, the rinderpest outbreak is depicted as particularly traumatic, both socially and economically. The plot of the novel revolves around the life and history of the Strie family. The Strie name holds power in the rural community: 'The narrow Strie farm was known in all of Salland; the name inspired awe in the richest farmers. Ancient myths resounded in the name; for hundreds of years the Stries had been dreamers and thinkers who had not associated with the others in the village. And this had increased their magical influence'.[42] The family are marked as outsiders to the community: the Stries are respected, but not well integrated in their village. Moreover, the family is haunted by

41 June Howard, *The Centre of the World. Regional Writing and the Puzzles of Place-Time* (Oxford: Oxford University Press, 2018), 42.

42 '[I]n geheel Salland was de smalle Strie-boerderij bekend; het ontzag voor den naam was in den rijksten boer. Oude mythen klonken in den naam; want door honderden jaren waren de Strie's droomende denkers geweest, die niet omgegaan waren met de anderen in het dorp. En dat had hun magischen invloed vermeerderd'. Cohen, *Ver van de menschen*, 166, published in parts in *Onze eeuw*, 10 (1910).

legend: 'there was a legend that the Stries would perish if they took part in village life too much; they were to avoid the inns, and wild company, and the game of cards. When a Strie sought after lower desires, our dear Lord burned his book of life in the high heavens, and wrote him off from the list of the privileged'.[43]

Ver van de menschen traces this family's downfall and survival, focusing primarily on Hannes Strie and his son, Peter, over the course of several decades. Throughout the novel, several disasters threaten the rural community. Cohen seems to be particularly interested in the rhetoric of blame that surrounds disaster in the rural community and in its effects on the dynamics between 'insiders' and 'outsiders'. The community is described as somewhat impenetrable: 'You had to have lived in Merloo from childhood to understand this quiet world far from the people'.[44] The region is thus described as isolated from the rest of society and difficult to understand for an outsider. The relationship between individual and community, and between local and foreign, is very important in this context. Tradition and legend – two important pillars of local identity – shape the way disasters are framed and interpreted by members of the local community.

Throughout the novel, legend and superstition are used to explain as well as predict disastrous events. For instance, the community is convinced that if the castle of the local baron catches fire, the whole town and even the entire province will burn.[45] The Strie legend in particular is seen as the cause of disaster. The family is often blamed when misfortune comes to the family or the community, even if it is explicitly caused by others. For instance, when Peter's sister Gerardien develops a relationship with a local nobleman, she is murdered by a jealous local youth. Gerardien is blamed for arson after the young man burns down the castle and even receives blame for her own death; the narrator claims that 'no human on earth but the legend had killed her' and that her murder is therefore 'no case for an earthly judge'. Meanwhile, the townspeople protect her murderer from

43 '[E]r was een legende, dat de Strie's te gronde zouden gaan, wanneer ze te veel deel zouden nemen aan het dorpsleven; ze moesten de herbergen mijden, en woeste gezelschappen, en het kaartspel. Als een Strie lagere lusten zocht, verbrandde in den hoogen hemel onze lieve Heer zijn levensboek, en schreef hem af van de lijst der bevoorrechten'. Cohen, *Ver van de menschen*, 167.

44 'Je moet van kind-af in Merloo hebben gewoond, om deze stille wereld ver van de menschen te kunnen kennen'. Cohen, *Ver van de menschen*, 169.

45 'Denk ie dan neêt an de legende? As het kasteel in brand vlug, geêt 't heele darp mee. Bedenk dat toch. Heel Overiesel verbrandt'. Cohen, *Ver van de menschen*, 177.

persecution.[46] When the family farm is burnt down by the jealous husband of Peter's childhood friend, the Strie family is, once again, blamed: 'who dared accuse Piet Nievelt, when it was so clearly a punishment for the misdemeanour, by which a Strie played with his family name?[47] Both events are explicitly the consequence of crimes committed by others.

However, blaming the Strie family for disaster has a clear social function. The Stries are essentially outsiders to the community, though at times respected and at times despised. Other members of the community escape culpability by blaming the Stries. It also makes disaster *explicable.* The narrator notes that 'the farmers found comfort in the idea that misfortunate had not come to their lands for no reason. It was because Peter Strie and Hilde Wolters had desired each other, and this has caused the catastrophe'.[48] By blaming individual actions and desires for disaster, the wider community can avoid investigating their own accountability. This thus serves as a coping mechanism for dealing with disasters.

Throughout the novel, Cohen draws parallels between the Strie family and other persecuted outsiders, particularly the Jewish population in the area.[49] As a boy, Peter Strie befriends a young Jewish man – an unnamed character consistently called 'the Jew' – who remains his loyal friend and ally throughout the story. The two support each other throughout, against the prejudice of others in town. Both are treated as 'foreign', and their existence is identified as a threat. Early in the novel the narrator explains that a Jew and his family had moved to the region in the fourteenth century. Twenty years later, when crusading armies returned from Jerusalem and spread leprosy in the community, the Jew was accused of poisoning the wells and causing the illness, after which he was brutally murdered by a local knight.[50] This illustrates, early on, the role of prejudice in explaining disease and disaster: communities irrationally place blame on an individual, with horrific consequences.

46 'Dat was geen zaak voor den aardschen rechter ... de legende [had haar] vermoord, geen mensch ter wereld'. Cohen, *Ver van de menschen*, 183.

47 'Wie durfde Piet Nievelt beschuldigen, nu het zoo klaarblijkelijk een straf was voor de euveldaad, waarin een Strie met zijn naam gespeeld had?' Cohen, *Ver van de menschen*, 340.

48 '[D]e boeren vonden troost in de gedachte dat het ongeluk niet zonder reden over de landen was gekomen. Het was, omdat Peter Strie en Hilde Wolters naar elkander hadden verlangd, en dat had het onheil gelokt'. Cohen, *Ver van de menschen*, 213.

49 Cohen was himself Jewish, potentially motivating his interest in superstition, prejudice, and the dynamics of 'othering' in the novel.

50 Cohen, *Ver van de menschen*, 4.

Figure 9.2 *The Cattle Plague: Another One of the Blessings of War*, cartoon in *De courant* (23 August 1920), 5. KB – National Library of the Netherlands, The Hague. Source: Delpher (accessed 21 April 2022)

The actual cause of this fourteenth-century outbreak of leprosy is identified by the narrator to be the return of the Crusaders; war is the real root cause of disaster. This applies to rinderpest as well: in the description of the outbreak, the narrator explains the causes to the reader and notes that 'as it has been since the Migration Period: where war has burnt not only men were killed due to battle, but cattle too'.[51] The narrator states that the cattle plague initially raged due to the war, and after a ship from one of the countries at war transported infected cows to England, thousands of English cows died. A diseased cow was then exported to Holland, bringing rinderpest to Dutch farms. While countries at war had made peace, England and Holland both continue to grieve for their lost cattle.[52] Cohen describes the consequences of war as far reaching, affecting more than the countries involved in battles, with ramifications that last beyond the timeframe of the war itself. Notably, the countries at war remain nameless in the novel, perhaps suggesting that Cohen himself has no interest in placing blame. War and conflict in general, rather than a *specific* war or nation, are the cause of this agricultural disaster.

Due to the magnitude of the losses, the rinderpest and loss of cattle are described as a collective regional trauma. The narrator mentions that an earlier rinderpest a hundred years prior had also robbed the farmers of their cattle and that the people in West Holland had forgotten, but that 'in the East, on the heathland, rinderpest was a repeat of a drama just experienced'.[53] Although the tragedy itself is national, and the 1865 rinderpest disproportionately affects Holland and Utrecht, Cohen suggests that the cultural memory and response vary between regions. This is in line with the previously mentioned notion that peripheral rural regions are more 'rooted' in tradition and the past: in the province of Overijssel, the tragedy feels more recent, suggesting a different relationship to time in West Holland compared to the eastern provinces. Likewise, it suggests that the east has a different relationship to memory. Pierre Nora has argued that industrial growth caused a 'fundamental collapse of memory', identifying 'peasant

51 'En zooals het al sinds de volksverhuizing is geweest: waar de oorlog brandde, werden niet alleen menschen gedood onmiddellijk door den strijd, maar middellijk de runderstapel'. Cohen, *Ver van de menschen*, 354.

52 'De veepest woedde in de stallen; een schip met besmette koeien uit een land, waarvan men op de heide nog nimmer gehoord had, bracht zijn gevaarlijke lading in Engeland, en bij duizenden stierven daar de koeien. En weer werd een ziek rund, temidden van gezonde, geëxporteerd naar Holland. En door de lucht woei dichtonzienbaar de smetstof, van stal tot stal. De vrede tusschen de twee mogendheden, die niets met Holland en Engeland gemeen hadden, werd gesloten. Wat deed het er toe, wie won en wie verloor? In Holland en Engeland was rouw op de akkers!' Ibid.

53 'in West-Holland, waar het groote leven heerscht, vergeet men ras zulke dingen. In het Oosten, op de heide, werd de veepest als de herhaling van een zoo juist-beleefd drama'. Ibid.

culture' as a 'quintessential repository of collective memory' due to the transmission and conservation of values and habits across generations.[54] *Ver van de menschen* similarly suggests that the rural east has a privileged relationship to memory, one that binds its inhabitants together. The novel thus establishes a relationship between the region and its outside and explicitly reflects on the importance of the past in the present.

After a brief historical introduction to rinderpest's history and causes, the novel details the effect of the most recent outbreak on the local community in this town in Overijssel:

> When the water began to run from the eyes of one of Isfordink's cows... and in the blink of an eye the breathing of the cattle accelerated feverishly, when the eyes were swollen and red and the animals were coughing, even the stupidest servant knew that the cattle plague had broken out. But because the marks of illness did not reveal themselves for about five days, everyone understood that the whole region must be contaminated with the pest. The cows also coughed in Piet Nievelt's stable, and the animals moaned in wild terror for lack of breath.[55]

The descriptions are graphic at times, illustrating the devastating effects of the illness on the cows. Although this passage suggests that people have some level of understanding of the progression of the disease – 'everyone understood' that the whole region must be contaminated, since it takes days for symptoms to reveal themselves – superstitions about the cause of the outbreak soon begin to shape the community's response. For instance, the local farmer Tubertink 'did not believe in contamination' and was convinced that the disease only affected 'unclean stables'.[56] For a while, his cows remain unaffected. The farmer is so sure of himself that he suggests that other farmers bring their diseased cows to stay amongst his own and that the animals will heal, after which an infected animal is taken to his stables.

54 Pierre Nora, 'Between Memory and History. Les Lieux de Mémoire', *Representations* 26 (1989), 7–24, at 7.

55 'Toen het water uit de oogen van een van Isfordink's koeien ging loopen ... en in een oogwenk de adem van alle runderen koortsig zich versnelde, de oogen roodgezwollen zagen en de dieren hoestten, wist de domste knecht, dat de veepest was uitgebroken. Daar echter de kenteekenen zich eerst na een dag of vijf vertoonen, begreep ieder, dat de geheele streek met den besmetting bevlekt moest zijn. Ook in Piet Nievelt's stal hoestten de koeien, en de beesten kreunden in wilden angst om gebrek aan adem'. Cohen, *Ver van de menschen*, 354–55.

56 'Jan Tubertink ... geloofde niet aan besmetting. Hij zei, dat de ziekte alleen maar kwam in onzindelijke stallen'. Cohen, *Ver van de menschen*, 25.

Unsurprisingly, the animal does not get better.[57] Then, 'a few days later Peter Strie passed his farm. And the next day Jan Tubertink noticed that his livestock was also affected by the disease. Now people on the heath understood: Peter Strie was bewitching the cattle. How could the ringing of the bells help to prevent evil when one of the curse-worthy Stries remained on the land?'[58] The only action people undertake to prevent the disaster is ringing the church bells, a traditional method for warding off disaster since at least the Middle Ages. This indicates, once more, the community's rootedness in tradition and the past, as well as, perhaps, their conviction that the rinderpest outbreak was an act of God. As in Maas's *Landelijke eenvoud*, the community places blame on an individual for causing misfortune and invoking God's wrath.

However, Cohen has noted previously that people are aware that symptoms take a while to appear. An infected cow was brought into the stables, other cows were likewise infected, and Peter Strie is still blamed for merely walking past. There is at least the suggestion that this is deliberate scapegoating by people who have enough information and resources to know better. In fact, by 1865, information about the disease, its causes and prevention, was widely accessible. Newspapers and periodicals across the country reported on the cattle plague, and many included detailed information about the best preventative measures, such as quarantining diseased cattle and prohibiting sales of meat or hides from affected animals.[59] The need to isolate infected cattle was already well understood in the eighteenth century.[60] Regional authorities likewise called upon farmers to isolate diseased cattle, and one local newspaper from Middelburg warned farmers in the region not only to avoid any contact with livestock from Holland but also to steer clear of any cattle dealers that had visited Holland and even any other labourer who had worked in the area, to ensure that the province of Zeeland would remain unaffected.[61] In spite of the widespread availability of information to the contrary that was widely accepted and put into practice by nineteenth-century

57 Ibid.

58 'Enkele dagen kwam Peter Strie zijn boerderij voorbij. En den volgenden dag bemerkte Jan Tubertink, dat zijn vee ook door de ziekte was aangetast. Nu begreep het de heide: Peter Strie behekste de runderen. Wat hielp het luidden van klokken, die men luidde, allerwege, om het onheil te bezweren, wanneer een der vloekwaardige Strie's op het land bleef?' Cohen, *Ver van de menschen*, 26.

59 See e.g. *Landbouw-courant* (7 September 1865). Similar claims are repeated in numerous local and regional papers.

60 Rommes, 'De runderpest in Utrecht'.

61 'De veepest!', *Middelburgsche courant* (29 August 1865).

farmers, the town nevertheless becomes entirely convinced that 'everything was Peter Strie's fault. He had infected the stable and sown disaster wherever he went. And he had forced many farmers to sell their farms and travel far from their homeland, to strange countries'.[62] Blaming Peter Strie is, perhaps, more comfortable than accepting that the farmers have no control over the disease once it starts to spread or acknowledging that Tubertink himself might be to blame for allowing a diseased cow to be brought amongst his healthy cattle. Cohen implicitly invites readers to critically consider the motivations for blaming others.

Peter tries to reason with people, asking them how they know that he brought the cattle plague to the area and pointing out that '[t]he plague is all over the country and in England – have I brought all that here, and there as well? I haven't done anything'.[63] Peter emphasises the scope of the disaster, noting that the spread of the disease is not limited to his region, but to no avail: he is forced to leave the area to escape the hatred of his community. After he is banished, disasters continue to strike. Farms burn down due to lightning, and after a few years, the region is affected by another outbreak of rinderpest. Like several other regional novelists, Cohen emphasises the precariousness of rural existence and its vulnerability to threats and natural events. The novel suggests that it is the community's inability to accept this precariousness that causes further problems. As Judith Butler suggests in *Precarious Life*: 'though for some, mourning can only be resolved through violence, it seems clear that violence only brings on more loss, and the failure to heed the claim of precious life only leads, again and again, to the dry grief of an endless political rage'.[64] It is precisely this cyclicity of violence that Cohen illustrates in his novel, relying on a narrative frame of recurrence: throughout the story, grief, loss, and jealousy bring about acts of violence, resulting in more loss.

However, it is not just this cycle of blame and violence that undermines the cohesion of the rural community. After the rinderpest outbreak, and accompanying losses of income, many locals are forced to move abroad or to the cities, while urban speculators buy up the farms for little money.

62 'Alles was de schuld van Peter Strie. Hij had de stallen gesmet, en, waar zijn voet ging, onheil gezaaid. En hij had vele boeren gedwongen, om hun hoeven te verkoopen en ver van hun vaderland te trekken, in vreemde landen'. Cohen, *Ver van de menschen*, 27.

63 '"[H]oa weêt-ie, dat ik de veepest ebracht heb. In 't heele land is de veepest – en in Engeland is de veepest – heb ik dat dan allemoale hier en gunder ebracht? Ik heb niks edoan."' Cohen, *Ver van de menschen*, 30.

64 Judith Butler, *Precarious Life. The Powers of Mourning and Violence* (New York: Verso, 2004), xix. Although Butler focuses on the aftermath of 9/11, their writing about vulnerability is relevant here.

When farmers are forced to sell, their servants lose their employment: 'this meant worry about daily bread, and misery for many'.[65] Cohen emphasises this domino effect, showing how rinderpest undermines community bonds, leaves farmers unable to support each other, and forces people out of employment and homes. Social stability is undermined by disaster even before people turn to prejudice and blame. The people who populate Cohen's novel are described as rooted to the heath, and the land will always have 'terrifying attraction of mystery for those who are born there'. When rinderpest forces them to move to the cities, they 'remain always homesick for nature ... they could not speak except when you talked to them of the land that they had left'.[66] Peter Strie describes the 'societal imbalance' caused by these migrations from the country to the city, separating people from the land and communities to which they are so strongly connected.[67] In the city the people from rural Salland become outsiders themselves. *Ver van de menschen* therefore illustrates that rural precariousness is not merely economic; the vulnerability of the countryside might also threaten its very identity, dispersing those people who share the connection to the land and the collective memory of the region. Identity, history, and tradition, elements so important to communities in regional literature, are thus always at risk – perhaps this also adds import to their preservation or memorialisation in fiction.

However, while identity and community sometimes devolve into xenophobic suspicion of outsiders, Cohen's novel consistently portrays the inadequacy of 'us versus them' thinking. By portraying the migration of people in the rural community to the city, Cohen also shows that anyone can become an outsider. Ultimately, it becomes clear that banishing Peter – the suspected cause of the community's misfortune – does not solve anything. The vulnerability of the countryside is partly innate to rural existence. Failure to accept the fact that misfortune does, in fact, strike the land 'for no reason' results in the community's attempts to blame individuals, mostly outsiders whose acceptance by the community is always already at risk.[68] Cohen suggests that misfortune cannot be avoided and that, in many cases, no one can be blamed. In fact, solidarity and community are suggested as partial solutions

65 'Dat beteekende zorg om brood, en ellende voor velen'. Cohen, *Ver van de menschen*, 28.

66 'De heide heeft op hen, die er geboren zijn, de angstwekkende aantrekkingskracht van mystiek en mysterie'. Cohen, *Ver van de menschen*, 37. '[I]n hen bleef altijd het heimwee naar de natuur... Ze konden niet praten behalve als je ze van het land sprak, dat ze verlaten hadden'. Cohen, *Ver van de menschen*, 51.

67 '[M]aatschappelijke evenwichtsverstoring'. Ibid.

68 Cohen, *Ver van de menschen*, 213.

to the precariousness of rural life: during Peter's banishment, sharing his longing for home with someone from the same region eases his suffering, and the Jew helps him buy back his farm from investors.[69] Solidarity is thus a more effective coping mechanism than blame, improving instead of undermining a community's resilience in times of crisis. Throughout the novel, bonds in the local community help people find jobs or escape financial ruin through loans. This does not prevent disasters from striking: after Peter Strie restores his family name, and relationships are repaired, the farmers continue to 'fight a heavy battle with the land', and failed harvests and diseased livestock continue to threaten their livelihood.[70] However, the focus is on cooperation and rebuilding as a community. The novel is persistent in its focus on the social and emotional impact of economic loss and on solidarity and nonviolence as ways of coping with it.

Conclusion

Like many regional novels, *Ver van de menschen* emphasises social relationships in a small rural community. This community is initially described as isolated and far removed from society, but Cohen persistently illustrates how it is shaped relationally, impacted by events both at home and abroad. Rural inhabitants move to the cities or abroad; urban speculators buy local farms; outsiders live and work in the local village. The rinderpest that plagues the community happens on a much larger scale and is caused by a foreign war. Grief over the lost cattle is shared between England and the Netherlands and extends beyond the end of the war. In Overijssel grief also spans across generations as memories of previous rinderpest outbreaks linger. Far from isolated, the small rural community is in fact deeply embedded in transnational and transgenerational processes. This interdependence is emphasised by the rinderpest, which has a clear historical referent: the 1865–67 rinderpest outbreak in the Netherlands. The novel thereby invites readers to imagine their connection to others, both past and present.

Besides being economically devastating, rinderpest undermines social stability, both because poverty and bankruptcy force people out of the region

69 e.g. Cohen, *Ver van de menschen*, 57, 61, 337.

70 'Het is een zware strijd, dien hij met den grond te voeren heeft. En het is niet waar, dat de boeren minder te werken hebben, nu er weer een Strie op de heide woont: kwade zomers, die den oogst op het land vernielden, zijn niet uitgebleven, noch ziekten onder het vee, noch ziekten in het gewas'. Cohen, *Ver van de menschen*, 355.

and because people resort to superstition and blame in explaining their misfortune. While the novel emphasises the importance of tradition and community, it negates the validity of thinking in terms of 'self' versus 'other' and implicitly criticises strategies of blame in the face of disaster. Cohen engages this past disaster to convey moral lessons to present-day readers.

Moreover, he challenges existing dynamics of exclusion and notions of belonging, within a genre that was deeply concerned with constructing the identities of rural regional communities in relation to the nation and beyond. The novel encourages audiences to imagine grief as shared across nations and generations, offering an expanded sense of a social collective. This ability to rethink social relationships and collective life in the aftermath of a (trans)national catastrophe makes literary fiction a useful resource for understanding the cultural impact of disaster and for memorialising events and communities that – as the novel repeatedly emphasises – are at risk of disappearing and being forgotten.

About the Author

Anneloek Scholten is a PhD candidate at Radboud University, where she is working on a dissertation on Dutch nineteenth-century regional literature. In particular, her research engages with the relationship between the local, national, and transnational in rural fiction and the genre's circulation across borders. This PhD project is part of the NWO-funded Vici project *Redefining the Region. The Transnational Dimensions of Local Colour.*

Part 3

Disaster and Time

10 Coping with Epidemics in Early Modern Chronicles, The Low Countries, 1500–1850

Theo Dekker

Abstract
Historians have hypothesised that the increase of medical knowledge in the early modern period led to a shift away from religious towards 'scientific' explanations and prophylactic measures. The writings of contemporaries belonging to the 'middling' ranks of society tell a different story. This chapter presents a long-term perspective on how 104 non-medical experts coped with and reflected upon epidemics in the Low Countries. By using the corpus of the *Chronicling Novelty* project, I demonstrate that the middling sort used both religious and non-religious practices side-by-side. I show that between 1500 and 1850, natural explanations became more detailed and complex, but they remained in service of, or subordinate to, divine explanations. Moreover, although the idea of an angry and vengeful God was never far away, the notion of a benevolent God gained prominence in the seventeenth century.

Keywords
epidemic disease – chronicles – Low Countries – middling sort – knowledge

When early modern societies were struck by deadly epidemics, they used various methods to try and deal with these diseases. The extent to which these methods changed has been the subject of debate. On the basis of printed ordinances, legislation, and medical treatises, scholars have argued that there was a shift away from religious towards prophylactic measures in the late seventeenth century. This coincided with the hypothesis that an

Asperen, van, Hanneke and Lotte Jensen (eds): *Dealing with Disasters from Early Modern to Modern Times. Cultural Responses to Catastrophes.* Taylor & Francis Group, 2023
DOI: 10.5117/9789463725798_CH10

inverse relationship between the use of religious explanations, on the one hand, and natural or scientific explanations for epidemics, on the other, exists.[1] In various case studies, scholars have recently been questioning this inverse relationship by claiming that both religious and non-religious practices were used side by side, usually without apparent conflict.[2] What is missing, however, is a long-term perspective on the development of the relationship between these explanations, especially among non-experts. The *Chronicling Novelty* project, supervised by Judith Pollmann and Erika Kuijpers, recently digitised 313 handwritten Dutch chronicles, produced by 235 authors from the middling sort and from local elites. Since they were written especially in periods of crises, disasters, and hazards, chronicles enable us to develop a new outlook on how farmers, local officials, and merchants reflected upon and coped with epidemics in the early modern Low Countries.

Chronicles were usually written by literate males who used them to gather what they deemed to be important and 'useful' information about the past and especially the present of their locality. They did so chronologically without creating a hierarchy between their entries. According to historian Judith Pollmann, early modern chronicles have long been dismissed as a distinctly old-fashioned and uninteresting type of text.[3] In contrast to their medieval predecessors, early modern chroniclers were less restricted to the conventions of the genre. Everything the author regarded as useful was included, which gave chronicles the dubious reputation of being 'subjective' and 'trivial' sources among historians. Moreover, chroniclers barely wrote about themselves and their feelings, which (supposedly) makes them less

1 The research for this article was done with funding from the NWO project *Chronicling Novelty. New Knowledge in the Netherlands, 1500–1850*. I would like to thank Judith Pollmann and Erika Kuijpers for their helpful comments on earlier drafts.

The debate is summarised in John Henderson, *Florence Under Siege. Surviving Plague in an Early Modern City* (New Haven, CT: Yale University Press, 2019); Frank M. Snowden, *Epidemics and Society. From the Black Death to the Present*, Open Yale Courses Series (New Haven, CT: Yale University Press, 2019); David K. Chester, 'Theology and Disaster Studies. The Need for Dialogue', *Journal of Volcanology and Geothermal Research* 146: 4 (September 2005), 319–28, at 320. For the Dutch case, see L. Noordegraaf and Gerrit Valk, *De gave Gods. De pest in Holland vanaf de late middeleeuwen* (Bergen, NH: Octavo, 1988); Jacoba Steendijk-Kuypers, *Volksgezondheidszorg in de 16e en 17e eeuw te Hoorn. Een bijdrage tot de beeldvorming van sociaal-geneeskundige structuren in een stedelijke samenleving* (Rotterdam: Erasmus Publishing, 1994); A.H.M. Kerkhoff, *Per imperatief plakkaat. Overheid en pestbestrijding in de Republiek der Zeven Verenigde Nederlanden* (Hilversum: Verloren, 2019).

2 Henderson, *Florence Under Siege*.

3 Judith Pollmann, 'Archiving the Present and Chronicling for the Future in Early Modern Europe', *Past & Present* 230: 11 (2016), 231–52, at 231.

useful to study emotional responses. Consequently, the genre fell between the two stools of historiography and autobiography.[4] Despite the lack of emotions and the heterogeneity of the genre's content, however, chronicles are a promising source when we study them comparatively and as collections of useful knowledge.[5] Through a combination of computational and traditional historical methods, and by analysing the causal relations chroniclers constructed between (disastrous) events, structural patterns can be discovered between the 'trivial' and 'subjective' entries. Moreover, in combination with the longevity of the genre, chronicles enable us to reconcile the tensions in scholarly debates between generalised knowledge claims and individual experiences, and thereby to understand more fully the elements of both change and continuity in the history of disease and disasters. In this article I aim to do so through an analysis of the causal relations made by chroniclers and their use of both religious and non-religious practices to demonstrate how they coped with, anticipated, and reflected upon epidemics.

Of the 104 chroniclers from the Northern and Southern Netherlands who discussed epidemics, this article uses a small number of significant cases to illustrate the individual experiences of the chroniclers and the elements of change and continuity between 1500 and 1850, with an emphasis on the latter. In order to do so, I will first show how sixteenth- and seventeenth-century chroniclers searched for patterns to make sense of the present and to anticipate the future. Secondly, I will argue that the increase of empirical research and statistics led to a more positive notion of God by eighteenth-century chroniclers. Thirdly, I will present how they combined medical ideas to make sense of epidemics in the first half of the nineteenth century.

Searching for Patterns to Make Sense of Epidemics

Most sixteenth-century chroniclers emphasised natural alongside divine explanations for epidemics in their chronicles. It did not matter whether the chronicler was a cleric, rhetorician (member of a literary club), artisan, or official. The natural causes they discussed were generally limited to sublunar phenomena, with a focus on the Hippocratic-Galenic non-naturals. They were behavioural and environmental and comprised air, sleep and waking, food and drink, rest and exercise, excretion and retention, and the

4 Pollmann, 'Archiving the Present', 232.
5 Pollmann, 'Archiving the Present', 233–34.

passions (including sex). For example, the artisan from Antwerp Godevaert van Haecht (1546–1599) and the local official Zegerus ter Stege (1535–?) from Steenwijk (Overijssel) pointed out how rotten bodies infected the air. The nun Maria Luyten (c. 1540–c. 1570) from Weert (Limburg) wrote how the spit of contagious persons infected a whole village, while others stressed how poor diet, soft fruits, and famine could imbalance the four humours.[6]

The knowledge of natural causes in early modern chronicles was probably derived from both local medical practitioners and the information that was expressed implicitly in government messages, such as plague ordinances and treatises. Although this paper is concerned with the relationship between religious and non-religious practices, some measures present a striking resemblance to those taken during the COVID-19 pandemic: isolation, social distance, trade restrictions, border control, closed schools, and many more measures were installed to avoid further contagion.[7] Ecclesiastical and secular authorities often worked together in fighting the plague, for instance through the organisation and composition of processions. Both Albertus Cuperinus (c. 1500–c. 1560) from 's-Hertogenbosch and Marcus Vaernewyck (1518–1569) from Ghent mentioned how the authorities worked together to avoid contamination between districts during processions with holy relics.[8]

6 Godevaert van Haecht, *De kroniek van Godevaert van Haecht over de troebelen van 1565 tot 1574 te Antwerpen en elders*, Uitgaven van het Genootschap voor Antwerpsche Geschiedenis, 2 (Antwerp: 1929), 150–51. Zijger ter Stege, 'Aentekening van verscheyden saken, welke in en omtrent Steenwijk sijn voorgevallen op het papier eertijds gestelt door mr. Zijger ter Stege in sijn leven secretaris deser stadt Steenwijck' (Steenwijk: [17XX]), 0263, kept in Historisch Centrum Overijssel, Vereeniging tot beoefening van Overijsselsch Regt en Geschiedenis (VORG), handschriftenverzameling inv. no. 947, unpaginated, 1580–81. Maria Luyten, *Kronijk uit het klooster Maria-Wijngaard te Weert, 1442–1587*, ed. Jos. Habets, Ch. Creemers, and A. Nieuwenhuizen (Stichting Historisch Onderzoek Weert; Ghent: Geschiedkundige Heruitgeverij, 2004), 159: 'In 't selve iaer 1563 in meert of april quamder een pest te Weert, hier en daer geheel de stadt door; daer storvender veel van en genaesender oock veel van. Deese pest quam eerst in de beeckstraat bij Billeken van Horne, met eenen man die daer in de herbergh was, daer het die dochter van cregh als sy opgeveght hadde daer hij gespouwt hadde, daer sij oock van storf; maer nergens storf het soo seer als in de langhstraet binnen en buijten de poort; want in sommige huijsen storvender wel 9 oft thien of soo veel als er in een huijs was'.

7 Beatrice de Graaf, Lotte Jensen, Rina Knoeff, and Catrien Santing, 'Dancing with Death. A Historical Perspective on Coping with Covid-19', *Risk, Hazards & Crisis in Public Policy* 12: 3 (2021), 346–67, at 350.

8 Adelbertus Cuperinus, 'Die chronicke van der vermaerder en de vromer stad van 's-Hertogenbosch etc', in Cornelis Rudolphus Hermans (ed.), *Verzameling van kronyken, charters en oorkonden betrekkelijk de stad en meijerij van 's Hertogenbosch*, vol. 1 of 3 vols ('s-Hertogenbosch: Stokvisch, 1847), 104; Marcus Vaernewyck, *Van die beroerlicke tijden in die Nederlanden en voornamelijk in Ghendt 1566-1568* (Leiden: DBNL, 2007), 233; F.G.V., *Antwerpsch chronykje, in het welk zeer veele en elders te vergeefsch gezogte geschiedenissen sedert den jare 1500 tot het jaar 1574*

Figure 10.1 *Comet in the Months of February and March 1744*, hand-coloured engraving, included in the manuscript chronicle of Albert Pietersz. Louwen, 'Kronijk der stad Purmerende bevattende der selver opkomste en voortgang, in handschrift door Albert Pietersz. Louwen, Deel 1', Purmerend, 1766, adjacent to p. 286. Noord-Hollands Archief, Haarlem, 143 – Collectie van A. Pietersz. Louwen, 18

Contrary to what one might expect, the chroniclers included in the *Chronicling Novelty* corpus did not mention that the planets and the stars in the sky could have any influence on human bodies. Moreover, they hardly ever referred to almanacs as a source of useful knowledge. Previous scholars claimed that almanacs offered access to the knowledge of a 'large, non-expert audience' in relation to planetary motions, but if we believe the Ghent official Philip van Campene, this was 'pagan knowledge', which he considered to be nothing more than old wives' tales.[9] Only nature and the

zoo in die toen ... vermaarde koopstad als de andere steden van Nederland, ed. Frans van Mieris and Gerard van Loon (Leiden: Pieter vander Eyk, 1743), 231–33.

9 *Dagboek van Cornelis en Philip van Campene: Behelzende het verhaal der merkwaardigste gebeurtenissen, voorgevallen te Gent sedert het begin der godsdienstberoerten tot den 5en april 1571*, ed. Frans de Potter (Ghent: Annoot-Braeckman, 1870), 343, 359: 'dat de meynsschen te zeer huerlieder gheloove ende betrauwen stellen upde ghemeene almanacken, inde welcke sommeghe bescriven uuter ghestaethede vande planeten ende sterren de toecommende veranderinghe vanden

word of God were the true sources of knowledge on which they could rely during miserable times.[10]

Although authors did not see a causal relation between super- and various sublunar phenomena, they were interested in the correlation between these events. Especially in the late sixteenth and seventeenth centuries, chroniclers frequently recorded the appearance of comets and eclipses, along with earthquakes and parhelia (multiple suns) in their quest for patterns.[11] In most instances chroniclers made only a short note, but others were more elaborate and interpreted extraordinary meteorological phenomena as prodigious signs. Luyten, Vaernewijck, and Augustyn van Hernighem (c. 1540–c. 1617) from Ypres believed that God often communicated through natural and supernatural phenomena, or human behaviour or shape, to warn those who paid attention to God's book of nature. This could take the shape of 'bloody signs' in the sky such as comets, but also through children who were playing at funerals.[12] As a result, most chroniclers ascribed an active role to God, who could send messages from the heavens but whose messages did not have any direct influence on earth. In fact, people hesitated to interpret these signs when they happened but constructed their meaning in hindsight and in relation to other phenomena.[13]

Unlike the study of heavenly signs, the weather was one of the most important indicators for predicting a potential epidemic, which authors based on their extensive records of weather over time. When we take a closer look at the chronicles, we see that they used both religious and non-religious explanations for disasters. Recently, historians of the modern period have described this as *parallel practice*, yet in the early modern period there was no clear distinction between natural, human, and divine explanations.[14]

weere, dwelck ons es ghecommen vande heyndenssche, die zulck betrauwen hadden, dwelck den prophete Hieremias seer es mesprijsende ... so wie ooc zijn gheloove stelt inde planeten ende sterren des luchts, als of die macht hadden jeghens de zielen der meynsschen, es een afgodisch'.

10 *Dagboek van Cornelis en Philip van Campene*, 342–43.

11 Pollmann, 'Archiving the Present', 246.

12 Luyten, *Kronijk uit het klooster*, 177–78; Vaernewyck, *Van die beroerlicke tijden*, 256. Augustyn Van Hernighem, 'Beschrijving der stad Yper', book 1 (1595), 162, kept in Rijksarchief te Kortrijk, Fonds Goethals-Vercruysse, Ms 296.

13 Justus Billet, 'Den Polytye Boeck, ... beginnende den 22sten augusto in tjaer ons heeren 1658 (1658–68), book 11 (Ghent: 1668), fol. 6v, kept in Ghent, Stadsarchief Gent, bibliotheek 1LF2 en lGDl, 529 (C. Handschriften).

14 David K. Chester, Angus M. Duncan, and Christopher J.L. Dibben, 'The Importance of Religion in Shaping Volcanic Risk Perception in Italy, with Special Reference to Vesuvius and Etna', *Journal of Volcanology and Geothermal Research* 172: 3 (2008): 216–28; David K. Chester and Angus M. Duncan, 'The Bible, Theodicy and Christian Responses to Historic and Contemporary Earthquakes and Volcanic Eruptions', *Environmental Hazards* 8: 4 (2009), 304–32.

Van Campene, for example, wrote in September 1568 that an epidemic and the violence of Alba's army were two punishments of God, referring to the Four Horsemen of the Apocalypse.[15] However, a month before his reference to the Book of Revelation, he claimed that the pestilence was caused by a period of extreme heat.[16]

In the seventeenth and eighteenth centuries, chroniclers made similar observations, but it seems that other factors, such as the level of education, someone's profession, and the distinctions between the city and the countryside influenced how chroniclers wrote about epidemics. The Ghent police officer Justus Billet (1593–1682) used the height of the growing grain to measure deviant weather patterns, which offers a wonderful insight into the practice of weather observation and its relation to diseases. On 9 October 1666, he wrote that:

> The weather was very good, sweet, warm, and quiet, as if it were May, because instead of frost there was dew in the morning. As if it were summer, herbs, flowers, and trees appeared young and green, similar as in the summer season. Also, the sown grains, which were many, are already grown a half of a *vierendeel* high. Whether this is a good or bad sign is only known by God. What is certain, is that this warm weather could spread a contagious disease.[17]

To put it another way, not only the dew in place of the frost caught Billet's attention, but also the appearance of vegetation. Moreover, the grain was already a *vierendeel* high, which could mean either an additional harvest or crop failure, because it made the plants more vulnerable to cold weather. Although the meaning of God's sign was yet to be revealed, other consequences of the weather aberration were abundantly clear. Warm weather in the autumn, and especially the winter, was conducive to the spread of diseases, which could lead to casualties and disturbances that immediately affected his work as a police officer.

15 *Dagboek van Cornelis en Philip van Campene*, 175.

16 *Dagboek van Cornelis en Philip van Campene*, 157.

17 'In desen tijdt. Soo was het een weder, soo goet, soet, waerme ende stille, alof het inden Meij hadde gheweest, want het daude des smorghens, in plaetse van Rijm, alof het inden Somer gheweest hadde gheweest, staende de Cruijen, de Blommen, ende de Boomen, soo Jeughdich ende soo groen, ghelijck men die siet in het Somer Saijsoen. Oock soo staen der ghesaijde graenen, heel veijl, wel een half vierendeel hooghe, ofte beth ghegroijt; Of dit nu goeden ofte quaede Teekenen sijn, dat is Godt alleene bekendt; Dan het is seecker, dat dit waerem weder, niet goet en is voor de contagieuse Zieckte'. Billet, 'Den Polytye Boeck', fols 5r–6v

Two months later little had changed, which increased Billet's concern about the consequences. He wrote:

> Wednesday, December 22 [1666]. Here in Ghent, and other places in Flanders, warm, pleasant, and damp weather, combined with clear sunshine and southern winds made it feel like summer, however, Christmas is upon us. Consequently, there was a common fear that a great disease would result in mortalities among people. Moreover, people were afraid of a bad harvest due to cold nights, because the grains were too high, since they still grew lustily on the fields.[18]

Four days later the weather became 'normal' again, leaving the community behind with mixed feelings.

> Sunday, December 26, the day after Christmas day. The sweet weather started to change. A fierce and biting cold with frost and ice caused great joy among most people, both for the sake of their health and that of the fertility of the grains in the fields. However, poor people, who had a hard time keeping themselves warm, expressed their concern and sadness, because they knew that they would suffer tremendously until Easter was upon them.[19]

These fragments demonstrate how different groups in society responded to deviant patterns. As already stressed by Maslow and other psychologists since the 1940s, this depended on the socio-economic position of the authors along with cultural factors.[20] People with little to no income were primarily concerned with their struggle to survive, while more affluent people could

18 'Woonsdach den 22en dito. Soo was het hier tot Ghendt, ende apparent in alle andere Plaetsen van Vlaendren, een soo schoon, soet ende Vochtich weder, ghemeinghelt met schoon Sonneschijn, ende Zuijdelicken winden ghelijk of het somer hadde gheweest daer wij nochtans bij de kersdaeghen waeren; oversulcx wasser een ghemeene Vreese, datter eenighe groote Zieckten, ende sterften onder de Menschen soude commen, alsmede een slecht ghewas vande Graenen, alsoo die te hooghe, ende te gulsich stonden en groijden op het Velt, faucte van Rijmeghe, ofte Vorstighe nachten'. Billet, 'Den Polytye Boeck', fols 31r–32v.

19 'Sondach den 26en xbre 1666 sijnde den 2en kersdach, soo begost het goet ende soet weder te veranderen, in een straffe ende bijtende caude, met Rheijm ende Vorst, daer over veel Menschen seer verblijdt waeren, soo om de ghesontheijt vande Menschen, als om de Vruchtbaerheijt vande Graenen op het Velt, doch de Aerme Lieden, die niet wel gheduffelt en waeren, bethoonden daerover druck en droefheijt, want sij wisten wel, dat sij veel souden lijden eer het Paesschen was'. Billet, 'Den Polytye Boeck', fol. 32v.

20 A.H. Maslow, 'A Theory of Human Motivation', *Psychological Review* 50: 4 (1943), 370–96.

literally afford to worry about more complex cultural phenomena and therefore constructed different patterns.

In addition to the socio-economic position of the authors, their profession and the direct impact of the weather on their welfare also played a significant role in how they associated meteorological phenomena with potential epidemics. Investigating the chronicles of the sixteenth-century Catholic brewer Willem Jansz. Verwer (c. 1533–1580) from Haarlem and a seventeenth-century farmer from the small village of Jisp (Holland), we see, unlike with Billet, a positive attitude towards a warm autumn.[21] The brewer emphasised the 'clean dry summer and dry winter which was beneficial for the growth of barley'.[22] The farmer, on the other hand, recorded for several years the date when his cows and calves were brought into the stables at the end of the year. On 14 December 1652 he wrote that: 'it was not without God that we put our cows in the stable with beautiful weather',[23] meaning that for the entire autumn, the weather had been good enough for the cattle to stay in the fields. Consequently, the farmer did not have to buy hay, which implied a significant cost reduction. In other words, having a relatively warm autumn was a godsend for a significant part of society, while others feared potential disaster.

What changed in the practice of chronicling epidemics among seventeenth-century chroniclers was the use of tables and lists. This elementary form of descriptive statistics was used by chroniclers to study the course of an epidemic, in contrast to recording only the total number of deaths. An early example can be found in the chronicle of the Protestant teacher Pieter van Godewijck (1593–1669) from Dordrecht, who was an acquaintance of the famous and influential contemporary physician Johan van Beverwijck (1594–1647). Godewijck published Beverwijck's pest treatise in rhyme.[24] Employing rhyme served, among other things, as a mnemonic device for those who could not read.[25] The message for the population was clear. On the one

21 Willem Janszoon Verwer, *Memoriaelbouck. Dagboek van gebeurtenissen te Haarlem van 1572–1581*, ed. J.J. Temminck (Haarlem: Schuyt & Co, 1973), 171.

22 'Een schoene droge zomer ende een droge winter, niemant mocht bedeincken van sulck lach water, alst dese zomer is geweest ... Die vruchten zijn redelick gewassen, als garst, cooren, hennip'. Verwer, *Memoriaelbouck*, 216.

23 'Niet sonder godt 1652- den – 12 – maent den – 14 – dach teweten in desember haelde wij onse koeije op het stal met moij weer droech weer'. 'Aantekenboek betreffende Jisp, met beschrijving, 1647–1716' (Jisp: 1716), unpaginated 14-12-1652, 0954, kept in Waterlands Archief, Collectie persoonlijke documenten en handschriften, 1518–1968, inv. no. 1.

24 Pieter van Godewijck, *Remedie voor de pest* (Dordrecht: pr. H. van Esch, 1636).

25 Robert Darnton, 'An Early Information Society. News and the Media in Eighteenth-Century Paris', *The American Historical Review* 105: 1 (2000), 1–35, at 19.

hand, the pestilence was caused by God as a punishment for the collective sins of His people. He operated through nature; therefore, southern winds and direct sunlight between sunrise and sunset needed to be avoided. On the other hand, the audience was offered a list of remedies, with a strong emphasis on prayer, purifying the air, and dietary recommendations to strengthen the body.[26] This combination of religious and non-religious elements did not disappear after the seventeenth century, as is often suggested, but changed in a way that was crucial for most contemporaries to make sense of an epidemic.

Closer to God with New Knowledge: Empirical Research and Statistics

From the second half of the seventeenth century, but especially since the eighteenth century, farmers, tax collectors, and merchants (similar to scholars and officials) started to use statistics to understand the course of an epidemic.[27] Consequently, two developments emerged. On the one hand, earlier explanations were refuted based on empirical and comparative research. On the other hand, statistics were used in the service of divine explanations to find a pattern and to make sense of God's divine providence. The first development is demonstrated in the chronicle of Protestant Jan Kluit (1722–1811) and the second in the chronicles of Albert Pietersz. Louwen (1722–c. 1791) and Lambert Rijckxz. Lustigh (1656–1727).

The tax collector Kluit wrote in 1748 about a 'mysterious' disease among a regiment of the Scots Brigade, which was stationed in Brielle during the War of the Austrian Succession. According to him, the high mortality rates

26 Godewijck, *Remedie voor de pest.*

27 L.F. Cattenbelt, 'Eenige gedenckwaerdige geschiedenissen raeckende meest het landt van Overijssel ende deszelfs steden? Kroniek van Overijssel en in het bijzonder Zwolle, geschreven door L.F. Cattenbelt, 1678–1681' (Zwolle: 1678–81), kept in Historisch Centrum Overijssel, Vereeniging tot Beoefening van Overijsselsch Regt en Geschiedenis (VORG), handschriftenverzameling, 0263, inv. no. 909; Albert Pietersz. Louwen, 'Kronijk der stad Purmerende bevattende der selver opkomste en voortgang, in handschrift door Albert Pietersz. Louwen, 18e eeuw, Deel 2, eerste gedeelte' (Purmerend: s.pub., 1791), kept in Noord Hollands Archief, Haarlem, 143 – Collectie van A. Pietersz. Louwen, inv. no. 19; Jan Kluit, 'Historische jaerboeken der stad Briel', book 1, parts 1–2 (Brielle: 1756), 501, kept in Streekarchief Voorne-Putten Rozenburg, inv. nos 1–2. Jessica Otis, 'By the Numbers. Understanding the World in Early Modern England', PhD dissertation, University of Virginia, 2013, 234–35; Mary Lindemann, *Medicine and Society in Early Modern Europe* (2nd ed., Cambridge: Cambridge University Press, 2010), 225–28.

among the Scottish soldiers demanded an explanation.[28] Some referred to the work of the sixteenth-century chronicler François le Petit (1546–?), who claimed that the air in the South Holland town of Brielle was miasmatic, while others suggested that it was not so much the air itself but the change of environmental conditions between Scotland and the Low Countries that made the Highlanders sick. According to Kluit, this was a plausible idea, but the empirical data suggested an alternative explanation that was more likely. It was widely believed that the cities of Breda and Gorinchem had the purest air of the Republic, and many soldiers died even there. To find the true cause for the high mortality rates, Kluit wrote that he started an investigation to prove that the cause of the disease was not related to the air.[29]

Firstly, Kluit refuted Petit's claim that the air in Brielle was miasmatic, since Petit also argued that the air was healthy in Maassluis, which is located across from Brielle. According to Kluit, the air quality could not change significantly over such a short distance. In addition, he argued that Petit had probably never been to Brielle and therefore could not offer a valid judgement based on empirical evidence.[30] Secondly, after observing and questioning the nutrition of the Highlanders, he concluded that the 'mysterious' disease was caused by the amount and the preparation of their food. After reporting his findings, petty officers surveyed the vegetable market and recommended that the Highlanders should cook their meals and eat in a 'civilised' manner. According to Kluit, these measures were effective, and the Scots became as healthy as the Dutch. Based on these observations, Kluit concluded that it was not the air that made the Highlanders sick but their way of life, and especially their food consumption.[31]

Eighteen years later a report was published on the demographic decline in Brielle between 1747 and 1766. The decline was explained by the authorities with reference to impure air, a theory which Kluit had refuted earlier. He wrote that he 'felt obliged to take away the misconceptions and to convince the world of the opposite, especially, since I am able to prove it by experiment'.[32] According to the report, 2,163 children were born, and 2,422 people died, which meant a demographic decline of 259 people in ten years.

28 Henk Brouwer, 'Malaria in Nederland in de achttiende en negentiende eeuw', *Tijdschrift voor sociale geschiedenis* 30: 9 (May 1983), 140–59, at 140–43. Brouwer convincingly demonstrated that it could not be malaria in this specific period.

29 Kluit, 'Historische jaerboeken', book 1, parts 1–2, 339–57.

30 Kluit, 'Historische jaerboeken', book 1, parts 1–2, 36–43.

31 Kluit, 'Historische jaerboeken', book 1, parts 1–2, 339–57.

32 'Maar ik ben verpligt tot wegneming van deze verkeerde begrippen de weerelt van 't contrarie te overtuigen, te meer, dewijl ik nu in staat ben om zulx proeff-kundig te kunnen betogen'. Jan

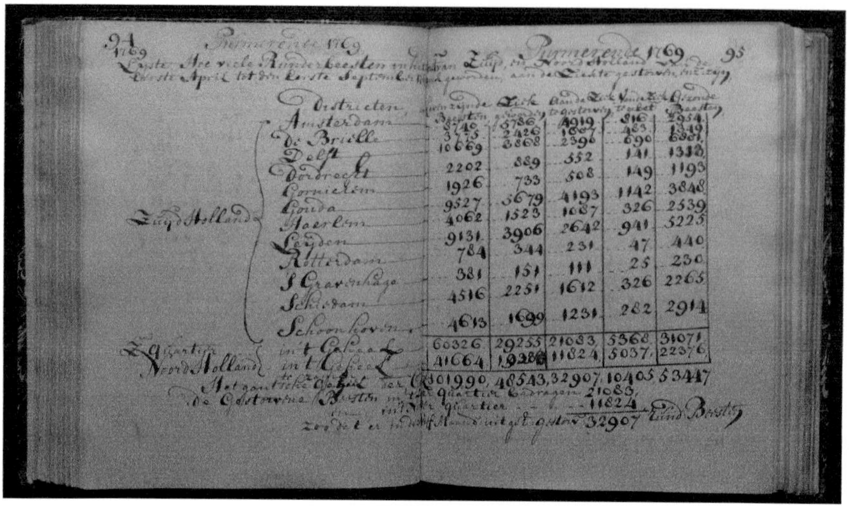

Figure 10.2 Table in the manuscript chronicle of Albert Pietersz. Louwen, 'Kronijk der stad Purmerende bevattende der selver opkomste en voortgang, in handschrift door Albert Pietersz. Louwen, Deel 2, eerste gedeelte', Purmerend, 1769, 94–95. Noord-Hollands Archief, Haarlem, 143 – Collectie van A. Pietersz. Louwen, 19

According to Kluit, these numbers could easily be explained. Moreover, he could provide evidence that the air of Brielle was among the purest of the Republic. According to Kluit, most deaths had occurred between 1747 and 1748, when there was a birth deficit of 247 people. Consulting the *Jaerboeken* ('Yearbooks'), a historical periodical published since 1747, he concluded that many Highlanders were quartered in that period and died because of their rough lifestyle. Discounting the soldiers, there were 67 more births than deaths, which meant that there was growth instead of a demographic decline. Based on his own experiment, by carrying out a census and comparing the numbers with the average number of deaths each year, Kluit concluded that the air of Brielle was comparable to that of North Germany, which was regarded as among the healthiest and purest in Europe.[33]

Even though this epidemic was mainly restricted to the Highlanders, it demonstrates the critical engagement of the chronicler with the causes of potential disasters. In a similar manner Kluit investigated the outbreaks of rinderpest, of which he experienced the direct consequences. By keeping a record of how many cows died, recovered, or were not infected, categorised

Kluit, 'Historische jaerboeken der stad Briel', book 2, parts 3–4 (Brielle: s.pub., 1767), 1397, inv. nos 3–4, Streekarchief Voorne-Putten Rozenburg.

33 Kluit, 'Historische jaerboeken der stad Briel', book 2, parts 3–4, 1400.

by age and location, he could study the course of God's judgement.[34] Kluit believed, as did Louwen from Purmerend, that as a good Christian he could discover a pattern in God's judgements. Louwen, who also used statistics to measure the severity of the rinderpest outbreak per city, wrote in 1775 that: 'one third of the sick cattle were cured, while two thirds died. This was truly a special predestination of divine providence'.[35] According to him, God spared enough cattle so that people could still maintain themselves, which was proof of His mercifulness.

This idea of a merciful or benevolent God was more frequently expressed by chroniclers since the eighteenth century. The farmer and alderman (*schepen*) Lustigh from Huizen (Holland) initially claimed that rinderpest was a plague sent by God as punishment for not living according to His commandments. However, he also believed that an epidemic was not only a punishment but also test. God provided humans with the knowledge to stop the plague.[36] Therefore, people should not only pray and repent but actively try to understand God's plan to guide them to the righteous path.[37] Taking the right measures and remedies in honour of God was rewarded, while a passive attitude or acting out of greed made the punishment more severe.[38]

Some scholars have argued that this notion – that God provided humans with the knowledge to explain His natural laws, which was related to the mechanisation of the early modern worldview – resulted in the idea that God was no longer the *causa proxima* but became the *causa remota*.[39] In other words, the increase of 'natural' knowledge resulted in a concept of a deity operating at a distance, replacing earlier ideas associated with an active God who regularly intervened. As a result, since the middle of the seventeenth century, both Catholics and Protestants increasingly tended

34 Kluit, 'Historische jaerboeken der stad Briel', book 2, parts 3–4, 217.

35 'Een derde van het ziek geworde rundvee gebeterd, en twee derde van dezelve gestorven zijn; waerlijk eene eenbijzondere voorbeschikkinge der Goddelyke voorzienigheijd'. Albert Pietersz. Louwen, 'Kronijk der stad Purmerende bevattende der selver opkomste en voortgang, in handschrift door Albert Pietersz. Louwen, 18e eeuw, deel 2, tweede gedeelte', (Purmerend: s.pub., 1791), 296, kept in Noord Hollands Archief, Haarlem, 143 – Collectie van A. Pietersz. Louwen, inv. no. 20.

36 Lambert Rijckxz. Lustigh, *Kroniek I van Lambert Rijckxz Lustigh* (Huizen: Drukkerij Visser, 1973), 2, 8–10, 21.

37 Lustigh, *Kroniek I*, 2, 10, 13–14, 21, 25.

38 Lustigh, *Kroniek I*, 11.

39 Jan Willem Buisman, *Onweer. Een kleine cultuurgeschiedenis 1752–1830* (Nijmegen: Vantilt, 2019), 84.

towards a view of nature as an expression of God's providence and His care for all creation, rather than as a theatre of His anger and judgement.[40]

Although historians demonstrated that this transition can be observed among early modern scholars, the behaviour and reflections of the chroniclers present a less clear image. In contrast to the scholars and experts who have been traditionally studied by historians of science, those who did not belong to the educated elite continued to perceive disasters more often as a direct intervention from God and more frequently used divine and natural explanations.

The chronicle of Lustigh, especially, clearly presents the combination and the hierarchy of various explanations. Citing 1 Kings 8:37–40, he argued that prayer and penance were the most important remedies to stop the cattle plague, but only side by side with quarantine measures and hygiene regulations.[41] The measures he prescribed were based on several observations in which he included the composition of the air and other meteorological factors to demonstrate their relationship with the rinderpest outbreak. Although Lustigh believed that God ultimately determined the composition of different types of air, his account at the same time bears great resemblance to the atomistic and corpuscular explanations of two well-known physicians: van Beverwijck and Willem Swinnas (1620–1672).[42]

According to Lustigh, God dropped the 'pestilential air' on the earth. Afterwards, it 'stuck' (*aenkleven*) to the water and grass, which became infected and contagious (*besmettelijck*). As a result, not only people who went outside with their bare legs to milk the cows became infected, but also the cows who breathed the air and consumed the water and grass.[43] It is particularly important that Lustigh remarked that 'bare legs … were contaminated, with great inflammation and pain'.[44] It was common knowledge that pestilential air could be contained and transmitted, for example by wood, cotton, woollen cloth, and the fur of cats and dogs, but in the case of Lustigh the composition of the air penetrated the skin and

40 Rienk Vermij, *Thinking on Earthquakes in Early Modern Europe. Firm Beliefs on Shaky Ground* (Abingdon: Routledge, 2021), 185–86.

41 Lustigh, *Kroniek I*, 8–10, 74.

42 Lustigh, *Kroniek I*, 2; Johan van Beverwijck, *Kort bericht om de pest voor te komen. Ten dienste van de gemeente der stadt Dordrecht* (1636), 10; Willem Swinnas, *De pest-stryt, beharnast met veel voor-treffelyke genees-middelen* (Leiden: Johan Le Carpentier, 1664), 18–19.

43 Lustigh, *Kroniek I*, 3.

44 'de menschen met hare bloote benen des morgens melckengingen, met groote vierigheijt en serigheijt besmet wierden'. Ibid.

caused a reaction, resembling the descriptions offered by van Beverwijck and Swinnas.[45]

To put it another way, Lustigh did not perceive divine and natural phenomena as separate realms. Although his chronicle testifies to an extensive knowledge of natural factors that, according to him, played a role in the spread of the epidemic, ultimately it was God who dropped the 'particles' on the earth and could remove the pestilential air with cold winds and lightning.[46] As a result, God was not only the creator of natural laws, but He continued to actively intervene. When two farmers dissected a sick cow out of curiosity in 1713, God punished them directly with severe pestilential air, from which they died within 24 hours.[47] This passage can be read as a warning for those who did not acknowledge God as the *causa proxima* and illustrates that God was not only the creator – His efficacy was still perceptible in the material realms.

There were few differences between Catholics and Protestants in this regard. Both tended towards a view of nature as an expression of God's providence and his care for all creation, but without discarding the belief that God could directly intervene and punish His people for their sins. Consequently, chroniclers more often referred to God as described by Louwen, as 'a merciful father, who indeed visits His children with His punishing rod, but nevertheless through the wrath and punishment wants to demonstrate his mercy and to remember His caring will'.[48]

Combining religious and non-religious practices to stop an epidemic remained common practice in the nineteenth century for diseases such as cholera, rinderpest, and smallpox. Even vaccination was initially perceived as an additional method to prevent sickness, but without causing a paradigm shift among the middling sort. Scholars have perceived the eighteenth and early nineteenth centuries as a transitional phase in which religious causes and Hippocratic-Galenic medicine were replaced by a theory of contagion and germs that became characteristic of 'modern' medicine.[49]

45 Vivian Nutton, 'The Seeds of Disease. An Explanation of Contagion and Infection from the Greeks to the Renaissance', *Medical History* 27:1 (1983), 1–34, at 25.

46 Lustigh, *Kroniek I*, 8–9.

47 Lustigh, *Kroniek I*, 13–14.

48 'eene Barmhartige Vaeder, die zijne Kinderen wel met zijne Tugtroede bezoekt, maar nochthans, in de Toorn en Straffe, zijne Genade wil bewijzen, en des ontferme wil gedenken'. Louwen, 'Kronijk der stad Purmerende, Tweede deel. Tweede stuk', 296.

49 Snowden, *Epidemics and Society*, ch. 12; Michael Worboys, 'Contagion', in Mark Jackson (ed.), *The Routledge History of Disease* (Abingdon: Routledge, 2016); Samuel K. Cohn, Jr., *Cultures of Plague. Medical Thinking at the End of the Renaissance* (Oxford: Oxford University Press, 2009), 139.

However, when studying the behaviour and reflections of our non-expert chroniclers, we see that they were willing to experiment with new ideas, without abandoning old ones. As we will see, both practices could coexist both within society and in one person without experiencing any tension. Chroniclers continued to pray for lightning to purify the air while they got their children vaccinated in the meantime.

New Ideas, Old Practices?

In April 1832 the Catholic chronicler Edouard Callion from Ghent reflected extensively on the advice and measures issued by the authorities to combat a cholera epidemic in cooperation with the city's medical committee.[50] After writing down all precautions and remedies, Callion continued with the religious practices. On 13 May St Bavo's Cathedral organised a procession of penance with the relics of St Macharius to stop the 'contagious disease'.[51] Despite the rain, the turnout was overwhelming. Several times, the procession had to stop because people desperately tried to touch the relics, according to Callion.[52] A month later the St Nicholas Church organised another procession whereby barrels with tar were lit to 'purify the air of evil miasmas', thus demonstrating a combination of medical and religious practices.[53]

Despite the similarities between nineteenth- and sixteenth-century practices, they were subject to change. As we have seen in Kluit, even the most trivial observations could lead to a form of cognitive dissonance, which preluded alternative explanations. This also happened in Ghent in June 1832. Callion wrote that one of the halls of the former Bijloke Abbey – which functioned at that time as a home for elderly men (*oudemannenhuis*) – was renovated according to the highest contemporary standards with an abundance of space and high ceilings. However, cholera also broke out in that hall, 'which proves that uncleanliness or unhealthy air do not necessarily cause cholera. The disease could also be caused by other factors which are not yet known to us'.[54]

50 Edouard Callion, 'Gentsche kronijke: 1525–1835', vol. 11: 1832–1835 (Ghent, 1835), 476–82, kept in Ghent University Library, BIB.G.014248 v.01.

51 Callion, 'Gentsche kronijke', vol. 11, 484–85, 503.

52 Callion, 'Gentsche kronijke', vol. 11, 485.

53 'heden des avonds wierden 'er pektonnen gebrand in het nieuw poortjen; om de locht te zuyveren van de kwaede dampen'. Callion, 'Gentsche kronijke', vol. 11, 494.

54 'Men heeft aldaer iets zeldzaem opgemerkt, het welk bewyst dat het niet altyd onreynigheyd of ongezonde locht is die den Cholera veroorzaekt; maer dat de ziekte aen oorzaken moet toegeschreven worden, die tot hier toe onbekend zyn'. Callion, 'Gentsche kronijke', vol. 11, 499.

Like Callion, the Protestant Frisian farmer Lieuwe Jans de Jong (1798–c. 1855) wrote about the cholera pandemic, but more interesting are his reflections on the smallpox vaccine and what contemporaries called the lung plague (contagious bovine pleuropneumonia). Besides being a devoted Protestant, de Jong was an advocate of the smallpox vaccine. Both his son and three daughters were vaccinated in 1841, and three of them again in 1850.[55] Like Lustigh, he believed that God punished His people for their sins, but He also provided humans with the knowledge to stop an epidemic. Only fatalists deny this and obstruct God's will, according to de Jong.[56]

By combining traditional methods with what he had learned from vaccination, de Jong tried to understand how he could protect his cattle. Besides the usual causes such as bad nutrition, miasmas, and meteorological phenomena, de Jong also questioned whether a contagious substance could cause lung plague by making a comparison with smallpox.[57] According to de Jong, the contagious substance came from infected mice, and the problem was made more severe by farmers who did not take care of their fields. They thought that flooding their lands would solve the problem, but in fact it made things worse. Storks do not eat dead mice, meaning that the putrefaction of their bodies contaminated the air and released a contagious substance harmful to livestock and perhaps also to humans, according to de Jong.[58]

Based on his own observations, autopsy, and other farmers' experiences, de Jong concluded that the lung plague expressed itself as an infection of the lungs accompanied by pulmonary oedema. This fluid was probably blood turned to water, according to de Jong, because sufferers had little blood when they died. He did not know how it worked exactly, but it could have been caused by the airflow.[59] As a result, he used a combination of (preventative) methods to save his cattle from the lung plague. These were not restricted to isolation, medicine, and bloodletting, as advised by veterinarians, but also included prayer to God, because only He could purify the air from the

55 *De dagboeken (1825–1855) van Lieuwe Jans de Jong, boer te Poppenhuizen onder Oldeboorn* (s.l., 1998), unpaginated, 13-03-1841, 10-04-1850.

56 *De dagboeken*, 11-1854.

57 *De dagboeken*, 1831, 02-1843, 04-1843, 07-1851: 'Zoude het vee ook een smetstof tot longziekte bezitten, even als men redeneert, dat alle menschen een pokstof bezitten, die door mededeling van deze ziekte, deze stof als ontwikkelt en uitbreidt en tot ziekte en dood brengt, al naar de meerder of mindere stof in het ligchaam of longen aanwezig?'

58 *De dagboeken*, 07-08-1847.

59 *De dagboeken*, 06-1851, 1854.

contagious substance with lightning and thunder and consequently absolve His people from their sins.[60]

Conclusion

Despite the heterogeneity of the corpus and the supposed lack of emotional expression by the authors, we have seen, by studying their work as collections of knowledge, a long-term development in how chroniclers coped with disasters. The increase of medical knowledge and the focus on prophylactic measures and hygiene regulations by experts and local authorities resulted in the idea that there was a shift away from religious towards prophylactic measures in the late seventeenth century. However, analysing the experiences of the chroniclers demonstrates that natural explanations became more detailed and complex, but almost always in service of, or subordinate to, the divine explanation. Moreover, besides the natural factors, the role of the divine changed as well. The notion of a benevolent and providential God gained prominence in the seventeenth and especially the eighteenth century, although the idea of an angry and vengeful God was never far away. In the first half of the nineteenth century, God played a less prominent role in comparison to the previous centuries. However, in the end, both Catholics and Protestants turned to traditional religious practices in desperate times, despite their differences.

Studying the reflections and behaviour of the middling sort and the connections they made, we see the boundaries of the analytic categories present in the historian's toolbox dissolve. While concepts such as divine and natural, or miasma and contagion, have long been used as binary categories by historians, we see that contemporaries in fact closely associated them with each other.[61] Focusing on medical and religious practices, we see a difference but no categorical distinction. Both practices complemented each other in the early modern period in various ways. For example, religious explanations were not only used to make sense of the phenomena that gave rise to disasters such as epidemics but could also function as a strategy to cope with apparently inexplicable human suffering.[62]

60 *De dagboeken*, 01-01-1846, 07-1847, 09-1847, 12-1853.

61 Lindemann, *Medicine and Society*, 53.

62 Chester and Duncan, 'The Bible, Theodicy', 198; Judith Pollmann, *Memory in Early Modern Europe, 1500–1800* (Oxford: Oxford University Press, 2017), 166–67; de Graaf et al., 'Dancing with Death', 352 and 357.

As a result, chronicles offer an additional perspective on how non-experts reflected upon the combination of prophylactic measures, medicine, and religious practices. They complement historical studies on early modern scholars, scientists, officials, and clerics by offering a window into how the middling sort coped with uncertainty and periods of crises. As we have experienced during the COVID-19 pandemic, dealing with disasters is inextricably connected to making sense of disasters. The explanations of modern medicine alone were not sufficient to deal with this epidemic. People used various coping mechanisms to deal with uncertain situations. Some started writing chronicles and diaries, others turned to their faith, while some constructed an alternative reality to get some sort of grip on their lives. In the early modern period this was no different. Then as now, it is important to take all these voices seriously in order to understand society.

About the Author

Theo Dekker is a PhD candidate at the Institute for History of Leiden University and funded through the NWO-funded project *Chronicling Novelty. New Knowledge in the Netherlands, 1500-1850*. Using more than 300 local chronicles, Dekker studies how the appropriation of new knowledge influenced ideas on causality and the appreciation of the new among the 'middling' ranks of the Low Countries. He is especially interested in how knowledge of nature and the use of numbers changed in the early modern period, and how chroniclers reflected upon epidemics, meteorological phenomena, and dearth.

11 Coverage in Dutch Newspapers of Earthquakes in Italy and Beyond before Lisbon 1755

Joop W. Koopmans

Abstract
This chapter discusses tidings about earthquakes retrieved from the digitised Dutch newspapers before the Lisbon Earthquake of 1755. It focuses on how these newspapers reported about previous earthquakes and how detailed their contents were. Other addressed questions: Which parts of the world did they deal with most frequently, did they explain the causes of earthquakes, and what type of media culture do they represent? This chapter demonstrates that readers of early modern Dutch newspapers were already familiar with earthquake reports long before 1755. These papers mainly treated data, although within the existing societal framework. They represented, as far as possible, a media culture of accuracy, although sometimes with a touch of sensationalism. Readers could read dominant opinions only between the lines and draw conclusions for themselves. The widely accepted idea was that earthquakes were punishments from God and that people should repent their sins.

Keywords
earthquakes – newspapers – early modern era – Dutch Republic – Europe – media history

Since 2003, the northeastern part of the Netherlands has been confronted with many earthquakes due to the natural gas extraction in the Groningen gas field. However, earthquakes as a natural phenomenon, not caused by human intervention, have always been rare in this country. Furthermore,

Asperen, van, Hanneke and Lotte Jensen (eds): *Dealing with Disasters from Early Modern to Modern Times. Cultural Responses to Catastrophes.* Taylor & Francis Group, 2023
DOI: 10.5117/9789463725798_CH11

they mostly took place in the south of the Netherlands, as in Roermond in 1992 (moment magnitude 5.3). Yet from time immemorial, people in the Rhine-Meuse delta have been aware that far more destructive earthquakes happened elsewhere. Before the spread of manuscripts, printed books, and media, they could conclude this predominantly from oral tradition. Bible verses and classical texts with stories about earthquakes will also have played an important role in the process of memory construction.[1]

This situation implies that when the first Dutch newspapers appeared, beginning in 1618, readers will not have been surprised to find reports about earthquakes in the columns. Of course, the supply and selection of such items determined for the greater part which earthquakes became known in the Dutch Republic via printed news media. Since early modern Dutch newspapers included mainly European news, it is obvious that they primarily helped to spread ideas about the circumstances in Europe. In other words, people's impressions never reflected every part of the world equally. Besides, only over time was it possible to develop specific notions about questions such as where earthquakes often occurred – at least when people combined their insights with what they heard from older generations. Furthermore, it is self-evident that the more impressive or unique the available news about certain earthquakes, the better it would be remembered.

Portugal's 1755 earthquake, which completely destroyed its capital Lisbon, is the best known early modern earthquake in Europe. Its effects, in particular a tsunami, were noticed over a much larger area than Portugal. News media published extensively about the Portuguese devastations. Readers were shocked by the reports about the numerous victims and long-lasting consequences. Therefore, it is no wonder that the 1755 Lisbon Earthquake is also one of the best researched early modern disasters in relation to European media.[2]

Until now, the information on earthquakes provided in news media before and after 1755 has been scrutinised much less – at least, not the

1 See e.g. Matthew 28:2, when an earthquake happens while an angel descends from heaven, or Amos 1:1 and Zechariah 14:5 about the earthquake during the reign of Judah's king Uzziah (c. 749 BCE); Richard C. Hoffmann, *An Environmental History of Medieval Europe* (Cambridge: Cambridge University Press, 2014), 304–13.

2 Concerning the Dutch media about this earthquake, see chapter 'The 1755 Lisbon Earthquake and Tsunami in Dutch News Sources. The Functioning of Early Modern News Dissemination', in my *Early Modern Media and the News in Europe. Perspectives from the Dutch Angle* (Leiden: Brill, 2018), 243–64.

Dutch context.[3] This chapter will be restricted to the situation before the Lisbon Earthquake of 1755, focusing on how early modern Dutch newspapers reported about previous earthquakes. How often did they include such reports, and how detailed were their contents? Which parts of the world did they address the most, and did they explain the causes of earthquakes in their accounts? In short: what perceptions and discourses can we gather from the early Dutch newspapers concerning earthquakes before November 1755, and what type of media culture do they represent?

This chapter will discuss tidings about earthquakes retrieved from the Delpher collection of digitised Dutch newspapers.[4] Many variants of the Dutch words *(aard)beving(en)* ('[earth]quake[s]') have been tested, which resulted in a set of circa 820 items.[5] About 50 of them appeared to be of less relevance, since they mainly concern reports in which extreme weather circumstances are compared with the effects of earthquakes, and advertisements for publications about earthquakes. Although these items are relevant for questions such as how often and in what ways Dutch people were confronted with the phenomenon of earthquakes, they are less suitable for topics related to numerical issues, for example, the question of which earthquakes Dutch *courantiers* (the newspaper publishers) covered the most.

3 The situation varies per country and per case. See e.g. Romano Camassi and Viviana Castelli, 'Looking for "New" Earthquake Data in the 17th–18th Century European "Newssellers" Network', *Journal of Earthquake Engineering* 8: 3 (2004), 335–59. See also Emanuela Guidoboni and John E. Ebel, *Earthquakes and Tsunamis in the Past. A Guide to Techniques in Historical Seismology* (Cambridge: Cambridge University Press, 2009), 41–146, concerning newspapers and other types of written historical sources about earthquakes. Another angle: Kerrewin van Blanken, 'Earthquake Observations in the Age before Lisbon. Eyewitness Observations and Earthquake Philosophy in the Royal Society', *Notes and Records. The Royal Society Journal of the History of Science* 76: 3 (2022), 27–48, https://doi.org/10.1098/rsnr.2020.0005. The most recent general research: Rienk Vermij, *Thinking on Earthquakes in Early Modern Europe. Firm Beliefs on Shaky Ground* (London and New York: Routledge, 2020).

4 On 30 March 2021, Delpher included 32,620 newspaper copies in the Dutch language, published in the period 1618 until 31 October 1755; https://www.delpher.nl/nl/kranten (accessed 30 March 2021).

5 Consistency of Dutch spelling did not exist during the seventeenth and eighteenth centuries. It was therefore necessary to try several variants. On 1 March 2021, the following variants led to results: *aardbeeving, aardbeevinge, aardbeevingen, aardbeving, aardbevinge, aardbevingen, aerdbeeving, aerdbeevinge, aerdbeevingen, aerdbeving, aerdbevinge, aerdbevingen, aertbeeving, aertbeving, aertbevinge*, and *aertbevingen*. Variants of the word *beving* led to a few other dozens of results, but they also brought many irrelevant items because of different meanings. Other search terms were not used to keep the research feasible. Furthermore, it is questionable whether they would have produced different conclusions. Variants, e.g. of the Dutch word *aardschudding*, led to only a few extra results in Delpher (accessed 31 March 2020): two in *Oprechte Haerlemse courant* (22 September 1740) and (19 March 1743), both Italy, and one in *Opregte Groninger courant* (7 August 1750), which had already been retrieved via the keyword *aardbeevingen*.

Furthermore, one must keep in mind that not all of the early modern Dutch newspaper copies have been preserved and not all the stored copies have been digitised.[6] It is nevertheless possible to provide general impressions, because the number of circa 770 remaining reports is not a small sample. We may assume that observations based on these data are both representative and exemplary for the news that all Dutch newspapers published about earthquakes before November 1755.

The most significant newspapers in this research are, in decreasing order of importance: *Oprechte Haerlems(ch)e courant* (est. 1656), *Amsterdamse courant* (est. 1670), *(Opregte) Leydse courant* (est. 1686), *Opregte Groninger courant* (est. 1743), *'s Gravenhaegse* (or *The Hague*) *courant* (est. 1708), *Leeuwarder courant* (est. 1752), and the Delft *Hollandsche historische courant* (est. 1721). Several other long-running newspapers were published in the Dutch Republic before 1755, notably *Utrechtse courant* (est. 1675) and *Rotterdamse courant* (1717–29; restarted 1738). Unfortunately, the Delpher collection included few copies of these titles during the time of research.[7]

Earthquake Areas in News Reports

Before November 1755 Dutch readers would not immediately expect tidings about earthquakes in Portugal, but rather from regions surrounding the Mediterranean Sea.[8] This assumption is based on the fact that around 50 per cent of the selected reports concerned earthquakes on the Italian peninsula.[9] Another 10 per cent dealt with quakes in the Ottoman Empire, the huge territory that stretched from the Balkans and Greece in Europe

6 See Arthur der Weduwen, *Dutch and Flemish Newspapers of the Seventeenth Century, 1618–1700*, 2 vols (Leiden: Brill, 2017) for insights into preserved Dutch newspapers before 1700.

7 The *Amsterdamse courant* was a merger of several titles. The two important predecessors are *Courante uyt Italien, Duytslandt, &c* (est. 1618) and *Tydinghe uyt verscheyde quartieren* (est. 1619). More information about the newspapers mentioned also in Esther Baakman and Michiel van Groesen, 'Kranten in de Gouden Eeuw', in Huub Wijfjes and Frank Harbers (eds), *De krant. Een cultuurgeschiedenis* (Amsterdam: Boom, 2019), 20–45; Joop W. Koopmans, 'Expansion and Restraint in the Dutch Newspaper Market, 1700–1795', in Arthur der Weduwen and Malcolm Walsby (eds), *The Book World of Early Modern Europe. Essays in Honour of Andrew Pettegree*, 2 vols (Leiden: Brill, 2022), vol. 2, 481–510.

8 Reports about three rather recent earthquakes in Portugal before 1755 had been published in *Leydse courant* (25 April 1734), *Oprechte Haerlemse courant* (22 August 1750), and *Opregte Groninger courant* (7 March 1752).

9 In this observation, the Italian peninsula is a synonym for the present state of Italy; also five reports about earthquakes on Malta have been included here.

to Syria, Palestine, and Egypt in the Middle East, and Algiers on the North African Coast. Approximately 30 of the remaining 40 per cent included a range of many European states, leaving another 10 per cent related to the rest of world. The last category – a result of the age of European travels – demonstrated that earthquakes were a global phenomenon.

When we stay in Europe, apart from Italy and the Eastern Mediterranean, Dutch reports covered – in decreasing order – earthquakes in the German states and Hungary (including Transylvania), on the British Isles, in France, Portugal, Switzerland, Spain, the Southern Netherlands, the Dutch Republic, Poland, Sweden and the rest of Scandinavia, and the Baltic region. These findings fairly correspond with the information provided by AHEAD, the digital European Archive of Historical Earthquake Data.[10] The long list of earthquakes between 1618 and 1755 in this archive, however, also makes it clear that the digitised Dutch newspapers do not mention every earthquake. On the other hand, it is peculiar that not all of the earthquakes mentioned in the Dutch reports can be found in the list of AHEAD.[11]

Moving from the state level to the local situation in Europe leads to more specified outcomes. The reports about Italian earthquakes show a great variety of cities and other locations. Together, they give the impression that earthquakes could happen anywhere in Italy. Nevertheless, most reports concern quakes that took place in the regions of first Naples and Sicily, followed by Rome, Tuscany, and Calabria. Here mainly the regions are mentioned, since a great number of accounts include several geographical names of towns and villages. This is logical because most earthquakes struck large areas, corresponding to geological and not political borders.

Concerning the rest of Europe, many regions and places were covered. Most German and Swiss cities that are mentioned in this respect fit in the earthquake zone of the Rhine Rift Valley, the northern part of the Alps and the surroundings of Leipzig. One example is the August 1728 earthquake that caused damage and victims in the Palatinate and the regions of Strasbourg (now France), Frankfurt am Main, Freiburg, and Basel (Switzerland).[12]

10 See European Archive of Historical Earthquake Data, https://www.emidius.eu/AHEAD/index_en.htm (accessed 3 April 2023).

11 Perhaps some early modern reports include errors such as incorrect place names, which could explain why they cannot be found in AHEAD, but this may also mean that not all the early modern news media have been seriously used for finding data about earthquakes so far.

12 *Amsterdamse courant* (10 August 1728) and (12 August 1728); *'s Gravenhaegse courant* (13 August 1728); *Leydse courant* (16 August 1728). For Leipzig, see *Oprechte Haerlemsche courant* (5 November 1711). Among other Swiss cities, Geneva was also mentioned, see *Oprechte Haerlemsche courant* (9 July 1716) and (22 March 1753).

Over the years, Dutch readers could also conclude from occasional news reports that earthquakes happened on the British Isles, particularly in England.[13] The same goes for France, Portugal, Spain, and the other European countries listed. Since news reports about earthquakes in those states were published irregularly, it must have been difficult for interested readers to discern specific geographical territories.

In the Turkish part of the Ottoman Empire, the cities of Smyrna (today Izmir) and Constantinople (present-day Istanbul) generated much earthquake news. The first extensive and also emotional accounts describe the July 1688 earthquake that destroyed a major part of Smyrna. 'It had pleased God' to turn the city upside down, but the Dutch merchants had been saved, according to one of the reports.[14] Another devastating earthquake covered in the Dutch papers occurred in both Constantinople and Cairo in 1754.[15] It would be excessive to mention all the years with Turkish earthquakes; nevertheless, it is clear that almost all generations of Dutch readers were confronted with trembling stories from this part of the world. Roughly speaking, this also applies to Ottoman Greece. The news about this country usually reached the Dutch Republic via Venice, which ruled several Greek islands.[16]

Reports concerning earthquakes in Latin America concentrated on the Caribbean Isles, Peru, and Chile. The 1650 report from London, mentioning an earthquake on Barbados, seems to be one of the earliest pieces of information about non-European earthquakes in the Dutch printed papers.[17] A subsequent message was delivered by a Dutch ship in 1669, stating that the islands of Martinique and Guadeloupe had been struck by several earthquakes in the previous year.[18] Jamaica is the most frequently specified

13 News about earthquakes on the British Isles before 1755 can be found in the Dutch newspapers of at least 1681, 1683, 1690, 1692, 1725–28, 1734, 1736, 1739, 1744, 1747, 1749, 1750, and 1753–55. Most of them were relatively light.

14 'Het heeft God belieft, op den 10 deses omtrent de middag door een sware Aertbeving de Stadt van Smirna het onderste boven te keeren'. *Oprechte Haerlemsche courant* (16 September 1688). See also *Amsterdamse courant* (16 September 1688).

15 See e.g. *Leydse courant* (18 October 1754), (21 October 1754), and (11 December 1754); *Opregte Groninger courant* (25 October 1754), (5 November 1754), (15 November 1754), (22 November 1754), and (10 December 1754); *Leeuwarder courant* (26 October, 2 November, and 14 December 1754); *Oprechte Haerlemsche courant* (2 November 1754).

16 See e.g. *Oprechte Haerlemsche courant* (17 November 1665), (24 July 1683), and (15 March 1692); *Amsterdamse courant* (30 January 1677) and (11 January 1681); *Leydse courant* (16 April 1723); *'s Gravenhaegse courant* (9 September 1729); *Opregte Groninger courant* (27 June 1752) (via Ancona).

17 *Courante uyt Italien, Duytslandt, &c* (2 July 1650). About other earthquakes on Barbados, see *Amsterdamse courant* (21 September 1702) and (13 November 1749).

18 *Oprechte Haerlemsche courant* (3 January 1669).

Caribbean island in the Dutch reports, with earthquakes in at least seven years. However, apart from 1692, the reports concerned only minor cases.[19] Cuba (1693 and 1724), Montserrat (1690 and 1735), and Saint-Domingue (or Hispaniola; 1723 and 1752) are other hotspots mentioned more than once. In 1752, for instance, the Haarlem newspaper reported that an earthquake had destroyed 397 of the 400 houses in Port-au-Prince, the capital of the present state of Haiti.[20]

The first Peruvian earthquake to be addressed, which damaged Lima severely in October 1687, was described in the Dutch newspapers of June 1688.[21] The October 1746 earthquake, completely destroying Lima and its port town, generated several items in the papers of June 1747.[22] The earlier January 1728 earthquake in Suriname must have drawn special attention in the Dutch Republic because of the colonial presence of Dutch plantation owners. The reports were surely a great relief to related news readers, since the quakes had caused only fear and no damage.[23] Information about earthquakes in North America was rare. This news arrived in the Netherlands generally via reports from London. It contained, for example, sentences about quakes that had been felt in Boston in 1727 and Nova Scotia in 1753.[24]

The islands along the northwest African coast also pop up a few times in the digitised papers. In 1705 the Haarlem newspaper mentioned a powerful earthquake on the Canary Islands. In 1718 it cited a quake that had been felt on several Cape Verdean and Canary Isles.[25] In 1731 quakes on Graciosa Island, north of Lanzarote, and in the city of La Laguna on Tenerife were reported. Both had taken place at the end of the preceding year.[26] A report

19 See e.g. *Oprechte Haerlemsche courant* (13 July 1688), (18 July 1690), (2 November 1703), (29 August 1750), and (2 March 1752); *Amsterdamse courant* (5 February 1695) and (5 April 1703); *Leydse courant* (9 September 1720); *'s Gravenhaegse courant* (26 October 1729) and (7 September 1750). Concerning the 1692 earthquake on Jamaica, see further on.

20 See e.g. *Oprechte Haerlemsche courant* (18 July 1690), (24 February 1693), and (30 May 1752, this earthquake occurred on 21 November 1751); *Leydse courant* (31 December 1723), (6 March 1724), (8 March 1724), and (28 February 1752); *Amsterdamse courant* (13 November 1749); *'s Gravenhaegse courant* (17 October 1735).

21 *Amsterdamse courant* (8 June 1688), *Oprechte Haerlemsche courant* (10 June 1688).

22 *'s Gravenhaegse courant* (19 June 1747) and (26 June 1747); *Amsterdamse courant* (20 June 1747); *Oprechte Haerlemsche courant* (20 June 1747) and (27 June 1747); *Leydse courant* (23 June 1747). Other earthquakes, also or only in Chile, were reported in 1731, 1740, and 1752.

23 *Amsterdamse courant* (22 May 1728); *Leydse courant* (24 May 1728).

24 *'s Gravenhaegse courant* (16 January 1728) and (12 May 1728); *Oprechte Haerlemsche courant* (13 May 1728) and (23 June 1753). See also *Amsterdamse courant* (9 March 1728) with an announcement of a day of prayer because of the earthquakes.

25 *Oprechte Haerlemsche courant* (30 April 1704) and (6 August 1718).

26 *'s Gravenhaegse courant* (7 March 1731).

about a major quake on Madeira in November 1731 reached the Dutch press in June 1732 by a circuitous route: through a letter from Philadelphia that had been sent to Dublin.[27] News about sub-Saharan Africa was scarce in the early Dutch newspapers, so it is not surprising that we lack any information about earthquakes from this continent.

The eight covered earthquakes in Asia, outside this continent's part of the Ottoman Empire, struck Iran (Mashhad, 1687; Kashan, 1755), Japan (1703), China (Macao, 1728; Beijing, 1730), India (Chandannagar near Kolkata, 1737), and the Dutch East Indies, now Indonesia (Java, 1699; Amboyna, 1755). Except for the Mashhad Earthquake, which was covered with just one sentence in the Haarlem newspaper of 1687, they did not appear in the digitised Dutch news columns before the eighteenth century.[28] The Japan earthquake must have been the Genroku Earthquake of 31 December 1703, which struck Edo (present-day Tokyo) heavily. This disaster led to fires and a tsunami, which were also both mentioned in the Dutch report that was retrieved from the returning ships of the Dutch East India Company, but not before 1705.[29]

Most of the reports – both European and non-European cases – were devoted to urban regions, while earthquakes also struck the countryside, wilderness, and the seas. However, few correspondents made an effort to write about uninhabited or less-populated territories. One of them wrote about the 1693 earthquake that destroyed one third of Sicily. It was said to have struck the island in such a way that previous generations would not have recognised it anymore.[30] Another example is a report about the 1726 earthquake in Hungary that split a mountain, half of which would have fallen into Danube River.[31]

Frequency of Earthquake Reports and the Shocks of 1692

The first digitised remark about an earthquake, which presumably happened in Austria in April 1623, was included in the Amsterdam newspaper *Tydinghe*

27 *'s Gravenhaegse courant* (19 June 1732). This must be the same earthquake that destroyed Santa Cruz de Aguer, the predecessor of present-day Agadir in Morocco. Madeira had also been struck by an earthquake in 1725, according to *Oprechte Haerlemsche courant* (9 June 1725).
28 *Oprechte Haerlemse courant* (1 March 1687), (30 July 1705), (16 August 1731), (18 August 1731), (25 October 1731), (5 June 1738), (10 June 1755), and (14 June 1755); *Nouvelles extraordinaires de divers endroits* (1 October 1699); *Leydse courant* (2 November 1729) and (9 June 1755); *'s Gravenhaegse courant* (15 August 1731); *Leeuwarder courant* (14 June 1755); *Opregte Groninger courant* (28 October 1755).
29 *Oprechte Haerlemse courant* (30 July 1705).
30 *Oprechte Haerlemse courant* (7 March 1693).
31 *Amsterdamse courant* (3 December 1726).

uyt verscheyde quartieren from 8 May 1623. Time would tell how significant it would be, according to the author.[32] Subsequent messages about earthquakes in the same newspaper concerned cases in 1629 (Tuscany), 1642 (Alsace), and 1646 (Leghorn). The 1629 earthquake had swallowed a complete village with all its inhabitants, which must have been horrifying news to read. The 1642 report is interesting because it is an early example in the Dutch digitised newspapers in which the author attributes the earthquake to 'natural causes' instead of supernatural powers, which were connected with blood-spitting fountains in the same report.[33]

In the subsequent years and before the 1680s, the digitised newspapers include just two dozen other reports about earthquakes. This is not surprising, since few Dutch newspapers were published in that period. Furthermore, the news supply from abroad was still in an early stage.[34] From the 1680s onwards, however, the selected newspapers include an increasing number of items about earthquakes, including them almost every year, with a varying number of cases per year. Nearly identical texts could be found in multiple papers, since they collected news via similar channels and copied items from each other. This means that the knowledge about – and probably the fear of – the effects of earthquakes was spread among a larger group of Dutch citizens during the eighteenth century.

The September 1692 earthquake in Western Europe is one of the first extensively covered earthquakes in Dutch newspapers. The *Oprechte Haerlemse courant* opened the reporting in its edition of 20 September 1692, with a message from The Hague dated two days earlier. The editor writes about a possible earthquake that had been felt in The Hague on 18 September, and in all places within a distance of 'three to four hours' – a typical contemporary way of measuring. However, he suggests that it could also be the effect of a firework depot explosion. That idea was immediately refuted in the same paper's next report, dated 19 September, which describes the earthquake as it had been experienced in Antwerp.

The Haarlem newspaper's next edition from 23 September put an end to all doubts as well, with an account from the Dutch province of Zeeland's capital of Middelburg, describing how the earthquake had been felt there. The report's first sentence is conspicuous, since it states that people had feared the same

32 The city archive of Linz (Austria) could not confirm this earthquake from any source.

33 *Tydinghe uyt verscheyde quartieren* (9 June 1629), (17 January 1643), and (12 March 1646). Cf. Vermij, *Thinking on Earthquakes*, 161.

34 Cf. Carlos H. Caracciola, 'Natural Disasters and the European Printed News Network', in Joad Raymond and Noah Moxham (eds), *News Networks in Early Modern Europe* (Leiden: Brill, 2016), 756–78.

Figure 11.1 Jan Luyken, *Earthquake that Destroyed Port Royal (Jamaica) in 1692*, etching, 10.9 × 15.6 cm, published in the second volume of Johann Ludwig Gottfried's *Historische kronyck*, 1698, cols 1615–16. Rijksmuseum, Amsterdam, RP-P-1896-A-19368-1584

fate as 'those from Jamaica'. This was an implicit reference to recent messages in the Dutch newspapers of 28 August 1692 about a severe earthquake on Jamaica in June 1692. That quake was said to have destroyed two thirds of the island's capital Port Royal within two minutes and had probably caused 2,000 fatalities, which in hindsight seems to be an accurate figure.[35]

The Amsterdam editions of 23 and 25 September 1692 were also full of reports about the nearby earthquake. They clarified that the region affected included Cologne and other towns across the German border, as well as Liège, Brussels, and Bruges in the Southern Netherlands; London, Portsmouth, and other cities in south-east England; and Lille and Paris in France, although this country's capital had felt it only very briefly. The 1692 earthquake caused much material damage, in particular fallen chimneys, according to the reports.[36] Scientist Christiaan Huygens felt the earthquake in his mansion Hofwijck near The Hague. He, too, at first thought that

35 See *Amsterdamse courant* (28 August 1692) with reports from both London and The Hague. *Oprechte Haerlemse courant* of the same date reported that three-fourths of Port Royal had been destroyed, and this paper estimated the number of fatalities between 1,500 and 2,000. Cf. US Geological Survey, 'Earthquake Hazards Program', https://web.archive.org/web/20091217203828/http://earthquake.usgs.gov/earthquakes/world/events/1692_06_07.php (accessed 5 May 2021).
36 See also *Oprechte Haerlemse courant* (25 September 1692) and (30 September 1692).

it was an explosion, but after a few days he found out – possibly via the newspapers – that it had really been an earthquake.[37]

At present, we can conclude that the many news items in the Dutch newspapers can be explained both by the 1692 earthquake occurring nearby and because it was one of the most powerful in Western Europe reported up to that time. It measured an estimated magnitude of 6.3 in the epicentre, which was the Walloon city of Verviers.[38] In the same period, the Southern Netherlands generated much news as well because they were one of the battlefields of the War of the League of Augsburg against France. Therefore, many people living there were hit particularly hard.

In 1692 Dutch readers were confronted with news not only about the Western European and Jamaican earthquakes but also with reports on rather harmless shocks in the cities of Nafplio in Greece, Genoa, Steinfeld in the German Eifel, and Cologne.[39] Earthquakes had become a recurring topic in the Dutch newspapers. They predominantly represented a culture of factual information that could be retrieved from and confirmed by other news media such as letters, sheets, pamphlets, and papers from abroad. Compared to scholarly pamphlets and magazines and extensive discourses and histories, they were not much focused on philosophical notions and learned explanation. But precisely what kinds of factual information did the newspapers provide?

Reporting Styles and Characteristics

Over the years, the content of earthquake reports in Dutch newspapers showed many similarities; in other words, their patterns did not change much. The news items usually indicated the earthquake's place or region, the moment at which it had occurred (day or date, sometimes with the exact hour), and its gravity (through an adjective), as the following short notice in the *Amsterdamse courant* of 23 June 1746 exemplifies 'Naples 31 May. Monday night, people felt a light earthquake here'.[40] More extensive messages could

37 Christiaan Huygens, *Oeuvres complètes*, 22 vols, vol. 19: *Mécanique théorique et physique 1666–1695*, ed. J.A. Vollgraff (The Hague: Martinus Nijhoff, 1937), 311.

38 About this earthquake: Pierre Alexandre et al., 'The 18 September 1692 Earthquake in the Belgian Ardenne[s] and Its Aftershocks', in Julien Fréchet, Mustapha Meghraoui, and Massimiliano Stucchi (eds), *Historical Seismology. Modern Approaches in Solid Earth Sciences* (Dordrecht: Springer, 2008), vol. 2, 209–30.

39 *Amsterdamse courant* (15 March 1692), (11 October 1692), and (1 November 1692).

40 'Napels den 31 May. Maendag nagt gevoelde men hier een ligte aerdbeving'. The duration could also be indicated as a long Ave Maria ('een groot Ave Maria') or half of the prayer of Our

A°. 1746. N°. 75

AMSTERDAMSE Donderdaegse Courant

Werd uytgegeeven by
Bezyden de Beurs,
Post-Comptoir;

HENDRIK LINSEN,
by het Antwerpsche
den 23 Juny.

I T A L I E N.

Napels den 31 May. Maendag nagt gevoelde men hier een ligte aerdbeving. Deze morgen zyn 2 vreemdelingen, die zig op de binnen-plaets van 't hof bevonden, in arrest genomen, niettegenstaende zy met paspoorten voorzien waren.

Parma den 7 Juny. Het werd bevestigd dat de gener. Berenclau Rivalta by verrassing heeft ingenomen; en over de 600 man so ruytery als voetvolk krygsgevangen gemaekt. De Koning van Sardaigne is aen 't hoofd van een talryke armée op marsch, met intentie om het beleg van Tortona te ondernemen.

Mantua den 8 Juny. Dewyl de Prins van Lichtenstein zig te Fiorenzuola onpaslyk bevind, heeft de generael Botta *per interim* het commando van 't leger op zig genomen.

G R O O T B R I T T A N N I E N.

Londen den 17 Juny. Met een expresse die dingsdag avond wegens den hertog van Cumberland arriveerde, heeft men dat Lochiel met 500 mannen de wapenen had neergelegt en zig aen zyn koningl. hoogheid onderworpen; dat de zoon des Pretendents zig nog in Schotland onthield, en dat de rebellen het fort Au-

Palfy over Bon in deeze stad gearriveerd. Eergisteren wierd aen de inwoonderen der dorpen rondom Siegburg aengezegt dat binnen 4 dagen tyds in die nabuurschap een corps Keyzerl. troupen op de Heyde van Hangelohr een camp zal komen formeeren. Den 13, wierden 250 recruten na het Arembergse regiment gevoert. Deezer dagen wierd de bisschop van Roermond alhier in de Dominikaner kerk door den Pausfyke legaet ingewyd.

Frankfort den 19 Juny. De nog alhier gehouden wordende Keur- en Opperrhynsche Kreytzvergadering is steeds met de marsch der Keyzerl. troupen na Braband beezig, om by derzelver doortogt alle excessen voor te komen. Prins Karel zal tegen het einde der aenstaende week hier zyn: de Keyzerl. armée zal tegen den 10 July haer zamelplaets by Coblenz hebben, en in het Luxemburgse trekken.

N E D E R L A N D E N.

Leuven den 16 Juny. De collonel Franquini die van Weenen gekomen is om de plaets van den baron Trenk te vervullen, gisteren met een corps van 7 of 800 Panduren en Croaten tot aen 't dorp Pelenberg genaderd zynde, gong de marquis d'Armentieres, daer van kondschap gekreegen hebbende, aen 't hoofd van 10 piquets

Figure 11.2 Part of the front-page of *Amsterdamse courant* (23 June 1746), with news about an earthquake in Naples. KB – National Library of the Netherlands, The Hague. Source: Delpher (accessed 22 April 2022)

include the duration of earthquakes in minutes, their numbers of shocks and aftershocks, and their natural consequences, such as flooding and landslides.

Most earthquakes were qualified either as light (*licht*) or heavy (*zwaar*), but other adjectives were also used, such as small (*klein*), great (*groot*), fierce (*hevig*), harsh (*hard*), strong (*sterk*), severe (*fel*), violent (*violent*), terrible (*verschrikkelijk*), horrible (*afgrijzelijk* and *gruwelijk*), dangerous (*vervaarlijk*), and dreadful (*allerijselijk*).[41] The 25–26 April 1687 earthquake in Naples was even classified as a 'very heavy earthquake' ('seer sware Aertbeving').[42] In the 1750s the Haarlem editor introduced the phrase 'harmless earthquake' ('onschadelyke aardbeving') for classifying it as a light case without serious damage.[43] Other

Father ('een half Pater-noster'). *Haerlemse courante* (10 July 1660); *Oprechte Haerlemsche courant* (25 September 1692).

41 See e.g. *Amsterdamse courant* (21 September 1702): 'kleyne aerdbeving' on Barbados, (4 August 1729): 'heevige aerdbeeving' in Velletri near Rome, (29 April 1730): 'felle aerdbeeving' in Massa and Carrara, Tuscany, (20 May 1747): 'afgryzelyke aerdbeving' in Foligno, etc.; *Haerlemse courante* (10 July 1660): 'harde Aertbevinge' in Bordeaux; *Oprechte Haerlemse courant* (24 July 1683): 'violente Aertbeving' in Vicentino region, (21 September 1683): 'groote Aerdbeving' in Ledbury (26 October 1683): 'schrickelijcke Aerdbeving' in Coventry (4 September 1734): 'stercke Aerdbeving' on Iceland; *'s Gravenhaegse courant* (13 August 1728): 'vervaerlyke aerdbeevingen' in the Abruzzo region; *Opregte Groninger courant* (6 June 1752): 'gruwelyke Aerbeving in Stavanger; *Leeuwarder courant* (26 October 1754): 'alleryzelykste Aardbeving' in Constantinople.

42 *Oprechte Haerlemsche courant* (24 May 1687).

43 See e.g. *Oprechte Haerlemsche courant* (10 October 1751): Nocera and Gualdo, Italy, (2 December 1751): Naples, (28 March 1752): Falun, Sweden, (29 June 1752, 10 August, and 21 October 1752):

rather vague descriptions were also used to indicate minor cases, such as 'some movements of an earthquake' ('eenige bewegingen van een Aertbeving').[44]

Before the introduction of seismic magnitude scales, such as the Richter scale in 1935, words were one of the best ways to provide the readers with impressions. The Cypriot earthquake of April 1735, for example, was described as the heaviest since time immemorial on the isle.[45] An earthquake's force could also be compared with a previous one. The 1703 earthquake in Rome, for instance, was considered stronger than the earthquake that had happened during Roman emperor Nero's reign.[46] Such comparisons were rare, however.

Longer reports included information on whether or not an earthquake had caused damage, and if so, what kind of damage, such as collapsed churches, houses, and other buildings. The same goes for numbers of fatalities that ranged from a few people to several thousands. For example, a report about the disastrous earthquake that struck Palermo in 1726 speaks about 3,000 people that had been found dead among the debris.[47] The earthquake in the Kingdom of Valencia of 1748 was said to have led to 5,000 people being buried alive, in conjunction with the collapse of many buildings, such as the medieval castle of Montesa.[48] A dramatic report about the earlier earthquake that destroyed the southern Italian city of Benevento in 1688 states that few of the 6,000 inhabitants had survived the disaster.[49]

The numbers of casualties in the accounts about non-European earthquakes will have particularly impressed Dutch readers; they were perhaps beyond their imagination. First reports about the Beijing Earthquake of 30 September 1730 include estimations of around 100,000 victims, a figure that was later adjusted downwards to 36,000, which is still a dramatic number.[50] Even more impressive must have been the number of 300,000 deaths, allegedly caused by a cyclone and earthquake in India's

Central Italy.

44 See e.g. *Oprechte Haerlemsche courant* (16 December 1690) and (19 December 1690).

45 *'s Gravenhaegse courant* (7 September 1735).

46 *Oprechte Haerlemsche courant* (27 February 1703).

47 *Oprechte Haerlemsche courant* (12 October 1726). A report in this newspaper of 10 October 1726 mentions the figure of even 8,000 victims.

48 *'s Gravenhaegse courant* (29 April 1748); Elisa Buforn et al., 'The 1748 Montesa (Southeast Spain) Earthquake – A Singular Event', *Tectonophysics* 664 (2015), 139–53. See also the extensive report in *Hollandsche Historische courant* (2 May 1748).

49 *Oprechte Haerlemsche courant* (3 July 1688). The present estimation of victims of this earthquake called Sannio is about 10,000, but this figure concerns the whole province.

50 *'s Gravenhaegse courant* (15 August 1731); *Oprechte Haerlemsche courant* (16 August 1731), (18 August 1731), and (25 October 1731).

Chandannagar (near Kolkata) on 11 October 1737, resulting in a 13-metre storm surge in the Ganges River. The account in the Haarlem newspaper of 5 June 1738 also mentions the drowning of many animals and the loss of thousands of small boats and one French, two Dutch, and eight English ships.

The information about the human victims was far from precise, however, as Roger Bilham argued in 1994. At the time, Kolkata had about 20,000 inhabitants and the number of fatalities was most likely not greater than around 3,000 people. Bilham even disputes the existence of an earthquake as such, since it is not mentioned in the British East India Company reports. However, Pitta Govinda Rao rejected that idea, stating that an earthquake did occur in the vicinity of Kolkata during the night of 11 October.[51] At the time, European contemporaries could rely only on the sensational news reports, which would be trusted until the twentieth century.

The reports rarely include names of victims, most of which were anonymous ordinary people. An exception is Gregorius (or Joris) Croock, who was on his way to Constantinople in 1667 to become a Dutch resident in the Ottoman Empire.[52] However, he never reached his destination because he lost his life during the destructive earthquake that ruined Ragusa (present-day Dubrovnik) in April. His travelling companion Jacob van Damme, the intended Dutch consul of Smyrna, survived the tragedy, and returned to The Hague via Venice, where he informed the States General about the circumstances.[53] The Ragusa Earthquake was one of the most devastating earthquakes in the region that is part of Croatia today, killing around 2,000 citizens.[54]

A frequently mentioned human reaction, especially during the start of an earthquake, was going outside, leaving houses and public buildings in a hurry. This could lead to congestion in the streets, which happened in

51 Roger Bilham, 'The 1737 Calcutta Earthquake and Cyclone Evaluated', *Bulletin of the Seismological Society of America* 84: 5 (1994), 1650–57; Pitta Govinda Rao, 'A Probe into the Calcutta Earthquake of 1737', *Current Science* 69: 5 (1995), 476–78.

52 Other exceptions are 'Princess Altieri', who was reported to have had a miscarriage in 1750 because of an earthquake, and the French ambassador Pierre Puchot des Alleurs, who passed away because of a stroke due to an earthquake in Constantinople. *Oprechte Haerlemsche courant* (10 March 1750); *Opregte Groninger courant* (10 January 1755).

53 *Oprechte Haerlemsche courant* (14 May 1667) and (31 May 1667); O. Schutte, *Repertorium der Nederlandse vertegenwoordigers, residerende in het buitenland, 1584–1810* (The Hague: Martinus Nijhoff, 1983), 99, 332–33.

54 Paola Albini and Andrea Rovida, 'From Written Records to Seismic Parameters. The Case of the 6 April 1667 Dalmatia Earthquake', *Geoscience Letters* 3 (2016), 1–9. In 1667 Van Damme's sad story and other information about the earthquake could be found in the annual *Hollandsche Mercurius*; see vol. 18, 43–50. According to *Oprechte Haerlemsche courant* (7 January 1681), the restoration of Ragusa went well.

Figure 11.3 Jan Luyken, *Earthquake that Destroyed Ragusa (Dubrovnik) in 1667*, etching, 10.8 × 15.5 cm, published in the second volume of Gottfried's *Historische kronyck*, 1698, col. 965. Rijksmuseum, Amsterdam, RP-P-1896-A-19368-1509

London in September 1692. That earthquake struck at the busiest moment of the stock exchange.[55] A great panic also occurred in Rome a week after an earthquake in February 1703, when malicious men called upon all people to leave their houses 'on the orders of the Pope' during the night. The urban militia had to refute the appeal and calmed the people by telling them that the Pope had never given such an order.[56]

Normally, many people fled to gardens in the cities or to the countryside because of the destruction of their houses, danger of further collapse, and fear of aftershocks.[57] Queen Christina of Sweden, who lived in Rome after her abdication, and other important persons decided to have tents put up in the gardens of their palaces in June 1668, after a rumour about an impending earthquake had been spread.[58] Poor people had to sleep in the open air or in tents and provisional barracks or sheds made from timber, if these were provided.[59] In 1737, when King Charles VII of Naples slept in his royal tent

55 *Amsterdamse courant* (25 September 1692).

56 *Oprechte Haerlemsche courant* (1 March 1703).

57 In 1702 it was problematic for prisoners to escape because they were chained on bars and hung in the air during an earthquake. See *Oprechte Haerlemsche courant* (22 April 1702).

58 *Oprechte Haerlemsche courant* (10 July 1688).

59 See e.g. *Oprechte Haerlemsche courant* (17 November 1703).

because he was afraid of new earthquakes, it was even newsworthy to mention that he had got a cold.[60] The elite could also afford to leave their home towns and go to country houses. For example, Elbert de Hochepied, ambassador of the Dutch Republic in the Ottoman Empire, swapped Constantinople for a village near the Black Sea because of ongoing earthquake danger in 1748.[61]

Some the stories were dedicated to governmental actions, such as organising shelter for survivors, reparations and their costs, and precautions and bans. In 1688 Spanish king Charles III cancelled bull fighting in Madrid because of the sad tidings about an earthquake in the Kingdom of Naples, which was one Charles's possessions. When in 1703 smoke from an earthquake would supposedly have caused brain damage in Rome, the authorities decided to bring three citizens to the madhouse and to imprison several others because of their 'strange opinions'.[62] In 1728 the Pope prohibited festivities during Carnival because of ongoing earthquakes, rain and floods, and in 1732 he banned the performance of comedies during Carnival out of respect for the earthquake victims in Naples.[63] Papal admonitions to support the poor with alms were also newsworthy, as were his orders for restorations such as the rebuilding of Norcia in 1703.[64] In 1751 an Italian bishop sold all his silverware to help people who were ruined by the latest earthquake in his territory.[65] Dutch readers will very likely have appreciated such forms of solidarity.

We can assume that most of the province of Holland's newspaper readers were Protestants.[66] Since most European earthquakes happened in Roman Catholic countries, these readers were regularly confronted with typical Catholic reactions to disasters. A few examples will suffice here. In 1724 the clerics of Faro (Portugal) organised a procession with a new statue of St Barbara, hoping that by doing this they would be spared from new

60 *Oprechte Haerlemsche courant* (7 March 1737).

61 *Leydse courant* (13 September 1748).

62 *Amsterdamse courant* (21 June 1703).

63 *Oprechte Haerlemsche courant* (19 February 1728) and (27 December 1732). Another ban of carnival in this newspaper of 14 February 1711.

64 See e.g. *Amsterdamse courant* (17 February 1703) and (24 February 1703); *Oprechte Haerlemsche courant* (14 January 1704). An example of societal support is a collection in the whole Kingdom of Naples for survivors of an earthquake on Sicily in 1751. *Opregte Groninger courant* (17 September 1751).

65 *Oprechte Haerlemsche courant* (30 September 1751).

66 Most newspapers were published and read in Holland, a province with 20 per cent Roman Catholics at most. Their percentage in the Dutch Republic in 1726 has been estimated at 34 per cent. Hans Knippenberg, *De religieuze kaart van Nederland. Omvang en geografische spreiding van godsdienstige gezindten vanaf de Reformatie tot heden* (Assen and Maastricht: Van Gorcum, 1992), 23.

earthquakes and other disasters.[67] In 1727 a Neapolitan report stated that after a stormy night with an earthquake the weather became quiet as soon as the city began its devotions. This piety was a reaction to the exhibition of St Januarius's head, which had been ordered by the archbishop. Furthermore, the viceroy of Naples had visited a church to reveal the 'miraculous Statue of the Crucifix'.[68] In 1728 the Pope admonished the people to gain an indulgence by praying for mercy because of the many disasters, including earthquakes.[69] In 1751 a Genoese reporter explained the lack of earthquake damage as a result of God's mercy and the protection of Maria.[70] The Dutch newspapers published such eyewitness accounts and digressions without comment or judgement. They only translated or copied the information they retrieved.[71]

Thanksgiving masses, soul masses for deceased earthquake victims and other – not exclusively Roman Catholic – prayers were also reported.[72] In 1705 the Pope had ordered prayers for the ailing Portuguese king in all the churches of Rome because the king had supported this city after its recent earthquakes.[73] Another characteristic Christian reaction can be found in a story about the 1731 earthquake in the southern Italian city of Foggia. The author described it as a 'terrible spectacle' that seemed to be the start of Judgement Day.[74]

A supernatural phenomenon was noted in 1733. People were said to have seen a bloody cross in the sky during an earthquake in the Kingdom of Naples.[75] When the local bishop referred to this during the consecration, the host appeared to be bloody as well. Protestants – who rejected the transubstantiation doctrine – must have frowned when they read such a sensational account, which was probably published only out of fascination. Dutch readers were also confronted with the measures taken by the Neapolitan viceroy against several monks in 1730. He threatened to expel them from all monasteries in his kingdom if they would continue their dispersion of false warnings regarding an expected earthquake. Many people had also been scared by the alleged sweating of a

67 *Leydse courant* (31 March 1724). A terrible earthquake had struck the Algarve at the end of 1722 according to *Leydse courant* (24 February 1723) and (26 February 1723).

68 *Leydse courant* (7 November 1727).

69 *Leydse courant* (30 January 1728).

70 *Oprechte Haerlemsche courant* (16 December 1751).

71 See e.g. Koopmans, *Early Modern Media*, 262–64.

72 See e.g. *Amsterdamse courant* (24 February 1742), (15 March 1703), (17 March 1703), (21 July 1729), and (8 February 1742); *Oprechte Haerlemsche courant* (26 February 1693), (17 March 1703), and (31 March 1703).

73 *Oprechte Haerlemsche courant* (7 April 1705).

74 *Oprechte Haerlemsche courant* (9 June 1731). The idea of Judgement Day is also present in the account of the 1667 Dubrovnik earthquake in *Hollandsche Mercurius* 18 (1667), 45.

75 *Amsterdamse courant* (10 January 1733).

statue of the Virgin Mary and had left their houses half naked during the night. Some of them had stopped carriages of 'decent people', had forced them to buy wax candles for the 'miraculous Crucifix' and plundered houses.[76]

In 1713 all Christians will have been worried about the news about an earthquake in Palestine that had greatly damaged the chapel of the Holy Sepulchre in Jerusalem and the 'Temple' in Bethlehem. The short report ended with the remark that the hatred and conflicts between the Greek Orthodox Christians and the Roman Catholics over the possession of the holy places in Palestine increased daily.[77] Although the report does not make the connection, Dutch readers might well have concluded that the earthquake reflected societal disharmony. Nonetheless, this disaster did not lead to more harmony between different Christian denominations.

Overall, the newspapers did not provide many indications regarding the causes of the earthquakes mentioned. In 1727, however, the Leiden paper stated that the earthquakes in the Algarve had without a doubt been caused by a huge fire between the Portuguese cities of Faro and Tavira.[78] It was difficult to contradict such an explanation in a period when scholars still discussed possible reasons predominantly within a religious framework, while religious authorities continued to tell people that it was God's will. Nevertheless, a story in the Groningen newspaper of 1750 is exemplary for the increasingly rational approach towards the phenomenon. When a young English lady, who denied all stories about quakes, suggested during an earthquake that she could cause a next shock by jumping, the Groningen newspaper characterised this as stupid behaviour of a person without knowledge.[79] It would be interesting to research whether this editorial reaction was unique or exemplified a new trend in the newspapers that were published after 1750. Anyhow, when geologists found out in the nineteenth century that earthquakes happened along fault lines, such forms of speculation could be rejected forever.[80]

Final Remarks

This chapter demonstrates that readers of early modern Dutch newspapers were already familiar with earthquake reports long before the

76 *'s Gravenhaegse courant* (27 April 1731).

77 *Oprechte Haerlemsche courant* (11 February 1713).

78 *Leydse courant* (26 February 1723). About this fire theory and other theories about causes of earthquakes before 1755, see Guidoboni and Ebel, *Earthquakes and Tsunamis*, 147–75.

79 *Opregte Groninger courant* (3 November 1750).

80 Vermij, *Thinking on Earthquakes*, 2.

devastating Lisbon Earthquake of 1755. Although these papers covered predominantly political news from Europe since their start in the late 1610s, earthquakes would soon be a part of the regularly discussed natural phenomena. This was not restricted to the most serious cases. On the contrary, even minor quakes, with no or little damage, were considered newsworthy. News correspondents deemed even shocks that had merely frightened people to be important enough to communicate. The very idea of sudden shakings of the ground that could destroy complete regions was extremely terrifying. Every new report about another earthquake will have revived or strengthened this fear. Besides, since reports of earthquakes came from a great variety of places in the world, Dutch readers could get the impression that this phenomenon seemed to happen everywhere.

While many news items about earthquakes remained short and did not exceed one or a few sentences, reports about serious cases became longer over the years. This was not so much the result of editorial choices but became possible because of the expanding news dissemination in Europe during the eighteenth century. Reports could increasingly include specific details about the earthquakes themselves, their damage and consequences, numbers of victims, and the reactions of survivors. And sometimes they also narrated trivial news. When Dutch people were involved in the covered earthquakes, that fact was particularly addressed. This practice can be considered an enduring journalistic characteristic: to report how many fellow citizens were involved in a disaster abroad and whether they passed away.

As long as people could not explain earthquakes in a satisfactory way, speculation remained all kinds of publications. Typically, however, the Dutch newspapers mainly treated facts and figures, although within the existing societal framework. They represented, as far as possible, a media culture of accuracy, although sometimes with a touch of sensationalism. Readers could read dominant opinions only between the lines and draw conclusions for themselves. The well-known religious (and widely accepted) idea was that earthquakes were punishments of God, and that people should repent their sins. This thought was neither denied nor questioned, nor openly proclaimed in the reports, but it was implicitly taken for granted.

Dealing with earthquakes was a matter of fate and acceptance. Dutch newspaper readers could only hope that next issues would bring more positive news that would correct earlier dramatic accounts and, in the meantime, that they themselves would be saved from earthquakes and other disasters.

About the Author

Joop W. Koopmans is Associate Professor of Early Modern History at the University of Groningen, with a special interest in the interaction between media and politics. He has written many books and articles in this field, among them *Early Modern Media and the News in Europe. Perspectives from the Dutch Angle* (Leiden: Brill, 2018). His latest monograph is *Het nieuws verbeeld. Oorlog en vrede in de titelprenten van de Europische Mercurius (1690-1750)* (Verloren 2021), a book about the frontispieces of the news periodical *Europische Mercurius*. Since 2020 he has served as chair of the Royal Frisian Society (est. 1827).

12 The Development of Disaster Prints and Publications in Japan, 1663–1923

Julia Mariko Jacoby

Abstract

Disasters such as earthquakes were an important topic of early modern Japanese print media because the Tokugawa Shogunate (1603–1868) censored political news. Beginning with the boom of the print market in the seventeenth century, visual representations, narrative genres, and tropes emerged in order to report, illustrate, and interpret disasters. They also persisted after print culture was transformed through the arrival of Western technology in the late nineteenth century. This chapter analyses the development of single-page disaster prints and three prominent examples of disaster publications – earthquake reports in book length – from their beginnings to the Great Kantō Earthquake, by tracing selected visual and narrative techniques and tropes: the catfish as a satirical symbol derived from folkloric religion, the dramatic visualisation of threat from earthquake and fire, visual and narrative representations of the spatial extent and degree of damage, and the stories of individual experiences told in increasingly standardised narrative genres.

Keywords

media history – Japan – *namazu* catfish prints – Great Kantō Earthquake – Ansei Edo Earthquake – memory

In 1855 a strong earthquake of estimated magnitude 7 hit Edo (present-day Tokyo; in this chapter, Tokyo before 1868 is referred to as Edo), the political capital of early modern Japan. As the earthquake occurred near the surface, it caused considerable damage through quaking and fire and killed between 7,000 and 10,000 people. Amidst the chaos of the earthquake's

Asperen, van, Hanneke and Lotte Jensen (eds): *Dealing with Disasters from Early Modern to Modern Times. Cultural Responses to Catastrophes.* Taylor & Francis Group, 2023

DOI: 10.5117/9789463725798_CH12

aftermath, hundreds of woodblock prints were circulated illegally, most of them depicting catfish (in Japanese *namazu-e*, literally 'catfish pictures'). At that time, the catfish was said to cause earthquake disasters in popular Japanese lore, and the pictures mostly showed anthropomorphised catfish symbolising the earthquake in a satirical manner: they were beaten up by victims, apologised for the damage they caused, or distributed wealth to the Edoites.[1]

The 1855 earthquake that caused this flood of satirical prints happened during a time of political turmoil: Two years prior, Commodore Perry had arrived in Japan with his 'Black Ships', modern American gunboats, to force Japan, which had mostly self-isolated from overseas trade since the seventeenth century, into an unequal trading relationship with the United States. In 1854 the Tokugawa Shogunate, which ruled Edo-period Japan from 1603 to 1868, signed unequal treaties with the US and Britain. Treaties with other European empires followed in 1858. As a result, Japan opened five ports for trade, exposing the country to the risk of colonisation or subjugation observed in East and Southeast Asia. This stirred political tension and discontent with Tokugawa rule.[2]

The general sentiment of instability caused by political crisis and the natural disaster erupted in the catfish woodblock prints. There were prints that connected the earthquake to the political crisis but did not necessarily depict it as a tragedy. In one example, the catfish looks like a whale, blowing out *koban*, Japanese oval gold coins, to the delight of the surrounding spectators. But in front of the whale, two small boats are depicted. Their occupants are seen attacking the much more powerful opponent in vain. As the large black catfish evokes contemporary depictions of Perry's Black Ships, it is safe to assume that this picture processes hopes and anxieties in Edo towards the Western powers through the earthquake disaster: both could mean ample business opportunities but also severe dangers.[3]

It is no coincidence that natural disasters were used to process and discuss the political situation: Japan was ruled in the early modern period by the

1 Gregory Smits, 'Shaking Up Japan. Edo Society and the 1855 Catfish Picture Prints', *Journal of Social History* 39: 4 (2006), 1045–78; Susanne Formanek, 'Japanische Reaktionen auf (Natur-) Katastrophen. Am Beispiel des großen Edo-Erdbebens von 1855 und seiner medialen Aufarbeitung', in Christa Hammerl and Ilja Steffelbauer (eds), *Naturkatastrophen. Dramatische Naturereignisse aus kulturwissenschaftlicher Perspektive* (Vienna: Mandelbaum Verlag, 2014), 48–63.

2 For an introduction to the mid-nineteenth-century crisis in Japan, see Elise K. Tipton, *Modern Japan. A Social and Political History* (New York: Routledge, 2016), 24–41.

3 Gregory Smits, *Seismic Japan. The Long History and Continuing Legacy of the Ansei Edo Earthquake* (Honolulu: University of Hawai'i Press, 2013), 147–53.

Tokugawa Shogunate, a military government that had risen to power after 150 years of civil war and censored all news coverage that could threaten their legitimacy. Natural disasters were among the few newsworthy topics that were not suppressed by the Shogunate. Thus, printers eagerly took up reporting on natural disasters and, as seen, sometimes even used them to hint at political events.[4] This is reflected in hundreds of surviving woodblock prints (*kawaraban*, literally 'tile-block prints') featuring natural disasters: in one of the most prominent *kawaraban* collections, they amount to approximately half of the prints.[5] As a consequence, a rich variety of disaster reporting and visual representation developed in Japan after print culture took off in the seventeenth century. They formed a range of tropes and narratives that proved remarkably persistent even when print culture was fundamentally transformed by the arrival of Western printing technology and photography.

This chapter presents and analyses the developments and transformations of the two main print mediums used for conveying catastrophe from the Edo period until the early twentieth century. First, there are single-page prints from the early modern and modern periods, which resembled Western pamphlets in function, and provided news reports, maps and illustrations, or satire. Second are book-length disaster reports, consisting of a variety of visual and literary genres, which I call disaster publications. Three prominent examples are presented here: *Kaname'ishi* ('Keystone') from 1663, *Ansei kenmonshi* ('Account of Seen and Heard Things of the Ansei Period') from 1855, and *Taishō daishinsai daikasai* ('The Great Earthquake and Great Fire of the Taishō Period') from 1923. This selection allows us to trace visual representations, narrative genres, and tropes employed to report and illustrate earthquake disasters in important stages of the genre in detail: from the initial appearance of disaster publications in the seventeenth century, their peak towards the end of the Edo period which manifested in a rich print production after the aforementioned Ansei Edo Earthquake of 1855, to their transformation through the arrival of Western technology in the second half of the nineteenth century and their emergence as part of the modern mass press in the aftermath of the Great Kantō Earthquake of 1923 (the worst earthquake disaster in Japanese history, which caused the death of over 105,000 people and destroyed two thirds of Tokyo).

4 Itoko Kitahara, *Saigai jānarizumu. Mukashihen* ('Disaster Journalism. Volume on the Past') (Tokyo: Rekishi Minzoku Hakubutsukan Shinkōkai, 2001), 82.

5 Kitahara, *Saigai jānarizumu*, 11–12; Hideo Ono, *Kawaraban monogatari. Edo jidai masukomi no rekishi* ('The Kawaraban Story. The History of Edo Period Mass Media') (Tokyo: Yūzankaku Shuppan, 1988). Of the 572 prints of the Ono collection, 258 deal with fires, earthquakes, storms, floods, and volcanic eruptions.

This chapter shows, in five parts, the continuities and changes in Japanese disaster prints and publications and their history, highlighting the development of selected visual representations, narrative genres, tropes, and motifs that not only conveyed reports on damages but also spatial imaginations of the disaster's scale and a variety of emotions and interpretations of disasters. In the first part, I explain the emergence of disaster prints and publications within the early modern print culture, their accessibility and consumption, and their transformation after the arrival of Western print technology and culture. The next two parts concern imageries of disaster. In the second part, I explore the catfish lore and its use as a satirical symbol of earthquakes. The catfish prints also give clues about how earthquake disasters were interpreted in Japan – not only as tragedies but also as a cosmic rectification or business opportunities. The third part traces the visual representation of threat experienced by the earthquake victims from the seventeenth to the twentieth century: earthquake victims were consistently depicted as mindlessly running in between buildings that were simultaneously burning and crumbling. The last two parts shed light on spatial representation and narrative genres. The fourth part analyses visual and narrative techniques used to represent the degree and spatial extent of damage. These included maps, lists, and descriptions of city tours. The last part is concerned with how the fates of individuals during disasters were conveyed. They were told in a variety of narrative genres such as tragic stories (*aiwa*), heroic 'beautiful tales' (*bidan*), and ghost stories (*kaidan*), which became increasingly standardised over time.

Japanese Print Culture and Disaster Reporting

In the Edo period, print culture emerged along with a flourishing urban consumer culture. Japan in the early modern period was marked by a long period of peace maintained by the Tokugawa Shogunate and enjoyed a steady population growth and an increase in urbanisation, production, and trade. This led to the rise of consumerism especially in Edo, which, with over one million inhabitants, surpassed European capitals in the early 1700s. Entertainment facilities and consumer products sprang up to satisfy samurai serving the Edo court as well as a growing number of wealthy merchants and city dwellers.[6] Consequently, print culture also experienced a boom. Prints began to enjoy wide distribution and differentiation beginning at the end of the seventeenth century. Instead of movable type, printers employed single woodblock prints,

6 For an overview of Edo-period Japan, see Tipton, *Modern Japan*, 1–23.

which made it easy to combine script and illustrations, thereby increasing their commercial appeal. Pictures were increasingly printed in colour and reached a high level of sophistication while remaining relatively cheap and accessible.[7] The Tokugawa Shogunate tried to control the flourishing printing industry by enacting strict censorship: The author and publisher had to be fully listed in the print and were held responsible for its contents.[8] Therefore, quickly produced single-page newsprints, the aforementioned *kawaraban*, were printed anonymously. However, prints were tolerated as long as the content was not offensive to the Tokugawa Shogunate, as was the case with natural disasters.[9]

As literacy in Japan was remarkably high, prints were accessible to a multitude of people.[10] In addition, the *kawaraban* were sold by so-called *yomiuri* (literally 'read-sellers'), who read the contents aloud when they sold their prints, thus also making printed content available to illiterate audiences.[11] *Kawaraban* were an important medium of disaster coverage, listing damage, providing maps illustrating the scope of the disaster, or offering satire. They were used as memorials: some disaster *kawaraban* were found included in individual diaries. Some people also used them to communicate their disaster experience or report their safety to their families in the provinces. There were disaster *kawaraban* printed that deliberately left space to write on.[12]

A large market for popular books intended to entertain also emerged, the so-called *kusazōshi* (literally 'grass books') that encompassed a multitude of genres. They were distinguished by the colours of their book covers and ranged from novels and travel guidebooks to pornography.[13] Conceptualised as consumer goods, they increasingly featured rich coloured illustrations.

7 On the publishing and reading culture in Edo Japan, see Mary Elizabeth Berry, *Japan in Print. Information and Nation in the Early Modern Period* (Berkeley: University of California Press, 2006); Peter F. Kornicki, *The Book in Japan. A Cultural History from the Beginnings to the Nineteenth Century* (Leiden: Brill, 1998).

8 Richard H. Mitchell, *Censorship in Imperial Japan* (Princeton, NJ: Princeton University Press, 1983), 3–5; Kornicki, *The Book in Japan*, 324–52.

9 Sepp Linhart, 'Die mediale Bewältigung von Erdbebenkatastrophen in Japans Geschichte. Flugblätter und Ansichtskarten', in Hammerl and Steffelbauer, *Naturkatastrophen*, 64–89, at 65–66.

10 Richard Rubinger, *Popular Literacy in Early Modern Japan* (Honolulu: University of Hawai'i Press, 2007).

11 Gerald Groemer, 'Singing the News. Yomiuri in Japan During the Edo and Meiji Periods', *Harvard Journal of Asiatic Studies* 54 (1994), 233–61. The largest newspaper in Japan is still named after the *yomiuri*, reflecting the premodern practice.

12 Ono, *Kawaraban monogatari*, 213–14; Kitahara, *Saigai jānarizumu*, 23–25.

13 Mitsutoshi Nakano, *Edo no shuppan* ('Publishing in Japan') (Tokyo: Perikansha, 2005), 49; Ekkehard May, *Die Kommerzialisierung der japanischen Literatur in der späten Edo-Zeit (1750–1868). Rahmenbedingungen und Entwicklungstendenzen der erzählenden Prosa im Zeitalter ihrer ersten Vermarktung* (Wiesbaden: Harrassowitz, 1983), 150–81.

Books were printed in the urban centres such as Edo, Kyoto, and Osaka, but with travelling book vendors they were also circulated in rural regions. Book lending and reading aloud were regular practices. Thus, books enjoyed broad popularity.[14]

Disaster publications, reports on individual disasters in book form, emerged as an independent genre with the beginning of print culture in the seventeenth century. The first known disaster publication devoted to an earthquake, *Kaname'ishi*, was published in 1663 and laid the foundations of the genre by providing a potpourri of information, including descriptions of earthquake damage, anecdotes of individual experiences, and explanations for the quake's causes.[15] Many publications followed this approach of providing an entertaining mix of information about a disaster, a genre that peaked in the nineteenth century. *Ansei kenmonshi*, which appeared after the 1855 Ansei Edo Earthquake, might be the best-known disaster publication from the Edo period because of its prominent author, Kanagaki Robun (1829–1894) and its famous illustrator, woodblock print master Utagawa Kuniyoshi (1798–1861). It is therefore also analysed in this chapter.[16]

These disaster publications appeared with a short gap of one to several months after the earthquake. This allowed for a first general assessment of the disaster event. Compiled into one or multiple book volumes, disaster publications provided a comprehensive narrative of a disaster that addressed both the temporal sequence of events and the spatial extent of the damage, as well as cosmological interpretations and historical background. Japanese disaster publications thus helped to structure disaster events and give them a coherent narrative.[17] As Ansgar Nünning has written regarding catastrophes and crises, eventful occurrences like natural disasters are told as stories with specific cultural tropes and plot patterns attached, which preconfigures the interpretation and specific meaning attached to a situation.[18] Following

14 See note 7 of this chapter.

15 Ryōi Asai, 'Kaname'ishi', in Masachika Taniwaki, Masahiko Oka, and Kazuhito Inoue (eds), *Kanazōshishū* (Tokyo: Shōgakkan, 1999), 11–83; Itoko Kitahara (ed.), *Nihon saigaishi* ('Disaster History of Japan') (Tokyo: Yoshikawa Kōbunkan, 2006), 239.

16 Robun Kanagaki, *'Berichte über Gesehenes und Gehörtes aus der Ansei-Zeit'. Kanagaki Robuns (1829–1894) Bericht über das große Ansei-Erdbeben 1855 als Repräsentant des Genres der 'Katastrophendarstellungen'*, ed. Stephan Köhn, 2 vols (Wiesbaden: Harrassowitz, 2002), vol. 1, 37–44. Stephan Köhn provides a detailed genre history for disaster publications which he calls 'Katastrophendarstellungen'.

17 Ryūichi Narita, 'Kantō daishinsai no metahisutorī no tameni. Hōdō, aiwa, bidan', in *Kindai toshi kūkan no bunka keiken* (Tokyo: Iwanami Shoten, 2003), 197–213.

18 Ansgar Nünning, 'Making Crises and Catastrophes. Metaphors and Narratives Shaping the Cultural Life of Crises and Catastrophes', in Carsten Meiner and Kristin Veel (eds), *The Cultural*

historian Reinhart Koselleck, disasters must first be identified as events by providing a beginning and an end in retrospect.[19] According to Hayden White's notion on 'metahistory', historical events are then moulded into stories using known narrative traditions and tropes.[20] Disaster publications played an important role shaping those narratives, while they were subjected to established narrative traditions and genre conventions. This, in turn, shaped disaster experiences. Disaster publications assessed and mapped damage; narrated individual experiences of the disaster as tragedies, heroic tales, and absurdities that emerged from disaster; and concluded on a positive note with first reports on reconstruction. They were consumer goods, sold primarily as entertainment and served a certain voyeurism. But they also aspired to help the readers comprehend the incomprehensible tragedy, come to terms with the disaster, and move on. They gave cosmological or scientific explanations for the disaster and contextualised it in the long history of similar events.[21] They not only provided a chronological narrative of the disaster but also laid out the spatial extent of the disaster down to the level of city blocks. Thus, victims could situate their individual experiences within the historical event. In addition, they also included dramatic tales of individual experiences, so readers who were not present during the event were afforded a way to witness and participate in the disaster. Disaster publications therefore also took part in moulding disasters into collective experiences beyond the local level.[22] They could serve as an individual *lieux de mémoire* of the disaster that could be kept on one's own bookshelf.[23]

As will be discussed in more detail later in this chapter, many traditions of disaster reporting survived the transformation of print culture by the arrival of Western technologies in the Meiji period (1868–1912). Following the overthrow of the Tokugawa Shogunate in 1868, the threat of colonisation by

Life of Catastrophes and Crises. Facts, Forms, Fantasies (Berlin: De Gruyter, 2012), 59–88.

19 Reinhart Koselleck and Wolf-Dieter Stempel (eds), *Geschichte – Ereignis und Erzählung* (Munich: Fink, 1973), 560–71. On disasters as events, see Christiane Eifert, 'Das Erdbeben von Lissabon 1755', *Historische Zeitschrift* 274 (2002), 633–64.

20 Hayden V. White, *Metahistory. The Historical Imagination in Nineteenth-Century Europe* (Baltimore, MD: Johns Hopkins University Press, 1975).

21 Kitahara, *Nihon saigaishi*, 239. In the last volume of *Kaname'ishi*, a list of devastating historical earthquakes and cosmological explanations of the earthquake are given to present the described earthquake as a natural occurrence; it finishes by explaining its title (*Kaname'ishi* is the mythical stone that controls the earthquake-causing catfish) as an expression for the hope that the book would serve as a capstone helping the reader to move on from the tragedy.

22 Narita, 'Kantō daishinsai no metahisutorī no tameni', 205–13.

23 On the concept of memory space or *lieux de mémoire*, see Pierre Nora, 'Between Memory and History. Les Lieux de Mémoire', *Representations* 26 (1989), 7–24.

the Western powers was met by the new Meiji government with fundamental reforms introducing Western technology and institutions. Their goal was to build a strong military supported by modern industry and to gain Western recognition by meeting their standard of 'civilisation'.[24] This included the introduction of a modern press. Beginning in the 1870s, newspapers following Western models swept the markets and helped foster political debates. However, many woodblock printers still prevailed until the end of the nineteenth century and continued to provide illustrations.[25] After the end of the Russo-Japanese War in 1905, the Japanese press followed their Western counterparts in establishing a mass press, which was aided by the introduction of the rotation press and the effects of compulsory schooling established in the 1890s.[26] Disaster publications continued to play an important role, even for the development for the modern mass press: In 1923 the disaster publication *Taishō daishinsai daikasai* achieved a print run of a record-breaking 500,000 copies, paving the way for the first popular magazine to reach a million copies two years later by the same publisher, Kōdansha.[27] Since then, disaster publications have developed into glossy photo magazines that regularly appear after disaster events.[28] Visual representations, narrative genres, and tropes are explored below in greater detail.

The Catfish as an Earthquake Symbol

In the aforementioned catfish picture prints of the 1855 earthquake, the catfish symbolised the earthquake both for satirical purposes and for conveying certain interpretations of the disaster. The use of animals in

24 See, for example, Mark Ravina, *To Stand with the Nations of the World. Japan's Meiji Restoration in World History* (New York: Oxford University Press, 2017).

25 Itoko Kitahara, *Media kankyō no kindaika. Saigai shashin o chūshin ni* ('The Modernization of the Media Environment. With Emphasis on Disaster Photography') (Tokyo: Ochanomizu Shobō, 2012).

26 On the development of the mass press in Japan, see James L. Huffman, *Creating a Public. People and Press in Meiji Japan* (Honolulu: University of Hawai'i Press, 1997); Gregory James Kasza, *The State and the Mass Media in Japan, 1918–1945* (Berkeley: University of California Press, 1993); Amy Bliss Marshall, *Magazines and the Making of Mass Culture in Japan* (Toronto: University of Toronto Press, 2019).

27 Katsumi Iwasaki, 'Kaigenreika no besutoserā "Taishō daishinsai daikasai". 72 nengo no shohyō' ('The Bestseller under Martial Law "The Great Earthquake and Great Fire of the Taishō Period". A Review 72 Years Later'), *Masukomi-jānarizumu ronshū* 4 (1996), 56–84.

28 For the 2011 disaster, for example, Asahi Shinbunsha (ed.), *Asahi gurafu Tōhoku Kantō daishinsai. Ōtsunami to genpatsu* ('Asahi Graph on the Great Northeastern Disaster. The Great Tsunami and the Nuclear Power Plant') (Tokyo: Asahi Shinbunsha, 2011).

satire had a long tradition in Japan and was especially popular in the late Edo period.[29] The catfish had been associated with earthquakes since at least the seventeenth century and became subsequently established as a popular symbol for earthquake disasters. Apart from satire, the catfish prints also served as coping mechanisms derived from popular lore and vernacular religious beliefs on disasters. The lore has invited several explanations: It is believed that the catfish developed from a mythological dragon, which is located at the world's foundation in Buddhist cosmology and amalgamated with traditional local serpent deities.[30] Another popular explanation states that catfish were so sensitive that they could sense anomalies in the earth and thus were more likely to be caught shortly before earthquakes. This explanation was already offered in the Edo period and points to a popular practice of looking for unusual animal behaviour to predict earthquakes.[31] In either case, the catfish was also part of the Japanese animistic folk religion, Shinto, that assigned spirits and deities to sites in nature and natural phenomena.

The *namazu* lore thus operated within the Shinto myths: the catfish caused earthquakes through its movements. This was usually prevented by a keystone, the *kaname'ishi*, which restrains the catfish. Because the 1855 earthquake had happened in November, it was said that the Shinto deity that usually guarded the *kaname'ishi* had been away attending the annual gathering of deities in Izumo. Many 1855 catfish picture prints thus depicted how the catfish was reigned in again by the deity Kashima, who placed the *kaname'ishi* on top of it, as seen in the anonymous print *Namazu e no korashime* ('The Punishment for the Catfish', Fig. 12.1). Kashima is the bearded figure on the right of the catfish. The stone he is carrying has *kaname'ishi* written on it. This print combines this image with another popular recurring motif: one or several catfishes are beaten up by earthquake victims. In other cases, the catfish apologise for the mess they caused or even help rescue victims. It can be assumed that these prints served as a coping mechanism to overcome disaster. Some prints were even sold as charms and advertised to keep further earthquake damage away.[32]

29 Monika Jankiewicz-Brzostowska, 'Depictions of Animals in the Satirical War Prints of Kobayashi Kiyochika', *Art of the Orient* 9 (2020), 134–54, at 134–36.

30 Anna Andreeva, 'The "Earthquake Insect". Conceptualizing Disasters in Pre-Modern Japan', in Monica Juneja and Gerrit Jasper Schenk (eds), *Disaster as Image. Iconographies and Media Strategies Across Europe and Asia* (Regensburg: Schnell & Steiner, 2014), 81–90; Gregory Smits, 'Conduits of Power. What the Origins of Japan's Earthquake Catfish Reveal about Religious Geography', *Japan Review* 24 (2012), 41–65.

31 Smits, *Seismic Japan*, 182–88.

32 Smits, 'Shaking Up Japan', 1052–58; Formanek, 'Japanische Reaktionen auf (Natur-)Katastrophen', 56.

Figure 12.1 *Namazu e no korashime* ('The Punishment for the Catfish'), 1855. The University of Tokyo General Library, Ishimoto Collection, I-02-012. Source: University of Tokyo General Library

More often, however, the catfish picture prints of 1855 convey a reading of the disaster rooted in Chinese cosmology. Natural disasters were the expression of heavenly discontent with the ruling dynasty, and therefore both Chinese and Japanese rulers rushed to help disaster victims and aid with reconstruction. When the world was out of balance, divine intervention rectified it through natural disasters. For the urban population of the Edo period, imbalance especially meant inequalities of wealth. Urbanisation and trade had given rise to a class of wealthy merchants, upsetting the Confucian status order of the four occupations prevailing in the Edo period, which placed the merchants as the lowest status group behind the samurai, peasants, and the artisans and craftsmen. Therefore, many catfish picture prints depict flows of coins rectifying wealth distribution, for example catfish throwing coins. There are also other kinds of prints that pitted groups consisting of professions that profited from the disaster – mostly carpenters and other artisans involved in building houses, sometimes also prostitutes – against professions that lost.[33] This motif is also present in *Namazu e no korashime*: two carpenters, who can be identified by their typical black clothing, and a monk (who profited from giving memorial services) try to defend or plead for the anthropomorphic catfish in the middle

33 Smits, 'Shaking Up Japan', 1057–61.

who is beaten up by a book vendor, a merchant with his abacus, a female with a pipe, and a ghost (who has a flame where his feet should be) (Fig. 12.1). Employing satire, the catfish prints critiqued Tokugawa rule and society.

The *namazu* did not vanish with the end of the Edo period. They reappeared again as a symbol for heavenly rectification after the Great Kantō Earthquake in 1923, which happened during a period when consumerism became ubiquitous in urban areas and became debated as a negative effect of Western modernism and individualism.[34] The earthquake consequently caused a debate over whether this disaster was heavenly punishment for a society that had degenerated through decadent Western influences. Therefore, in cartoons that criticised consumerism the catfish appeared as a rectifying force threatening figures dressed in Western-inspired fashionable clothing pursuing frivolous activities.[35] Today catfish are enjoying a revival as government mascots, used to make disaster education more appealing.[36]

Visualising Threat

To visualise the experience of threat during an ongoing disaster, the early modern Japanese disaster prints offered dramatic and dynamic depictions of the ongoing disaster. These followed established visual tropes. Since fires were a typical consequence of earthquakes in Japan's crammed cities, victims running away from the flames became one major way of visualising earthquake disasters. An early example can be found in a disaster publication about one of the most notorious urban fires in early modern Japan, the Great Meireki Fire of 1657, *Musashiabumi* (literally 'Stirrup of Musashi', 1661) by Asai Ryōi (1612–1691) which was illustrated with uncoloured woodblock prints. In almost every image, people are depicted fleeing from the fire that engulfs the surrounding buildings. They are running agitatedly, some carrying luggage with them, which is also on fire. In

34 Harry D. Harootunian, *Overcome by Modernity. History, Culture, and Community in Interwar Japan* (Princeton, NJ: Princeton University Press, 2002).

35 J. Charles Schencking, *The Great Kantō Earthquake and the Chimera of National Reconstruction in Japan* (New York: Columbia University Press, 2013), 116–52; Gennifer Weisenfeld, *Imaging Disaster. Tokyo and the Visual Culture of Japan's Great Earthquake of 1923* (Berkeley: University of California Press, 2012), 178–89.

36 On the use of mascots and anthropomorphism in Japanese government institutions and companies, see Iain Macpherson and Teri Jane Bryant, 'Softening Power. Cuteness as Organizational Communication Strategy in Japan and the West', *Journal of International and Advanced Japanese Studies* 10 (2018), 39–55.

Figure 12.2 Kanagaki Robun, *Ansei kenmonshi* ('Account of Seen and Heard Things of the Ansei Period') (Edo, 1855), vol. 1. National Archives of Japan, Tokyo, 166-0416

the first disaster publication on earthquakes *Kaname'ishi* (1663), by the same author, architectural structures can be seen collapsing, for example a bridge which takes two people down with it. There is no fire present in the illustrations, probably because Kyoto had a tight-knit social fire control which prevented major fires.[37]

By the nineteenth century, the two depictions of disasters – collapsing building structures that bury people under them and people running from fires – were combined to create a dramatic effect. These include several such depictions by Utagawa Kuniyoshi in *Ansei kenmonshi* (1855) (Fig. 12.2). The illustrations are in colour, and the fires are depicted in bright red. Here, the spectator sees buildings shaken up and collapsing, indicated by flying debris around them. People can be seen buried in or narrowly escaping the rubble. The houses are burning and people are fleeing. Everything is happening simultaneously, knit together in a crowded, lively illustration.

37 Toshiaki Maruyama, *Kyōto no machiya to hikeshishū. Sono hataraki kishin no gotoshi* ('The Town Houses in Kyoto and the Fire Fighters. Their Deeds Resemble Gods') (Kyoto: Shōwadō, 2011).

The illustrations tell individual stories as well: on the right, a man is running out of the house naked, and another is carrying his elderly mother. These share similarities with the popular stories about individual experiences discussed below. This aggregation of the main earthquake destruction and fires was not realistic, as fires take some time to break out.[38] Nonetheless, this became a standard depiction of earthquake disasters.

The dramatic simultaneous depiction of earthquake and fire disaster was such a powerful and expected trope that it prevailed despite the problems the introduction of photography posed. Photography arrived with other Western technology in Japan in the mid-nineteenth century. Soon studios run by foreigners and Japanese sold photographs as souvenirs abroad and as attractions domestically.[39] However, as the long exposure times of early photography made it difficult to capture movement, woodblock prints still continued to flourish, as they allowed dynamic illustrations. This was especially true for depicting disasters. For example, after the 1891 Nōbi Earthquake, most pictures were taken for scientific purposes. A photo book of the earthquake was compiled with the help of John Milne, a leading British seismologist at the time, and mostly sold abroad.[40] At the same time, woodblock prints of the earthquake also circulated. In a woodblock print by Utagawa Kunitoshi (1847–1899), from the same artisan school as the illustrator of *Ansei kenmonshi* (1855), the familiar dynamic of buildings in the process of collapsing, bright red fires (now even more lively because of the use of aniline colours), and panicked people can be seen. Especially striking in this woodblock print is the emphasis on the collapse of modern architectural structures adopted from the West: on the right, a brick factory chimney, a symbol of modernity, is seen breaking off and crumbling, while a train is heading towards a collapsed iron bridge in the background.[41] As Gregory Clancey has elaborated, pointing to the failure of Western architecture was

38 From pre-war experiments, it is known that traditional Japanese wooden houses take approximately twenty minutes to be noticeably on fire. Uchida Yoshibumi, *Kenchiku to kasai* ('Architecture and Fires') (Tokyo: Sagami Shobō, 1942), 44. Theoretically, the scene could depict an aftershock, when already destabilised structures collapse, but this picture seems to show the main earthquake. For the study of historical seismology and aftershocks, see Emanuela Guidoboni and John E. Ebel (eds), *Earthquakes and Tsunamis in the Past. A Guide to Techniques in Historical Seismology* (Cambridge: Cambridge University Press, 2009).

39 Anne Tucker, Kōtarō Iizawa, and Naoyuki Kinoshita, *The History of Japanese Photography* (New Haven, CT: Yale University Press, 2003).

40 Kitahara, *Media kankyō no kindaika*, 40–55; John Milne and W. K. Burton, *The Great Earthquake in Japan, 1891* (Hong Kong: Lane, Crawford & Co., 1892).

41 Kunitoshi Utagawa, 'Gifu shigai ōjishin no zu' ('Figure of the Great Earthquake in Gifu City'), 1891. University of Tokyo Library, Ishimoto Collection II-02-009.

popular after the 1891 earthquake and led to a new appreciation of traditional Japanese artisanry.[42] Mirroring this, the traditional woodblock print was proven to be more efficient in depicting earthquakes in a dynamic manner than the then cumbersome Western photography.

By the time the Great Kantō Earthquake occurred in 1923, photography had mostly taken over as the main visual medium in the press. Photographs featured prominently in newspapers, magazines, and photo books. However, the Great Kantō Earthquake had caused damage to most of Tokyo's publishers or had at least toppled the type cases. Thus, news coverage stopped for several days. In this situation, reporting on disasters fell back to premodern patterns using a new medium: Postcards were printed in large quantities as they were cheap to produce. They were circulated widely and anonymously, as their premodern predecessors had been. This time, they dramatically showed the destruction and devastation of the earthquake and its aftermath, including piles of corpses.[43] Despite the advances in photographic technology, problems with capturing events in motion persisted. Therefore, illustrated prints and paintings remained popular ways to depict the Great Kantō Earthquake.[44] When photographs were used, they were manipulated and montaged, according to the publication's needs. For example, a quite peaceful scene of pedestrians was entitled 'People fleeing mindlessly while chased by the fire'. That photo was probably taken a few days after the disaster, when the smoke was still hanging over the city.[45] In other cases, the smoke was either montaged or painted in (in bright red) afterwards. In a postcard of the collapsed Western-style brick tower, the *Asakusa Jūnikai*, the first skyscraper in Japan and then a well-known landmark, a photograph of a row of people walking into the background was probably inserted on the right side (Fig. 12.3). Fire and smoke are painted in, while crumbling brick is added to the broken-off tower, reminiscent of the woodblock print illustrations of collapsing buildings. To deliver the visual cues for disaster established in the early modern period, 1923 postcard vendors had to combine photographs and illustration. When photography technology further improved, illustrations lost their importance. In 2011 the ubiquity of smart phones allowed for wide

42 Gregory Clancey, 'The Meiji Earthquake. Nature, Nation, and the Ambiguities of Catastrophe', *Modern Asian Studies* 40 (2006), 909–51.

43 Linhart, 'Die mediale Bewältigung', 76–84.

44 Weisenfeld, *Imaging Disaster*, 82–118.

45 Kōdansha (ed.), *Taishō daishinsai daikasai* (Tokyo: Kōdansha, 1923), 8 of the photography section.

Figure 12.3 Tatsuo Kimura and Toshio Ishi'i (eds), *Ehagaki ga kataru Kantō Dashinsai. Ishi'i Toshio korekushon* ('The Great Kantō Earthquake Told by Postcards. The Ishi'i Toshio Collection') (Tokyo: Tsuge Shobō, 1990), 11. Source: Tsuge Shobō, Tokyo

circulation of dramatic videos showing the ongoing disaster and people fleeing the approaching tsunami.

Spatial Distribution and Extent of Damage

One of the central roles of disaster prints was to give an orientation of the spatial extent of the disaster and to provide descriptions of how much damage had occurred. Descriptions of cities and tour guides were a popular genre in Edo literature, which is not surprising given the mobility between the capital and the provinces.[46] Disaster publications often adapted their tropes and techniques, making them disaster 'tour guides'. The three main techniques employed were visual representation on maps, lists of damage sorted by location, and descriptions structured as city walks.

The easiest way to visualise damage was to indicate it on maps. Since the Edo period, which was plagued by frequent urban conflagrations in its

46 Berry, *Japan in Print*, 139–84; Evelyn Schulz, *Stadt-Diskurse in den 'Aufzeichnungen über das Prosperieren von Tōkyō'* (*Tōkyō-hanjō-ki*). *Eine Gattung der topografischen Literatur Japans und ihre Bilder von Tōkyō* (*1832–1958*) (Munich: Iudicium, 2004).

crammed wooden cities, there had been a tradition to blacken or redden burnt areas on maps. Presumably, this developed from a pragmatic technique that allowed a quick distribution of disaster maps: woodblock printers simply printed with red over existing city maps that were often sold to allow visitors to orient themselves in the crammed chaotic neighbourhoods.[47] This form of visualising burnt areas was carried over to the modern period. The 1923 disaster publication *Taishō daishinsai daikasai* even utilises the early modern form: all inserted maps are printed in blue or black and the burnt or damaged areas are printed over them in red.[48] As Western disaster cartography employed colouring in a similar way, this form of visualising destruction is still prominently used today.[49]

The most common way to report on the intensity of damage and its spatial distribution was the use of lists. These were most often employed on *kawaraban* reporting on natural disasters. They listed damage by location. Japanese disaster historian Kitahara Itoko suspects they were leaked from official records and assessment of damage.[50] These damage lists were also printed in modern disaster publications, which reported disaster counts from the authorities: numbers of victims killed and destroyed houses, sorted by total and partial destruction. In *Taishō daishinsai daikasai*, the numbers are broken down to the district level, giving a very detailed account of the distribution and severity of destruction in the capital.[51]

The most popular way to give an impression of the damage within the urban space was to provide the narrative of a city walk. These were most common in disaster publications, which in the Edo period were structured as city walks. This reflects the connection between the act of speaking and measuring space through walking that historian Michel de Certeau has pointed out: The perception of space is structured by the pacing, trajectories, and places the walker ascribes meaning to (more popularly referred to as landmarks). This process echoes the act of speaking, which is also a tool of appropriating or constructing space in everyday life.[52] The narrated city

47 Kitahara, *Saigai jānarizumu*, 22–23.

48 Kōdansha, *Taishō daishinsai daikasai*, first map.

49 For example, an official map of the 1995 Kōbe Earthquake: Bosai Literacy Research Laboratory, 'Kōbeshi shigaichi chūshinbu de no higai', http://bosailiteracy.org/literacy/earthquake/house-damage/ (accessed 3 April 2023).

50 Kitahara, *Saigai jānarizumu*, 33–35.

51 Kōdansha, *Taishō daishinsai daikasai*, 1–27.

52 Michel de Certeau, *The Practice of Everyday Life* (Berkeley: University of California Press, 1984), 91–110.

walks in the disaster publications thus provided a spatial imagination of the disaster in which the reader could immerse themself.

Landmarks (*meisho*, literally 'famous places') also played an important role in Japanese urban literature, which ascribed stereotypical properties to places, popularising them in the process. As such, they became commonly known tourist destinations.[53] Disaster walks had their own landmarks, either as especially memorable places that suffered devastating losses or already famous landmarks that suffered damage or miraculously survived. The contrast between the existing image of a popular landmark in the minds of readers and how it was affected by the disaster provided a powerful indication of the severity of the disaster, although contrasting before and after depictions do not seem to have been popular before the twentieth century.[54]

The first known disaster publication on an earthquake, *Kaname'ishi* (1663) already had parts structured as a city walk. In its first volume, each chapter was anchored in a district of earthquake-stricken Kyoto, moving from northwest to southeast. The second volume surveys the damage in regions surrounding Kyoto before returning to southeast Kyoto. Each chapter provides loose descriptions of earthquake damage and stories of individual ordeals. These stories often take place at famous temples and landmarks that are known to the readers. Only the third volume alters the format to provide cosmological explanations and reactions of religious leaders. The later *Ansei kenmonshi* (1855) was more intricately structured. The title (literally 'a record of seen and heard things from the Ansei period') suggests it was conceptualised as a reportage. It is again structured by districts and results in a city walk starting from the hard-hit east to northeast of Edo, northeast to northwest in the second volume, and west to south in the last volume. It provides a greater variety of information, which was in large portion taken from *kawaraban* published on the earthquake.[55] *Ansei kenmonshi* provides lists of damage suffered in the respective districts and praises financial contributors to reconstruction. In between, there are collections of stories as well as a large variety of curious facts such as the workings of an earthquake-detecting machine.

In the twentieth-century disaster publication *Taishō daishinsai daikasai*, the structure anchored in space has given way to a structure dictated by narrative genres. However, the city tour is still an important structuring

53 Schulz, *Stadt-Diskurse*, 63–68.

54 *Taishō daishinsai daikasai* only shows one example of a tourist landmark before and after. Kōdansha, *Taishō daishinsai daikasai*, 74 of the photography section.

55 Kanagaki, *'Berichte über Gesehenes'*, vol. 1, 273.

element of these narratives. The city tour also got its own narrative genre: it entered a text genre comparable to Anglo-Saxon New Journalism, reportage narrated from the author's subjective point of view.[56] The city tour not only contains descriptions of the damage and of some individual tragedies but is also supported by the emotions expressed by the author. *Taishō daishinsai daikasai* includes a reportage that takes the reader on a disaster city tour to infamous disaster landmarks of mass death, such as an evacuation space in eastern Tokyo where 6,000 evacuees died in a firestorm.[57] Another mourns the lost famous landmarks of Tokyo.[58]

Narrating Fates of Individuals

The emotional core of disaster publications consisted of narrating the fate of individuals experiencing the disaster, the examples varying from tragic mothers losing their children to filial sons rescuing their fathers to miraculous survivals due to a supernatural phenomenon. In the Edo period, such narratives were sprinkled among other bits of information on a certain place within the disaster publication. Thus, the narratives were part of an entertaining variety of texts which represented the disaster story of a particular place. The stories were always tied to a specific space. Originally, they were probably orally transmitted (with questionable authenticity) and were collected to fit a certain expectation of narratives.[59] Over time, the stories were structured in increasingly standardised narrative genres. These were not exclusive to disasters but interacted with an emerging culture of telling and collecting stories in Edo Japan.[60]

These stories were used not only to evoke compassion by conveying individual tragedies and the catastrophes suffered by the victims, although this was maybe the most prominent genre (*aiwa*, literally 'sad stories'). There were also exemplary stories to praise morally upstanding or heroic acts (*bidan*, literally 'beautiful tales') which were intended to be morally

56 Matthew C. Strecher, 'From Kawaraban to Reportage. Toward a Theory for Japanese Literary Journalism', *Ritsumeikan Kokusai Kenkyū* 21: 3 (2009), 491–507.

57 Kōdansha, *Taishō daishinsai daikasai*, 33–46.

58 Kōdansha, *Taishō daishinsai daikasai*, 207–12.

59 In *Ansei kenmonshi*, some of the stories are given authenticity by claiming they were told by acquaintances of the author. In 1923 some magazines explicitly invited readers to send in *hiwa* or tragic stories. Narita, 'Kantō daishinsai no metahisutorī no tameni', 213–14.

60 Noriko T. Reider, '"The Emergence of 'kaidan-shū". The Collection of Tales of the Strange and Mysterious in the Edo Period', *Asian Folklore Studies* 60: 1 (2001), 79–99.

uplifting. Ghost stories and accounts of supernatural occurrences (*kaidan*, literally 'strange stories') were not only 'scary' but also provided closure by having ghosts take care of unfinished business. Finally, there were curious and peculiar stories which mainly provided entertainment and humour, such as miraculous survivals. In the following paragraphs, the development of the narratives is traced in the three aforementioned disaster publications, *Kaname'ishi* (1663), *Ansei kenmonshi* (1855), and *Taishō daishinsai daikasai* (1923).

In the seventeenth-century *Kaname'ishi*, the stories were distributed throughout the entire text. They culminated in short poems that were claimed to have been written by the stories' protagonists and often served as punch lines. The stories already featured typical tropes of the later genres but fluctuated between different kinds of emotional impacts, whereas the later standardised narratives served a single purpose. For example, tragic deaths in *Kaname'ishi* are always framed as deaths of beloved family members, a fixture in the moving *aiwa*. However, one *aiwa* also clearly fed voyeurism: a graphically described story about a woman dying while having her baby ripped out of her womb.[61] In another story, a child is crushed by a stone lantern in a temple, which is framed as tragic but also as heavenly punishment. The child had stolen food and behaved inappropriately in the temple before.[62] Other stories present entertaining curiosities, such as survivals: two people survive the disaster on the five-storied pagoda in Yasaka, which sways but does not collapse, but become very nauseous in the process.[63]

In all cases, the story is anchored in a specific place, in this case the intact pagoda as a disaster landmark. Many stories report strange or supernatural occurrences connected with the earthquake, such as animals changing behaviour or sightings of peculiar balls of light. In one third of the stories, the superstitiousness of the disaster victims or the helplessness of religious figures is pointed out as a curiosity. For example, because a shrine is rumoured to having been spared shocks during the earthquake, people in the vicinity rush to rip grass from the shrine's lawn as a talisman until the lawn is stripped bare.[64] *Kaname'ishi* already shows many themes regularly taken up by later disaster publications. It also uses a typical narrative structure leading towards a moralising conclusion or a punchline which was intended

61 Asai, 'Kaname'ishi', 21–22.
62 Asai, 'Kaname'ishi', 18–21.
63 Asai, 'Kaname'ishi', 30–32.
64 Asai, 'Kaname'ishi', 56–59.

to emphasise the exceptionality and curiosity of the occurrences witnessed during the disasters.

In *Ansei kenmonshi* of 1855, the stories are told in a more choreographed way and appear in loosely grouped collections inserted among the damage reports. The stories are still anchored in particular places and thus are not strictly sorted by genre, but it seems that one story was deliberately moved from the second volume to the third, where it starts a collection of stories highlighting good and bad human behaviour after the disaster.[65] About two thirds of the stories accordingly appear assembled in vaguely themed collections. The first volume lists several stories on miraculous survival, while the second volume has a collection of individuals that received warnings and survived as well as some stories of tragic deaths. The third volume assembles stories of strange supernatural occurrences and *bidan* about exemplary individuals displaying filial piety or loyalty to their superiors, disregarding their own situation. The stories are clearly intended for entertainment. There are *aiwa* reporting mass deaths or a story of a mother buried with her child with dramatic illustration. In *bidan*, although the protagonists are lauded as exemplary, the curiosities or ironies in the stories are highlighted. In one story, a man rescues his father buried by a collapsed house by cutting off his arm. What would normally be considered a crime is now an act of filial piety and rewarded with the pair's survival.[66] About one third of the stories concern apparitions and individuals predicting the earthquake, which leads to the survival of those who receive them. There are also ghost stories that follow typical patterns of *kaidan*: A stranger encounters a woman who entrusts him with a message for her father. Later, she turns out to be a ghost.[67] *Ansei kenmonshi* lists the stories concerning supernatural phenomena as equally authentic as other 'heard' stories, often citing acquaintances of the author as sources.

In the modern disaster publications circulating after the Great Kantō Earthquake of 1923, the narratives of individual fates were firmly grouped into their genres, and collections of those stories were titled with their genre names.[68] *Aiwa* collected emotional stories of torn-apart families but also mass death sites. The most important genre for the 1920s were the *bidan*, the 'beautiful tales', depicting exemplary individuals acting heroically during the disaster or fulfilling their professional duties. They often cross the genre

65 The story is announced in the table of contents for volume 2, which covers the Ueno district where the story is said to have taken place. Kanagaki, *'Berichte über Gesehenes'*, vol. 2, 26–27.

66 Kanagaki, *'Berichte über Gesehenes'*, vol. 2, 134–35.

67 Kanagaki, *'Berichte über Gesehenes'*, vol. 2, 162–67.

68 Kōdansha, *Taishō daishinsai daikasai*, 107–98.

boundaries to *aiwa*, because these heroic acts involve a disregard for individual safety.[69] For example, there is a story about a police officer who selflessly fulfils his duty while his entire family dies somewhere else. In the end, he dies heroically while saving children.[70] Others are exemplary in filial piety, rescuing elderly parents, or in their duty towards the emperor by sacrificing themselves while rescuing the emperor's portrait, which was venerated in schools in the context of State Shintō emperor worship. Given the values the *bidan* reflected, it is not surprising that they were later systematically collected and used in schools to educate children in moral values, supporting the ultra-nationalist values of the state, as historian Janet Borland has analysed.[71]

Various forms of entertaining stories were still present in *Taishō daishinsai daikasai*, which had a strong mass appeal but did not feature in other magazines which preferred *aiwa* and *bidan*. In *Taishō daishinsai daikasai*, the term *ibun* (literally 'strange hearsay') was invented to denominate them, but the stories show a greater variety. This indicates that they never formed a genre as specific as *aiwa* and *bidan* were. Nonetheless, the *ibun* show a greater continuity with Edo-period narratives because they often share an ironic or humorous punch line at the end. There were stories of miraculous survival. For example, a person survived the collapse of the Asakusa Jūnikai skyscraper because his shirt got stuck on a neon sign. Stories of seemingly heavenly intervention are told as curiosities. People declared dead scare their families who believe they are encountering ghosts but turn out to be very much alive. Another popular type of story included descriptions of business opportunities and professions profiting from the earthquake disaster, reminiscent of the Edo-period view that disasters rectified monetary imbalances. Still, ironies were important in these stories. For example, one describes how a temple that survived the fire made a lot of money from charm buyers while a water god temple nearby had burned down miserably.[72] While in the Edo period the explanation of supernatural phenomena was left ambiguous, the 1920s publication followed a scientific explanation of the disaster. Thus, the 'supernatural' stories in *Taishō daishinsai daikasai* were reduced to their humorous effect.

A variety of narratives were told about earthquake disasters in Japan, which were remarkably consistent over the centuries in subject matter and narrative structure. They aimed not only for compassion but also for

69 Narita, 'Kantō daishinsai no metahisutorī no tameni', 217–19.

70 Kōdansha, *Taishō daishinsai daikasai*, 141–44.

71 Janet Borland, 'Stories of Ideal Japanese Subjects from the Great Kantō Earthquake of 1923', *Japanese Studies* 25: 1 (2005), 21–34.

72 Kōdansha, *Taishō daishinsai daikasai*, 170–98.

catharsis. Some of them even professed educational aspirations or told morally uplifting tales. Many included ironies and humorous punch lines to provide relief from tragedy or voyeuristic entertainment. To the present day, collections of disaster stories can be bought as books, especially *aiwa* and *bidan*, as well as *kaidan*, the ghost stories.[73]

Conclusion

Disaster prints and publications in early modern Japan flourished due to the harsh censorship of the Tokugawa Shogunate on political discourse and developed a variety of visual representations, narrative genres, and tropes. There were several types of single-page woodblock prints, for which visual representations of the disaster played an important role. Most memorably after the Ansei Edo Earthquake in 1855, the earthquake was depicted as a catfish for satirical purposes and served as a coping mechanism rooted in folk religion. Illustrations of the disaster showed dramatic scenes of destruction and escape. The spatial extent of disasters was conveyed using maps and the degree of destruction using lists.

Tying those elements together, disaster publications in book form provided a first comprehensive assessment and narrative of the disaster. They were illustrated and structured as walks through the city destroyed by the earthquake and provided both a quantitative assessment of damage and narratives of individual experiences. Visual representations and narratives were tied to specific places and featured landmarks that were either already known to the readers or newly emerged as landmarks of the disaster. The narratives followed increasingly standardised patterns with different purposes: *aiwa* presented individual tragedies, *bidan* recounted exemplary deeds of morally upstanding individuals, and others were primarily entertaining, most notably (while other stories reported curiosities and absurdities caused by the disaster), tales of miraculous escapes or supernatural occurrences, or *kaidan*, ghost stories.

The Edo-period disaster narratives and motifs and their role in society were remarkably persistent beyond the introduction of Western thought and technologies in Japan. Most notably, the motif of people fleeing mindlessly

73 e.g. *Shinsai de hontō ni atta nakeru hanashi* (*Tear Inducing Stories That Really Happened During the Earthquake*) (Tokyo: East Press, 2011); Iku Asō, *Mae e. Higashinihon Daishinsai to tatakatta mumei senshitachi no kiroku* (*Move Forward. The Record of the Anonymous Fighters Who Fought with the Great East Japan Earthquake*) (Tokyo: Shinchōsha, 2011); Keisuke Udagawa, *Shinsaigo no fushigi na hanashi. Sanriku no 'kaidan'* (*Strange Stories after the Earthquake. The 'kaidan' from Sanriku*) (Tokyo: Asuka Shinsha, 2016).

from earthquake and fire depicted in dramatic interpretations was so powerful that until 1923, visualisations fell back on older illustration technologies or manipulated photographs to achieve the same effect. When modern printing technology failed, postcards were used in a similar way as the single-page woodblock prints in the Edo period. Other motifs and narratives also morphed into modern forms. In the case of city tours, the modern form of the reportage was employed. The narratives of individual experiences became even more differentiated and recognised as their own form of literary genres, such as *aiwa*, *bidan*, and *kaidan*.

The disaster prints served several purposes. They were commercialised entertainment, but they were also designed to narrate a comprehensive story of the disaster, to provide closure, to help victims see themselves as part of history, and to serve as *lieux de mémoire* for individual remembrance. The narratives also hint at various forms of perceiving and interpreting disasters. They were not only individual catastrophes but also macabre opportunities for entertainment or satire. In premodern Japan, earthquakes were also a cosmic rectifying force that balanced (monetary) injustice in society and punished evil acts. Although continuities from the premodern period survived Meiji modernisation, the Great Kantō Earthquake of 1923 marked a turning point: the earthquake was discussed as a punishment for an immoral westernised and commercialised society for a last time, but in its aftermath, disaster prevention by technology and scientific disaster education became ubiquitous, shifting the main interpretation of disasters to human failures.[74]

About the Author

Julia Mariko Jacoby is a historian of modern Japan specialising in environmental history and history of science. She earned her master's degree in modern and contemporary history, geology, and Latin at the University of Freiburg, where she also completed her PhD on 'Disaster Prevention in Japan 1885-1978. Natural Disasters, Scientific Expertise, and Global Transfers of Knowledge' in 2021. Her research has led her to the Universities of Osaka and Tokyo and to the Max Planck Institute for the History of Science in Berlin. She currently is a postdoc at the project *Cultures of Compromise* at the University of Duisburg-Essen.

74 Schencking, *The Great Kantō Earthquake*, 116–52; Julia Mariko Jacoby, 'Disaster Prevention in Japan 1885–1978. Natural Disasters, Scientific Expertise, and Global Transfers of Knowledge', PhD dissertation, Albert-Ludwigs-Universität Freiburg, 2021.

13 Breaking the Cycles of Catastrophe

Disaster, Time, and Nation in Dutch Flood Commemoration Books, 1757–1800

Adriaan Duiveman

Abstract

Historians argue that the eighteenth-century Dutch interpreted disasters in an overarching decline narrative. As such, catastrophes were understood as signs of an escalating political, economic, and moral crisis. However, this declensionist narrative was not the only interpretative framework that people could employ for dreadful events. This article traces the temporalities in four contemporary commemoration books on major Dutch flood disasters in 1757, 1775, and 1799. Their authors located recent inundations in time by comparing them to past catastrophes and imagining a future in which floods may or may not recur. The writers of two commemoration books recognised regular cycles of catastrophe, while the authors of the other two titles discerned an increase in the number of and the damage caused by inundations. Nonetheless, most authors provided their readers hope. Through either morality or technology, the writers asserted, people could prevent future catastrophes.

Keywords

floods – temporality – time – decline – memory – commemoration books

On the night of 14–15 November 1775, disaster came with the northwest wind. The combination of storm and spring tide led to a large-scale flood disaster around the Dutch South Sea (*Zuiderzee*). Houses, cattle, and people were taken by the water; many lives were lost. In the following year, two extensive commemoration books were published to memorialise the catastrophe. The *Historisch verhaal der overstroomingen in de Nederlanden* ('Historical

Asperen, van, Hanneke and Lotte Jensen (eds): *Dealing with Disasters from Early Modern to Modern Times. Cultural Responses to Catastrophes.* Taylor & Francis Group, 2023
DOI: 10.5117/9789463725798_CH13

Story of the Floods in the Netherlands') by anonymous authors and Johan Hendrik Hering's two-volume *Bespiegeling op Neêrlandsch Watersnood* ('Reflection on the Dutch Flood') collected stories from newspaper reports and private letters on the flood of 1775. On the basis of these documents, the authors reported in great detail on the recent events in the many places that had been struck. Nonetheless, the authors did not write about 1775 flood exclusively. Both commemoration books started with a long enumeration of flood disasters that had preceded the latest storm surge. Although the 1775 flood was exceptionally catastrophic, these books located the event in a long history that characterised the Dutch nation.

In this article I analyse four commemoration books from the second half of the eighteenth century, including the two I just mentioned. The two other commemoration books, dedicated to floods in 1757 and 1799, took a very different approach. The authors also referred to the history of floods, but they did not emphasise the stable recurrence of inundations. Instead, they argued that people could learn from disasters and that humans had the agency to prevent new catastrophes by employing technological measures. In the various commemoration books, I argue, one can find different temporalities. In addition, these sources provided an alternative to the decline narrative that was prominent in the eighteenth-century Dutch Republic.

Historians Petra van Dam and Harm Pieters provide valuable insights into the ways eighteenth- and nineteenth-century commemoration books contributed to the cultural memory of flood disasters. They show that the genre emerged in the eighteenth century, when authors compiled extensive books on a recent disaster to memorialise these events.[1] Nonetheless, van Dam and Pieters pass over fact that these books also referred to earlier inundations, thus locating the latest events in a longer history of calamities and risk. The history of temporalities, I contend, provides helpful analytical tools for understanding the remembrance of flood disasters in commemoration books.

Thus far, the historiography on early modern experiences of time has predominantly dealt with political discourses and actions.[2] Early modern environmental events have not yet been investigated in the context of

1 Petra J.E.M. van Dam and Harm Pieters, 'Enlightened Ideas in Commemoration Books of the 1825 Zuiderzee Flood in the Netherlands', in Pepijn Brandon, Sabine Go, and Wybren Verstegen (eds), *Navigating History. Economy, Society, Knowledge, and Nature* (Leiden: Brill, 2017), 275–97, at 275–76.

2 e.g. Tony Claydon, *The Revolution in Time. Chronology, Modernity, and 1688–1689 in England* (Oxford: Oxford University Press, 2020); Peter Fritzsche, *Stranded in the Present. Modern Time and the Melancholy of History* (Cambridge, MA: Harvard University Press, 2004); J.G.A. Pocock,

temporalities. Phenomena associated with the seasons, like snow and the harvest, fit a cyclical time frame.[3] However, in cultural landscapes like those in the Dutch Delta, the environment was partly made by its human dwellers. As we will see, some authors of commemoration books also recognised this fact. Furthermore, the majority of early modern Europeans interpreted nature-induced disasters as divine punishments, caused by the sins of people.[4] Seasonal high water was expected as part of the annual cycle – especially in the river areas. However, many eighteenth-century commentators explained exceptionally severe inundations as events that were caused by moral failings. Whether via technology or via religion, early modern Dutch authors recognised the environment as a 'space for human action'.[5]

During the eighteenth-century, Dutch moralists argued that their country experienced general decline.[6] Economic malaise, political unrest, and divinely ordered punishments were seen as its symptoms, but at its core, eighteenth-century Dutch moralists argued, lay a deep moral crisis.[7] Historian Adam Sundberg and literary scholar Lotte Jensen find this decline narrative in texts on disasters in the eighteenth century.[8] With regard to

The Machiavellian Moment. Florentine Political Thought and the Atlantic Republican Tradition (Princeton, NJ: Princeton University Press, 1975).

3 For the perceived 'timelessness' of nature, see Sabine Hofmeister, 'Nature's Temporalities. Consequences for Environmental Politics', *Time & Society* 6: 2–3 (1997), 309–21, at 312.

4 Adam D. Sundberg, 'Claiming the Past. History, Memory, and Innovation Following the Christmas Flood of 1717', *Environmental History* 20: 2 (2015), 238–61, at 242–45; Jan Willem Buisman, *Tussen vroomheid en verlichting. Een cultuurhistorisch en -sociologisch onderzoek naar enkele aspecten van de Verlichting in Nederland (1755–1810)*, vol. 1 of 2 vols (Zwolle: Waanders, 1992), 225–65.

5 Sara Miglietti and John Morgan, 'Introduction. Ruling "Climates" in the Early Modern World', in Miglietti and Morgan (eds), *Governing the Environment in the Early Modern World. Theory and Practice* (Abingdon: Routledge, 2017), 1–21, at 4.

6 This decline narrative does not fit solely into one of the two temporalities, linear or cyclical. The rise and fall of states was regarded by some authors as an inevitable cycle, albeit on a different time scale than the changing of seasons. However, a contrasting optimism about the ability of humans to break the cycle revealed a linear experience of time. See Jan Rotmans, 'Circles of Desire and the Corruption of Virtue. The Historical Thought of Cornelis Zillesen (1736–1828)', *De achttiende eeuw* 46: 2 (2014), 79–105.

7 Ernst H. Kossman, 'The Dutch Republic in the Eighteenth Century', in Margaret C. Jacob and Wijnand W. Mijnhardt (eds), *The Dutch Republic in the Eighteenth Century. Decline, Enlightenment, and Revolution* (Ithaca and London: Cornell University Press, 1992), 19–31.

8 Adam D. Sundberg, *Natural Disaster at the Closing of the Dutch Golden Age* (Cambridge: Cambridge University Press, 2022), 196–97, 203, 242; Lotte Jensen, '"Disaster upon Disaster Inflicted on the Dutch". Singing about Disasters in the Netherlands, 1600–1900', *BMGN – Low Countries Historical Review* 134: 2 (2019), 45–70, at 65–66.

Hering's commemoration book, Pieters argues that the author interpreted the recent disaster as a symptom of this imagined deterioration of the Republic.[9] Although this was a popular analysis at the time, I am convinced that the decline was neither the leading story arc of Hering's book nor that of the other authors. When they mentioned general decline of the Dutch nation, they did so only in passing.[10] Despite the lack of a decline narrative, some authors did imagine a Dutch nation with a history and a future. The 'fatherland', as they called it, existed in time.

My argument unfolds in five sections. In the first, I dwell briefly on the two strands of literature I engage with: history of temporalities and, in relation to that, the study of national identities. In the next, I analyse the two commemoration books written by so-called civil scientists. As I will show, these books operated in a linear temporality. The third and fourth sections investigate the two commemoration books on the 1775 flood. They provided readers with a cyclical history of flood disasters, presenting severe inundations as events that repeated every century. In the final section, I will analyse how the various commemoration books dealt with the concept of nationhood. Commemoration books connected the suffering of ancestors, contemporaries, and posterity.

Time, Nation, and Continuity

The flood commemoration books engaged with contemporary notions of time and nation in various, even contradictory, ways. In order to grasp this dynamic, we have to take a closer look at two strands of theory: the work on temporalities – i.e. experiences of time – and the research on nations as imagined communities.

Scholars have argued that the European experiences of time shifted during the eighteenth century.[11] Before then, Europeans regarded history

9 Harm Pieters, 'Herinneringscultuur van overstromingsrampen. Gedenkboeken van de overstromingen van 1775, 1776 en 1825 in het Zuiderzeegebied', *Tijdschrift voor waterstaatsgeschiedenis* 21: 1–2 (2012), 48–57.

10 Cornelis Zillesen, *Beschryving van den Watersnood van 't jaar MDCCXCIX., in verscheidene gedeelten van ons Vaderland door ysverstoppingen veroorzaakt, met Kunst-plaaten* (Amsterdam: Johannes Allart, 1800), vii–viii; Johan Hendrik Hering, *Bespiegeling over Neêrlandsch watersnood, tusschen den 14den en 15den Nov. MDCCLXXV*, vol. 1 of 2 vols (Amsterdam: Wed. Loveringh en Allart, 1776), 66; Lidia van der Souw-ten Hoven, 'Beschouwing over den zwaaren storm en hoogen watervloed', in Hering, *Bespiegeling*, vol. 1, 33–38, at 33.

11 e.g. Reinhart Koselleck, *Futures Past. On the Semantics of Historical Time*, trans. Keith Tribe (New York: Columbia University Press, 2004), 266–69; Eric Hobsbawm, *On History* (New York: The

predominantly as cyclical. Just like the natural passing of seasons, what happened in human societies in the past would just happen again. Changes occurred, of course, but they were never deemed radical or unprecedented. Because of that, people could project the past onto the future. In the eighteenth century, this sense of time moved to the background. Confronted with fundamental ruptures in all aspects of society – science, religion, politics – Europeans adopted an understanding of history as a linear timeline on which radical change was possible.[12] French authors of 'philosophical histories' – a typical Enlightenment genre – therefore constructed linear timelines on which civilisations and nations experienced periods of decline yet predominantly progress.[13] Writers claimed that, since people's future was open, humans could choose what it held.

Despite the radical ruptures that caused it, the shift of conceptions of time was a subtle one. Historian Judith Pollmann convincingly argues that even before the eighteenth century, Europeans were well aware of historical breaking points.[14] In the sixteenth century, she shows, people also perceived and discussed discontinuities. Likewise, a stable, cyclical understanding of time did not disappear in the eighteenth century.[15] Cyclical and linear notions of time coexisted, sometimes even in a single text or image.[16]

Furthermore, the distinction between the two temporalities is not always clear-cut. The general decline narrative that entertained the enlightened minds of the Dutch Republic, for instance, combined elements of both understandings of time.[17] While the downturn was viewed as a cyclical

New Press, 1998), 25; Lynn Hunt, *Measuring Time, Making History* (Budapest: Central European University Press, 2012), ch. 2: 'Modernity and History', 47–91, at 51–53, http://books.openedition.org/ceup/820. Historian Peter Burke locates the change in the seventeenth century. See his *The Renaissance Sense of the Past* (London: Arnold, 1969), 149–50.

12 Burke, *The Renaissance Sense*. Such a linear understanding of time was also present in Judeo-Christian texts, stretching history from the Fall to the unavoidable Apocalypse. Nonetheless, Renaissance authors and thinkers emphasised cyclical understandings of time. See the same, 87–89.

13 Siep Stuurman, 'Tijd en ruimte in de verlichting. De uitvinding van de filosofische geschiedenis', in Maria Grever and Harry Jansen (eds), *De ongrijpbare tijd. Temporaliteit en de constructie van het verleden* (Hilversum: Verloren, 2001), 79–96.

14 Judith Pollmann, *Memory in Early Modern Europe, 1500–1800* (Oxford: Oxford University Press, 2017), 51–72.

15 Claydon, *The Revolution in Time*, 14.

16 Brecht Deseure and Judith Pollmann, 'The Experience of Rupture and the History of Memory', in Erika Kuijpers, Judith Pollmann, Johannes Müller, and Jasper van der Steen (eds), *Memory before Modernity. Practices of Memory in Early Modern Europe* (Leiden: Brill, 2013), 315–29, at 318; Helge Jordheim, 'Against Periodization. Koselleck's Theory of Multiple Temporalities', *History and Theory* 51: 2 (2012), 151–71.

17 Rotmans, 'Circles of Desire', 79.

return to the period before the seventeenth-century Golden Age, authors also argued that the future was open for progress.[18] In other words: the cycle was not inevitable. These nuances show that the two temporalities are ideal types. They are scholarly constructs that help us to grasp an observable shift in the ways eighteenth-century people understood time. Yet the muddied thinking of contemporary writers seldom operated in one temporality only.

Interestingly, the notion of co-existing temporalities has also been picked up by academics with another specific research topic: national identities. Several scholars assert that the concept of a nation exists by the grace of a linear temporality.[19] National myths refer to a lost golden age or a glorious future. Geographer Nuala Johnson, however, argues that nations, as imagined communities, can build on both 'linear and recurrent time formulations'.[20] What matters to the members of nations is a sense of continuity: the idea that a community moves as one entity through time.[21] This continuity can be found in a transformation narrative but also in the stability of tradition. Recent studies by social psychologists confirm Johnson's claim. They establish that there are two ways that members of groups envision their collective continuity: cyclical and permanent, on the one hand, and linear and transforming, on the other.[22] These two temporal perspectives on the Dutch nation can also be traced in some of the commemoration books.

Learning from Past Disasters: Pierlinck and Zillesen

Jacob Pierlinck was a captain-lieutenant and engineer. When a flood hit in 1757, Pierlinck lived in 's-Hertogenbosch, a city in the south of the Dutch Republic. In the days that followed, victims from the surrounding areas

18 Eleá de la Porte, 'Polite Batavians. The Uses of the Past in Late-Eighteenth-Century Dutch Spectators', in *Discourses of Decline. Essays on Republicanism in Honor of Wyger R.E. Velema*, ed. Joris Oddens, Mart Rutjes, and Arthur Weststeijn (Leiden: Brill, 2022), 87–88, https://doi.org/10.1163/9789004470651.

19 Hakkı Taş, 'The Chronopolitics of National Populism', *Identities* 29: 2 (2020), 127–45, at 127–35; Carsten Humlebæk, 'National Identities. Temporality and Narration', *Genealogy* 2: 4 (2018), 36–53, at 40–41.

20 Nuala C. Johnson, 'From Time Immemorial. Narratives of Nationhood and the Making of National Space', in John May and Nigel Thrift (eds), *TimeSpace. Geographies of Temporality* (London and New York: Routledge, 2001), 89–105, at 92.

21 Benedict Anderson, *Imagined Communities. Reflections on the Origin and Spread of Nationalism* (London: Verso, 2006), 11–12.

22 Fabio Sani et al., 'Perceived Collective Continuity. Seeing Groups as Entities That Move through Time', *European Journal of Social Psychology* 37: 6 (2007), 1118–34, at 1119–20.

sought refuge in the town. Pierlinck felt sympathy for the wretches. The presence of all this suffering in his city and its surroundings led Pierlinck to produce the commemoration book, he claimed. The events had 'evoked in me the zeal – which I am indebted to this number of unfortunate people amidst I live – to investigate all these disasters in their origins'.[23] For Pierlinck, by knowing these 'origins' – the causes – people could act to prevent such events from happening again.

Tax officer, inventor, and hydrologist Cornelis Zillesen made similar claims in his commemoration book on the inundations of 1799. '[T]here is no effect without cause', he wrote, 'so the investigator of nature should, as much as possible, trace how causes and effects relate. ... A superficial historical narrative is of no use in this regard, but a story that benefits the experts of nature would be [helpful]'.[24] Zillesen's and Pierlinck's aim – tracing the natural circumstances that created river floods – manifests in the information they provided. Both writers tried to figure out, for instance, where water swelled in rivers. For this reason, both commemoration books discussed at length the water levels at various places along the rivers at various moments leading up to dyke breaches.

Pierlinck and Zillesen fit the category of eighteenth-century 'civil scientists': amateur scholars who wanted to serve society by studying practical topics and producing useful inventions.[25] Their publications, accordingly, engaged with a technocratic dialogue on floods.[26] Yet the books of both authors were not just dry enumerations of data. They also tapped into the genre conventions of commemoration books and appealed to the audience that wanted vivid descriptions of destruction. To illustrate the disastrousness of the recent events, Pierlinck and Zillesen told the personal stories of survivors. Anecdotes appeared often in early modern literary descriptions of floods and representations of disasters from that period in general. Literary

23 'hebben my, uit een yver, welken ik meen, aan zulken getal van ongelukkige Menschen, in welker midden ik woone, verschuldigd te zyn, opgewekt, om alle die Rampen in hunnen oirsprong na te vorschen ...' Jacob Pierlinck, *De verschrikkelyke watersnood, langs de Rivieren de Waal en de Maas, voorgevallen in de Maand February des Jaars 1757* (Amsterdam: Isaak Tirion, 1757), 2.

24 Free translation from the original: 'want is er geen uitwerkzel zonder oorzaak, zo diend de natuuronderzoeker, zo veel mogelijk op te speuren, hoe oorzaaken en uitwerkzelen in hun verband werken ... Een oppervlakkig Geschiedverhaal kan ons daarom in dezen van geen nut zijn, maar wel een zodaanig verhaal, waar uit Natuurkenners nut trekken kunnen'. Zillesen, *Beschryving*, 43.

25 Ad Maas, 'Civil Scientists. Dutch Scientists between 1750 and 1875', *History of Science* 48: 1 (2010), 75–103, at 75–76; Rotmans, 'Circles of Desire', 84.

26 Sundberg, 'Claiming the Past', 241, 249–50.

scholar Françoise Lavocat argues that these mini-narratives functioned as *pars pro toto*: they captured a large, destructive event in the specific experiences of one individual or a couple of people.[27] At the same time, these stories evoked pity in the readers. Zillesen introduced the dramatic story of a father who lost a child in the waves as a 'heart-rending case'.[28] Evoking emotion was a conscious goal.

The emotionally charged anecdotes contrasted stylistically with the scientific disquisitions on water levels.[29] However, in the two commemoration books, the touching stories strengthened the technological argument. Pierlinck argued that he chronicled all the misery to show to 'all our compatriots in which dreadful situation their fellow brethren' were brought by the floods.[30] Subsequently, he stated that he also wanted to tell how these calamities could be prevented. Emotional anecdotes evoked thus a sense of urgency for Pierlinck's technological plans. Zillesen does not make this connection so explicitly, but his moving stories would have had a similar legitimising effect.

Another way that Pierlinck proved the urgency of technological interventions was by referring to the past. He included a list with major floods in the river area since 1653. The inundation of 1757 was the twelfth time that such an event had occurred, Pierlinck claimed. 'Who does not fear for more severe consequences', the author asked rhetorically, 'after paying only little attention to this short register in which one sees in at once that – especially since the year 1740 – this corrupting malady increases?'[31] Here Pierlinck drew a rising line of disastrousness.[32]

Zillesen also referred to previous disasters. Like Pierlinck, he included a short list with historical floods along the major Dutch rivers, although he went further back in time. The first flood in his list was one in 1573.[33] And like Pierlinck, he saw a worrying rise in the number of disasters. 'It is highly remarkable', he concluded, 'that the [number of] ice congestions and

27 Françoise Lavocat, 'Narratives of Catastrophe in the Early Modern Period. Awareness of Historicity and Emergence of Interpretative Viewpoints', *Poetics Today* 33: 3–4 (2012), 254–99, at 265.

28 Zillesen, *Beschryving*, 69.

29 For Zillesen's use of literary techniques, see Lotte Jensen, 'De schoonheid van een ramp. De culturele verbeelding van de watersnood van 1799', *Vooys* 39: 1 (2021), 6–17, at 9–10.

30 Pierlinck, *De verschrikkelyke watersnood*, 33.

31 'Wie vreest niet voor droeviger gevolgen, die zynen aandacht, maar eenigzins, op dit klein Register vestigt; waar uit men met een opslag ziet, dat deze zoo bederflyke kwaal, hoe langer hoe meer, toeneemt, byzonderlyk zedert den Jaare 1740'. Pierlinck, *De verschrikkelyke watersnood*, 34.

32 See also Pierlinck, *De verschrikkelyke watersnood*, 39.

33 Zillesen, *Beschryving*, 38–41.

inundations that were caused by them, has increased in this century'.[34] The increase was the result of obstacles in the river, Zillesen speculated. One of the barriers he listed were groynes (*kribben*), man-made structures that prevented rivers from eroding their banks.[35]

Zillesen and Pierlinck dealt with time in a similar way. Although Zillesen went further back in time than Pierlinck, the long history of floods receives little attention in either commemoration book. When Zillesen compared the latest disastrous water levels to those of previous floods, he referred most often to recent inundations: 1784 and 1795.[36] Similarly, Pierlinck consistently compared the disaster of 1757 to a flood in 1740. The timespan in which the latest disaster fell was, in both commemoration books, relatively short.

Regarding the future, both authors attributed agency to humans. They showed that disasters were preventable and that the future was therefore malleable. As has been mentioned before, both commemoration books are relatively secular. Zillesen denied even explicitly that floods were the result of divine intervention. According to the engineer, river floods were 'not originally the judgements of God'.[37] Furthermore, if people would allow rivers to inundate their lands every winter, the fields would actually become more fertile.[38] By keeping the water out at all costs by building dykes and constraining the space of the river, Zillesen argued, people were actually creating the danger of floods.[39] In a sense, the engineer already envisioned ways of dealing with high water that are now applied in the Dutch water management project *Space for the River* (*Ruimte voor de rivier*).[40] In Zillesen's view, it was human agency that created catastrophe, and it was human agency that could solve the problem.[41]

Like Zillesen, Pierlinck was also sceptical about the preventive qualities of dykes.[42] People should always be on guard for floods, he wrote, and

34 Zillesen, *Beschryving*, 41.

35 On this groynes problem, see Toon Bosch, *Om de macht over het water. De nationale waterstaatsdienst tussen staat en samenleving, 1798–1849* (Zaltbommel: Europese Bibliotheek, 2000), 22; Alex van Heezik, *Strijd om de rivieren. 200 jaar rivierenbeleid in Nederland of de opkomst en ondergang van het streven naar de normale rivier* (Haarlem and The Hague: Van Heezik Beleidsresearch, 2008), 46–47.

36 Zillesen, *Beschryving*, 78–79, 98, 148, 160, 167–68, and 271.

37 Zillesen, *Beschryving*, xiii.

38 A similar argument can be found in Pierlinck, *De verschrikkelyke watersnood*, 47.

39 Interestingly, Hering also acknowledged these man-made risks for floods but did not elaborate on them. See his *Bespiegeling*, vol. 1, 43.

40 These proposals contrast with the projects to 'tame the rivers' in Germany in the late eighteenth and early nineteenth centuries. See David Blackbourn, *The Conquest of Nature. Water, Landscape and the Making of Modern Germany* (London: Jonathan Cape, 2006), chs 1 and 2.

41 Cf. Heezik, *Strijd*, 48–50.

42 Pierlinck, *De verschrikkelyke watersnood*, 8–9.

they should repair and maintain dykes and dams. However, the recent disaster showed that these measures were not enough. With regard to the area around 's-Hertogenbosch, he argued for the creation of an emergency spillway (*overlaat*) that would reduce high water levels.[43] The spillway would redirect excess water to the Biesbosch, a large wetland in the southwest of the Netherlands. In spite of his scepticism towards dykes and dams, the author did provide a technological solution that could prevent misery in the future.

The commemoration books of Zillesen and Pierlinck operated in a linear temporality. Their past was a time of change, and their future was open and malleable. The emphasis on scientific investigation and the technological prevention of disasters was not unique to the Enlightenment, as some scholars have shown.[44] Furthermore, a secular understanding of floods which emphasised natural causes also existed before the eighteenth century.[45] Yet, I think it is undeniable that Zillesen and Pierlinck manifested a mindset that historians associate with the era: rational, curious, and forward thinking. However, between the publication dates of these two commemoration books, two other commemoration books were written which took a completely different approach. Instead of forward, they looked backward in time.

Catastrophic Cycles in the *Historisch verhaal*

The *Historisch verhaal der overstroomingen in de Nederlanden* (hereafter *Historisch verhaal*) was sold in two volumes. While the publisher distributed the first, the anonymous authors – they referred to themselves as 'we' – were writing the second.[46] In the preface the authors thanked the people who had already sent them 'various neat and extensive messages from many places'.[47]

43 Pierlinck, *De verschrikkelyke watersnood,* 40.

44 Adriaan M.J. de Kraker, 'Two Floods Compared. Perception of and Response to the 1682 and 1715 Flooding Disasters in the Low Countries', in Katrin Pfeifer and Niki Pfeifer (eds), *Forces of Nature and Cultural Responses* (Dordrecht: Springer Netherlands, 2013), 185–202, at 199–200; Marie Luisa Allemeyer, *'Kein Land ohne Deich ... !' Lebenswelten einer Küstengesellschaft in der frühen Neuzeit* (Göttingen: Vandenhoeck & Ruprecht, 2006), 371–83.

45 Raingard Esser, '"Ofter gheen water op en hadde gheweest". Narratives of Resilience on the Dutch Coast in the Seventeenth Century', *Dutch Crossing* 40: 2 (2016), 97–107, at 99.

46 Pieters argues that the second volume was never published. See Pieters, 'Herinneringscultuur', 49. Nonetheless, I found a book in the KB – National Library of the Netherlands in The Hague (KW 656 M 57) that includes both volumes.

47 'verscheidene nette en uitvoerige berigten uit vele plaatsen ontvangen', in *Historisch Verhaal der overstroomingen in de Nederlanden, byzonder op den 14 en 15 van Slagtmaand des Jaars 1775 voorgevallen* (Amsterdam: G. Warnars and P. den Hengst, 1776), ch. 'Voorberigt'.

The writers asked other readers to send in reports on the inundations at their locations, so they could write an even more extensive second volume.[48] By collecting the events of the night of 14–15 November 1775 in one book, the authors emphasised the geographical scale of the disaster. The dyke breaches, the drowning cattle, the stories of families that sought refuge on their roofs – these events happened at various places but all at the same time. The storm surge did not hit the whole Dutch Republic. Because of their geographical location, provinces like Groningen were spared.[49] Nonetheless, the authors of the *Historisch verhaal* claimed that the disaster hit 'the fatherland'. The specific stories of families on roofs and other anecdotes that took place simultaneously functioned as *pars pro toto* of the storm surge that hit the whole nation.

Besides the geographical scale, the authors of the *Historisch verhaal* also stretched the disaster's temporal scope. The book did more than just recapitulate recent news. The full title translates to 'Historical Story of the Floods in the Netherlands, in Particular [the One] on the Fourteenth and Fifteenth of the Slaughter Month of the Year 1775'. The words 'in particular' were telling. Although the authors dedicated most space to the recent storm surge, they also extensively discussed earlier flood disasters. The first section (*afdeeling*) of the *Historisch Verhaal* was an enumeration of flood disasters in the Dutch delta since 806, when sea water flooded the fields of Frisia on St Thomas Day.[50] Unlike Pierlinck and Zillesen, who mainly mentioned recent disasters, the anonymous authors covered no fewer than nine centuries of inundations. The overview of historical floods ended in the year 1747 – 28 pages later – with a storm that 'gave birth to very large calamities in our fatherland'.[51] In between, the authors provided their readers with a canon of major flood disasters.

A canon implies selection.[52] The authors wrote that they had to pick events, since they could not include all these catastrophes. For instance, when discussing the thirteenth century, the authors asserted:

> Although our space does not make it possible to remember [*op te halen*] all the exceptional drawbacks which we experienced because of the water

48 Ibid.

49 *Historisch Verhaal*, 230–31.

50 *Historisch Verhaal*, 4.

51 *Historisch Verhaal*, 30–31. After the enumeration, the author closes the section with a couple of pages with geographical information.

52 On flood disaster canons, see also John Emrys Morgan, 'Understanding Flooding in Early Modern England', *Journal of Historical Geography* 50 (2015), 37–50, at 46; Pieters, 'Herinneringscultuur', 49.

in one or the other part of our fatherland, we cannot silently pass over the Marcellus Flood that particularly hit Frisia.[53]

The St Marcellus Flood was considered to be the most severe flood disaster of the century. However, the authors did not have objective measures for severity, especially not when they described disasters from centuries ago. They had to rely on collective memory, such as earlier disaster canons. It is therefore no surprise that the authors included the St Elisabeth Flood of 1421. Modern-day research shows that the flood was not the sudden, catastrophic event as it was remembered in later centuries.[54] The authors of the *Historisch verhaal* nonetheless claimed that it had 'surpassed most other severe floods'.[55]

The last quote shows that disaster canons enabled comparisons between floods.[56] For instance, in their description of a storm surge in March 1625, the authors of the *Historisch verhaal* argued that the water in Amsterdam rose one *duim* – a old measure of length which compares to an inch – above the level of the All Saints Flood of 1570.[57] This is one of many comparisons to earlier inundations that were made. In their introduction the authors argued that weighing severity was the main reason to include a historical overview. The information on previous flood disasters enabled their readers to 'compare this [recent] flood disaster with those of other epochs'.[58]

The authors deemed the catastrophe of 1775 exceptionally severe. Nonetheless, the recent catastrophe was embedded in the past. As noted before, a twenty-eight-page list with historical inundations preceded the reports on the storm surge. After that, the authors also referred to a village or town's experiences with previous flood disasters when describing recent events. For instance, when discussing the impact of the storm surge in Scheveningen, a coastal village near The Hague, the authors started with a description of a plaque in the local church that commemorated the 1570 All Saints Flood.[59] Hereby, they showed that it was not the first time that Scheveningen was hit by a storm surge.

53 'Schoon ons bestek het niet mogelyk maakt alle byzondere nadeelen, welke wy door de wateren, nu in het eene dan in het andere gedeelte van ons Vaderland, veroorzaakt vinden, op te halen, kunnen wy egter den Marcellus Vloed, door welken Friesland in het byzonder is getroffen geworden, niet met stilzwygen voorbygaan', in *Historisch verhaal*, 9.

54 Gerrit Jan Schiereck and Paul Visser, 'De Sint-Elisabethsvloed. Feiten en fictie', in Hanneke van Asperen, Marianne Eekhout, and Lotte Jensen (eds), *De grote en vreeselike vloed. De Sint-Elisabethsvloed 1421–2021* (Amsterdam: De Bezige Bij, 2021), 139–54.

55 *Historisch verhaal*, 12.

56 de Kraker, 'Two Floods Compared', 198.

57 *Historisch verhaal*, 19.

58 'dezen Waternood met dien van andere tyden te vergelyken'. *Historisch Verhaal*, 3.

59 *Historisch Verhaal*, 43–44.

In the *Historisch verhaal*, flood disasters are recurring events, with major inundations happening several times in a century. Because of that, the danger of the water is a stable threat in the lives of the Dutch. The authors did not present a story arc in the historical overview nor in the reports that followed it. Floods differed in severity, but the authors did not draw rising or declining lines. In other words: the authors did not claim that floods were becoming less or more catastrophic over time. The anonymous authors presented disasters thus as cyclical events. When discussing the dyke breaches in the polder of Dussen, they noted that here 'one was afraid that the despair of the year 1421 – when everything was inundated – would be reborn'.[60] According to the authors, the inhabitants of the polder feared for a flood disaster because of a similar catastrophe over three centuries ago. They feared repetition.

The authors wrote about the distant past and the recent events, but they said little about the future. Nonetheless, in the postscript they do envision the time to come. Just like the past, this future was strikingly cyclical. The authors closed the book with 'the whole-hearted wish that these provinces [*Gewesten*], by the protection of Divine Providence, may be protected against these disasters for long'.[61] The words 'for long' imply that the authors did not think that the inhabitants of the Dutch Republic would ever be safe from flood disasters. Even though God could postpone catastrophes, it was only a matter of time before the next inundation arrived.

Hering's Monument for Posterity

In many ways, the *Historisch verhaal* and Hering's *Bespiegeling* are similar. Just like the anonymous authors, he preceded the recent events with an extensive historical overview, and he also links the stories from various places to their local histories. And like the anonymous authors, he portrayed the disaster as an event that hit the whole 'fatherland'. However, Hering's commemoration book diverged in two key ways from the *Historisch verhaal*: it put the disasters in a far more explicitly religious frame, and it adopted a more literary style. Hering regarded his book as a literary work of art.

60 'Men was bedugt dat hier de ellenden van het Jaar 1421, toen 'er alles onder liep, zouden herboren worden'. *Historisch Verhaal*, 166.

61 'met den hartelyken wensch dat deze Gewesten, door de bescherming der Goddelyke Voorzienigheid, tegen dergelyke en andere onheilen lang mogen beveiligd blyven'. *Historisch Verhaal*, 231.

Johan Hendrik Hering was the editor of the newspaper *Amsterdamsche courant*.[62] In addition, he wrote books on topics like harsh winters and the island of Curaçao. The first volume of his *Bespiegeling*, counting almost 300 pages, described the storm surge and past floods. The second volume was an extensive appendix with sources that Hering used. Among them were letters from people in the affected areas. Although these reports were often rather plain and factual, the author presented the stories they conveyed in a vivid, even dramatic manner. Hering complained that his 'primary objects of contemplation are air and water' and that these led to similar stories from the various places.[63] Because of that, the author feared that he could not distinguish the various events sharply enough.[64] In order to solve this problem, he employed 'visual descriptions and jewels of art'.[65] With the latter, Hering meant literary devices. Flowery prose dominated the book, an aspect that was criticised by contemporaries.[66]

In his descriptions of the unfortunate night of the storm surge, Hering aimed at portraying events as vividly as possible. Like the other commemoration books, he told the personal stories of survivors and appealed to the emotions of his readers. Another way in which he made the descriptions lifelike was by situating himself, as an all-seeing narrator, reporting in the present tense.[67] This narrator travelled from disaster area to disaster area. He heard the wailing of the victims, felt the dykes crumble, and even saw a ship smashing on a dyke.[68]

Hering wanted to introduce a similar vividness to the historical overview with which he starts his book. Grandchildren listen to the life stories of their 'grey grandfather' with much dedication, the author wrote.[69] Plain history writing could not evoke this level of engagement, Hering argued. Because of that, the author introduced his readers to a personification of history: 'Dutch Antiquities' ('Neêrlandsch Oudheidkunde'). This female figure spoke directly to the audience and told it how she accompanied the first

62 Eco O.G. Haitsma Mulier, G.A. van der Lem, and P. Knevel, *Repertorium van Geschiedschrijvers in Nederland, 1500–1800* (The Hague: Nederlands Historisch Genootschap, 1990), 180–81.

63 Hering, *Bespiegeling*, vol. 1, vii.

64 Ibid.

65 Ibid.

66 Pieters, 'Herinneringscultuur', 52.

67 For a similar narrator in eighteenth-century disaster literature, see Peter Altena, '"Menschlievendheid schikt hulp van boven". Evert Schonck, Lambert Stoppendaal en de Nijmeegse watersnood van 1784 en 1799', in Rick Honings, Lotte Jensen, and Gert-Jan Johannes (eds), *Jaarboek Bilderdijk 2021. Rampen in de tijd van Bilderdijk* (Hilversum: Verloren, 2021), 31–50, at 42.

68 Hering, *Bespiegeling*, vol. 1, 192, 117.

69 Hering, *Bespiegeling*, vol. 1, 19.

inhabitants of the Netherlands. Dutch Antiquities regarded these previous inhabitants as her 'old friends'. During her narration, she even shed tears for their losses during flood disasters.

Hering's use of personification sets his historical overview apart from the one in the *Historisch verhaal* and other contemporary publications on floods. Nonetheless, the function of this literary device was not unique. Page after page, Dutch Antiquities listed chronologically the most severe flood disasters in the history of the Dutch delta. Hering's Dutch Antiquities noted that 'my old friends endured many severe floods, which have – from time to time – increased in weight, and I will touch upon the most severe ones'.[70] The description of these historic events did not follow a rising line of increased impact, and neither do the disasters decline in impact. Instead, major floods recurred 'from time to time'. This illustrates that Hering, like the authors of the *Historisch verhaal*, adopted a cyclical view of flood disaster history.

A major difference between the two books is the way they dealt with the future. Hering was far more interested in the times to come than the anonymous authors. In various parts of the books, he referred to posterity and how it would reflect on the 1775 flood disaster. One of the regions that endured the most damage was the province of Overijssel. Hering: '[M]any dungeons of bitterness open here in front of us, [those] which will let the heart[s] of the progeny shiver from fear'.[71] A couple of pages later, the author praised the generosity and hospitality of the citizens of the town of Zwolle, a city in Overijssel. They gave refuge to flood victims. The author wished that the citizens would receive a 'hereditary blessing over your lineages'.[72] Remarks like these show that Hering had the future generations of Dutch people in mind while writing his book. Furthermore, he imagined his book to be part of that future. Because of his publication, Hering wrote, the 1775 storm surge would become part of the collective memory of coming generations.

In one dramatic fragment, the author made his future-focused goals most explicit. In the middle of the first volume, Hering confessed to his readers that he could not bear to write more about all these sad events.[73] Then he addressed himself:

70 'Menige zwaare watervloeden hebben myne oude vrienden moeten doorstaan, die van tyd tot tyd in gewicht zyn toegenomen, en waarvan ik u de voornaamsten zal aanstippen', in Hering, *Bespiegeling*, vol. 1, 41.

71 'veele kerkers van bitterheden worden hier voor het gezicht geopend, welke het hart der nakomelingschap nog van angst zullen doen beven'. Hering, *Bespiegeling*, vol. 1, 240.

72 Hering, *Bespiegeling*, vol. 1, 241–42.

73 For a similar case, see Altena, 'Menschlievendheid', 40–41.

> Maybe you [Hering] will, at one moment, impress the insensitive hearts of their progeny. It would be unfair to let the generation that still have to be born be ignorant of this that happened these days.[74]

According to Hering, he had a responsibility not only to his readers but also to their children. The Dutch people of the future should know about the night of 14–15 November 1775.

Hering's ambition to educate the next generations is also reflected in the analogies he used to describe his *Bespiegeling*. He wrote that his two volumes were intended to be a 'building of letters' and a 'memorial column'.[75] One time, he used another word. '[I] would like to do complete justice to the historical [events], and, at the same time, by my labour erect an EBEN-EZER, to honour Him, who punishes and blesses'.[76] The term 'Eben-Ezer' refers to the biblical monument from 1 Samuel (4:1, 5:1, 7:12) built to thank God for a military victory. The fact that Hering compared his books to this mythical memorial column highlights two things. Firstly, it illustrates his fear for the forgetfulness of future generations. Secondly, it shows that he deemed remembering to be a divine mission.

Hering perceived collective amnesia as a constant threat for the Dutch. With regard to the eleventh century, his Dutch Antiquities told the readers that

> [s]ince then, for around half a century, the wind and the sea, tired from the previous havocs, hid in their holes, or they at least did not do noticeable damage. The previous fatalities were almost forgotten ... Some of your countrymen probably imagined that they, by taking careful measures, did not have anything to fear.[77]

74 'Misschien zult gy te eeniger tyd, op de ongevoelige harten van hun nakroost, indruk maken; 't zou derhalven onbillyk zyn, de Geslachten, die nog staan gebooren te worden, onkundig te laten van het geen, deezer dagen, gebeurd is!' Hering, *Bespiegeling*, vol. 1, 98.

75 Hering, *Bespiegeling*, vol. 1, vi, 13, 245. See also Michael Kempe, '"Mind the Next Flood!" Memories of Natural Disasters in Northern Germany from the Sixteenth Century to the Present', *The Medieval History Journal* 10: 1–2 (2007), 327–54, at 350.

76 'ik wilde gaarne aan het Historische alle mogelyke volledigheid geven, en te gelyk door mynen arbeid een EBENHAËZER oprichten, tot eer van HEM, die straft en zegent'. Hering, *Bespiegeling*, vol. 1, vi.

77 'Zedert verliep omtrent eene halve Eeuw, dat wind en baren, vermoeid door de vorige verwoestingen, zig in hunne hoolen schuil hielden, of ten minsten geene merklyke schaden aanbragten. De vorige noodlottigheden waren meest vergeeten ... Mogelyk verbeeldden zig zommige uwer landsgenoten, dat men, door voorzichtig genomen maatregels, diergelyke gevaren niet meer te duchten had ...' Hering, *Bespiegeling*, vol. 1, 46–47.

After a period with less severe floods, Dutch Antiquities asserted, people forgot about the threat of the water and God's divine punishment. Then the St Marcellus Flood (1219) hit the Netherlands. The personification of history showed that it was naïve to think that the nation was ever safe from the destruction of the waters.

Not only the water proved to be a constant threat, Hering and the personification implied. Forgetting the threat was a danger in itself. To understand the author's fear for forgetfulness, it is essential to look at the causes Hering sought for the disasters. As is already implied in the quote, the author considered 'careful measures' insufficient to counter future disasters. Dykes and other types of envirotech alone would not help the Dutch. Hering wrote:

> [I]t seems to be a general shortcoming in human nature, to lack esteem for the endured disasters, and always push away the fear for the future [catastrophes], especially when one starts to regard them as the natural [*gewoone*] effects of this or that cause, to which one can protect oneself.[78]

Blind trust in hydrological interventions rested on the assumption that floods were only a natural phenomenon. Here, Hering claimed, people made a costly mistake. Sins were the real cause of floods, and only conversion could prevent disasters.

In general, early modern Europeans followed the logic of the 'economy of sin': because of the sins of individuals – input – God punished communities with disasters – the output.[79] Even though Hering never pointed to specific sins or sinners, he also applied the economy of sin as an explanation of the 1775 storm surge.[80] His account therefore fits the wider, eighteenth-century understanding of disasters. Although there were secular understandings, like the ones discussed in this article, the punishment-for-sin interpretation remained dominant.[81]

78 'het schynt een algemeen gebrek in de menschelyke natuur, de doorgestaane rampen niet te achten, en de vrees voor de tookomende, altoos verre te stellen, voornamentlyk, wanneer men die begint te beschouwen als gewoone gevolgen van deeze of geene oorzaak, waar tegen men zig op de beste wyze zoekt te beveiligen, en 'er verder niet aan denkt, dan wanneer het gevaar vergroot'. Hering, *Bespiegeling*, vol. 1, 40–41.

79 Wolfgang Behringer, *A Cultural History of Climate*, trans. Patrick Camiller (Cambridge: Polity, 2010), 133; Adriaan Duiveman, 'Praying for (the) Community. Disasters, Ritual and Solidarity in the Eighteenth-Century Dutch Republic', *Cultural and Social History* 16: 5 (2019), 543–60, at 545–46.

80 Also noted in Pieters, 'Herinneringscultuur', 51.

81 Buisman, *Tussen vroomheid en verlichting*, vol. 1, 143, 217.

Sundberg has convincingly shown that remembrance was an important goal of eighteenth-century poets and preachers in their representations of flood disasters.[82] For these authors, keeping alive the memories of past disasters was a religious obligation. Since disasters were regarded as part of a divine plan, authors asserted that recent and past catastrophes taught their victims and others moral lessons. Forgetting disasters, and thus forgetting divine punishments that were disasters, was a sin in itself.[83] Hering also believed that amnesia would invoke God's wrath again.

Just as with the *Historisch verhaal*, Hering closed his book by envisioning the future. Hering, however, was more optimistic. He concluded:

> Now, nothing is in the way from … wishing you, beloved compatriots for who this memorial column has been raised, that our behaviour will not invoke any more revenge from heavens [*Hemelwraak*]![84]

In the case of the *Historisch verhaal*, the authors deemed it inevitable that a next flood would occur. They believed a merciful God would postpone the next flood but not stop inundations altogether. Hering, to the contrary, thought that people could learn from disasters, repent, and hence prevent new floods from happening. For Hering, the agency of people was not technological or environmental, as Pierlinck and Zillesen argued, but religious and moral.

Whose Past, Whose Future?

As argued earlier, national identities presuppose collective continuity. One of the two ways to establish such a sense of communal persistence involves the idea that the imagined community has a stable essence. Groups can search for this essence in traditions and mentalities. The other way to construct collective continuity is by stressing the phases of change that a group went through. This process makes the group. In Zillesen's and Pierlinck's commemoration books, the past is a time of transformation. This change is one for the worse, since they both see a rise in disastrousness.

82 Sundberg, 'Claiming the Past', 243.

83 Ibid.

84 'Thans blyft my niets overig, dan … u, BEMINDE LANDSGENOOTEN, voor welken deeze GEDENKZUIL gestigt is, toe te wenschen, dat onze gedragingen, geen Hemelwraak meer wekken!' Hering, *Bespiegeling*, vol. 1, 245.

Nonetheless, if people would learn from this past, the authors explicitly and implicitly claimed, their future was malleable. But whose past and whose future was this?

Pierlinck appeals to the idea of the 'fatherland' (*Vaderland*) – i.e. *patria* – a couple of times. At the beginning of the eighteenth century, authors did not use the word 'fatherland' exclusively for the whole Dutch Republic.[85] Some authors employed it to point only to their own province, like Holland or Guelders, or it even referred only their own city. Later on, the notion of the fatherland 'nationalised', as did the disasters that hit it.[86] Pierlinck also employed the term in its national sense. He asserted that other people with ideas on how to prevent river floods should share this knowledge with the whole nation, since this would benefit the whole of the Netherlands.[87] Yet the Dutch nation does not seem to be the entity that owns the past and the future of the river area. There is not a clearly defined 'we' in Pierlinck's commemoration book.

Zillesen's book also lacks a clear collective protagonist. This is striking, since a declining Dutch nation is central in other works he produced.[88] Zillesen's flood commemoration book was published five years after the Batavian Republic replaced the Dutch Republic. Whereas political power was held by many local authorities in the latter, the new government centralised the state. Part of this programme was the introduction of a new, central agency for water management.[89] Zillesen expected that the new agency could prevent river floods and the nation-*state* would be the actor of his imagined future.[90] When it comes to the past, however, Zillesen did not perceive the Dutch nation as a single character that struggled through centuries of floods.

The commemoration book of the anonymous authors addressed, as its title claimed, floods in the whole of the Netherlands. The historical disasters listed in the beginning took place in both coastal and riparian regions of the Republic. The authors also referred to the fatherland as the place where

85 Eco O.G. Haitsma Mulier, 'Het begrip "vaderland" in de Nederlandse geschiedschrijving van de late zestiende eeuw tot de eerste helft van de achttiende', in Niek van Sas (ed.), *Vaderland. Een geschiedenis van de vijftiende eeuw tot 1940* (Amsterdam: Amsterdam University Press, 1999), 163–79, at 169–71.
86 Jensen, 'Disaster upon Disaster', 65–66.
87 Pierlinck, *De verschrikkelyke watersnood*, 43.
88 Rotmans, 'Circles of Desire'.
89 Bosch, *Om de macht over het water*, 44–52; Bert Toussaint, 'Eerbiedwaardig of uit de tijd? De positie van de waterschappen tussen 1795 en 1870', *Tijdschrift voor Waterstaatsgeschiedenis* 18: 2 (2009), 40–50.
90 Zillesen, *Beschryving*, xv–xvi.

recent disasters took place. Readers could read the commemoration book as an account of the flood and storm disasters that hit the whole Dutch nation. The repetition of suffering connected contemporaries with the previous generations.[91]

More than the authors of the *Historisch verhaal*, Hering explicitly framed the Dutch nation as his protagonist. Thanking God for His mercy, the author claimed that 'the Netherlands was preserved this time; the Netherlands has been spared again!'[92] The nation that the author imagined was based on the cyclical recurrence of flood disasters. The author sought, I argue, the essence of the Dutch in the stability of experiences. This is illustrated by the fact that Hering's Dutch Antiquities did not distinguish between contemporary readers and their imagined forebears. She spoke about the readers' 'ancestors' and 'your fathers', but when discussing the impact of a flood in the ninth century, she remembered the following:

> when sea and Rhyne again, with equal drift, attacked your possessions ... salt and sweet water united to [cause] your deterioration, played a most dreaded role over the surfaces of your formidable lands, and caused a great maceration of the [number of] inhabitants and cattle.[93]

Here Dutch Antiquities addressed the readers in 1776. Yet, while she talked about a flood from eight centuries ago, she mentioned 'your possessions', 'your deterioration', and 'your formidable lands'. In Hering's envisioning of the past, there was no difference between the people who inhabited the Dutch delta then and those who lived there during the 1775 storm surge. The experience of flood was the same, and those who experienced it were the same.

In Hering's book, posterity was just as much part of the Dutch nation as the ancestors. However, while the connection with the past was the repetition of disasters, the future of the Dutch nation was open for Hering. He regarded the Dutch as a single moral community that could change its

91 This contrasts with the proud remembrance of water management successes described in Lotte Jensen, 'Floods as Shapers of Dutch Cultural Identity. Media, Theories and Practices', *Water History* 13: 2 (2021), 217–33, at 228–30.

92 'NEÊRLAND werd ditmaal nog behouden; NEÊRLAND is ditmaal wederom gespaard!' Hering, *Bespiegeling*, vol. 1, 14.

93 'wanneer de zee en rhyn andermaal, met gelyke drift, op uwe bezittingen aanvielen ... ; zout en zoet water thans tot uw bederf vereenigt, speelden eene allergeduchtste rol over de oppervlakte uwer heerlyke landsdouwen, en maakten eene groote vermagering aan inwooners en vee'. Hering, *Bespiegeling*, vol. 1, 43.

future for the better.[94] To put this in terms of temporalities: while the past had been cyclical, the future was linear. According to Hering, the Dutch could break the cycles of catastrophe.

Conclusion

In this article I compared four flood commemoration books. The two books written by Pierlinck and Zillesen operated in a linear temporality. They looked at recent floods, counted frequency, and looked at water levels to conclude that the disastrousness of floods rose with time. Furthermore, they argued that one could learn from the past. By learning from history, people were able to prevent potential floods. Pierlinck pleaded for an emergency spillway, Zillesen argued that people should give rivers more space and that they should benefit from inundations instead of creating disasters. This illustrates that both authors attributed agency to humans. People could do something about the floods and their future was malleable.

Pierlinck's and Zillesen's books provide a stark contrast to the two other commemoration books that are central in this article: the *Historisch verhaal* by anonymous authors and Hering's *Bespiegeling*. These two commemoration books paid far more attention to past disasters than Pierlinck and Zillesen did. In addition, they went back further in time, also discussing medieval floods. The lists with flood disasters stressed that the latest catastrophe was severe but still far from unique. Instead, the recent events were included in a pattern that stretched across centuries. Finally, the anonymous authors and Hering show more interest in the idea of the Dutch nation and its experiences through time, whereas Pierlinck and Zillesen do not introduce such a clear collective protagonist.

Nonetheless, the *Historisch verhaal* differs from Hering's *Bespiegeling* in the ways the anonymous writers described the future. The *Historisch verhaal* paid little attention to the time to come. When the authors reflected on it, they projected the past onto the future. In the logic of the cyclical temporality, they assumed that a subsequent, severe canon flood was simply inevitable. Hering, to the contrary, was far more interested in the future. He even regarded his own book as part of it. Claiming that he produced a 'memorial column', he wanted to educate the coming generations so that they could learn from the dreadful experiences of the victims of the 1775

94 Peter van Rooden, *Religieuze regimes. Over godsdienst en maatschappij in Nederland, 1570–1990* (Amsterdam: Bert Bakker, 1996), 89–95.

storm surge. If the readers would repent, they could prevent the repetition of the *Hemelwraak* – revenge from heavens.

Historians have shown that the decline narrative was a prominent national story in the eighteenth-century Netherlands. Scholars have convincingly argued that disasters, as divine punishments, were interpreted as symptoms of the general decline. Nonetheless, the commemoration books in this article add nuances to this generalisation. The publications operated in different temporalities, yet not one emphasised the decline narrative. On the one hand, the books that emphasised a cyclical recurrence of disasters showed that several severe floods hit the Netherlands every century. There was not a sudden rise of disastrousness caused by a moral collapse but a strict repetition of a historical pattern. On the other hand, the books that found that the number and severity of floods had increased over time did not explain this development by referring to the decline narrative. Instead, Zillesen and Pierlinck pointed to humans' mismanagement of their environment.

In the commemoration books, the tragedies of flood disasters connected readers with past generations. With exception of the *Historisch verhaal*, the commemoration books nonetheless provided readers with hope. Pierlinck, Zillesen, and Hering attributed agency to humans. Through technology or religion, the inhabitants of the Netherlands could prevent future disasters.

About the Author

Adriaan Duiveman is a historian and was PhD candidate at Radboud University in Nijmegen. As a member of the research project *Dealing with Disasters. The Shaping of Local and National Identities* (1421–1890), he investigated the cultural and social impact of catastrophes in the eighteenth century. In 2019, Duiveman was selected as one of the young 'Faces of Science' of the Royal Netherlands Academy of Arts and Sciences (KNAW). The Society of Dutch Literature (MdNL) and Royal Netherlands Historical Society (KNHG) awarded him the Elise Mathilde Essay Prize in 2021.

14 Disaster Memory and ‘Banished Memory’

General Considerations and Case Studies from Europe and the United States (19th–21st Centuries)

Christian Rohr

Abstract

In 1981 medievalist and cultural historian Arno Borst presented the thesis that today’s European societies have largely eliminated dealing with natural disasters from everyday life and that they have become a ‘society of banished memory’. This contrasts with premodern societies, which integrated the risk of natural disasters far more into everyday life. By ‘taming’ natural hazards through river straightening and various protective structures, especially since the nineteenth century, small and medium-sized events have generally been avoided, but serious events became even more devastating when hitting the unprepared population. A prolonged absence of extreme events, a ‘disaster gap’ (Christian Pfister), could thus significantly increase the catastrophic nature of a new event. This chapter analyses selected flood, avalanche, storm, and earthquake events from Europe and North America (late 19th c.–present) to show which factors might have contributed to reshaping memory cultures after catastrophic events and which encouraged banishing memory against better judgement.

Keywords

floods – avalanches – earthquakes – memory – Europe – United States

In 1981 medievalist and cultural historian Arno Borst provided an early case study of an extraordinary natural disaster in his essay ‘The Earthquake of

Asperen, van, Hanneke and Lotte Jensen (eds): *Dealing with Disasters from Early Modern to Modern Times. Cultural Responses to Catastrophes.* Taylor & Francis Group, 2023

DOI: 10.5117/9789463725798_CH14

1348', published in German in the renowned journal *Historische Zeitschrift*.[1] In this article he put forward the thesis that today's societies in Europe have largely eliminated dealing with natural disasters from their everyday lives and that they have become a 'society of banished memory' (in German: *Katastrophenverdrängungsgesellschaft*). He contrasted this situation to premodern societies, which had integrated the risk of fires, natural disasters, and weather anomalies far more into everyday life.[2]

This chapter deals with memory cultures in premodern and modern societies in Central Europe and in the United States. In a first step, I will examine the parameters involved in perceiving a natural hazard as a disaster. These considerations include the assumption that a vivid memory culture will raise risk awareness and contribute to a higher level of resilience, two basic features of 'cultures of risk management'.[3] On the other hand, societies with 'banished disaster memory' or at least 'neglected disaster memory' might be more surprised by and unprepared for extreme events and therefore more vulnerable. In a second step of this article, I will analyse historical cultures of risk management in Central Europe against river floods and avalanches. What had just been daily life to those societies, which events became extraordinary or even catastrophic, and why? Which prevention and coping strategies had been applicable and how were they influenced by a vivid memory culture? We will see how risk awareness changed after the straightening of riverbeds in the second half of the nineteenth century and how new types of awareness and memory cultures had to be reinstalled after destructive flood and avalanche disasters around 2000. In a final step, I will undertake a comparison with the USA. This section asks whether or not a 'rise up and never look back' mentality in dealing with disasters, as can be reconstructed for the 1906 San Francisco Earthquake and Fire, can also be generalised for recent events such as Hurricane Katrina in 2005 and its aftermath.

1 Arno Borst, 'Das Erdbeben von 1348. Ein historischer Beitrag zur Katastrophenforschung', *Historische Zeitschrift* 233 (1981), 529–69.

2 'Erdbeben als dauernde Erfahrung der Gesellschaft und der Geschichte anzunehmen, widerstrebt dem modernen europäischen Selbstgefühl zutiefst. Es isoliert Katastrophen in der Gegenwart und eliminiert sie aus der Vergangenheit, weil sie die Zukunft nicht definieren sollen'. Borst, 'Das Erdbeben von 1348', 532.

3 The term 'cultures of risk management' seems to me more applicable than 'cultures of disasters' as introduced by Greg Bankoff, *Cultures of Disaster. Society and Natural Hazard in the Philippines* (London: Routledge, 2003). By raising risk awareness and resilience, the perception of many hazards will be less catastrophic.

General Considerations on Disaster Perception

'Only human beings can recognize catastrophes, provided they survive them. Nature recognizes no catastrophes'.[4] When the Swiss author Max Frisch composed these words for *Man in the Holocene*, one of his later volumes, he could not have foreseen that they would still be continuously cited in studies on historical natural disasters decades later. As Frisch highlights, talking about disasters and catastrophes depends on the perspective and on human perceptions in general.

The study of natural hazards from a cultural history perspective has become very popular in the last two decades.[5] These studies tend to focus on the perception, interpretation, (risk) management, and commemoration of events by those affected and by human societies in general. In this context, the term 'natural hazard' is taken to mean the physical event itself, which in some instances impacts on the human environment, whereas 'natural disaster' is used to denote the perception of such an event by those involved.[6] Several factors are necessary for a natural hazard to be considered a natural disaster. Not all of them must necessarily be relevant at the same time, but at least three or four should be applicable[7]:

4 Max Frisch, *Man in the Holocene. A Story*, trans. Geoffrey Skelton (London: Methuen, 1980), 102, originally published in German as *Der Mensch erscheint im Holozän* (1st ed., 1979).

5 See, for instance, Christian Rohr, *Extreme Naturereignisse im Ostalpenraum. Naturerfahrung im Spätmittelalter und am Beginn der Neuzeit* (Cologne, Weimar, and Vienna: Böhlau, 2007); Christopher M. Gerrard and David N. Petley, 'A Risk Society? Environmental Hazards, Risk and Resilience in the Later Middle Ages in Europe', *Natural Hazards* 69 (2013), 1051–79; Thomas Labbé, *Les catastrophes naturelles au Moyen Âge, XIIe-XVe siècle* (Paris: CNRS Éditions, 2017); Gerrit Jasper Schenk (ed.), *Historical Disaster Experiences. Transcultural Research – Heidelberg Studies on Asia and Europe in a Global Context* (Cham: Springer, 2017).

6 See, among others, Enrico Louis Quarantelli, *What Is a Disaster? Perspectives on a Question* (London: Routledge, 1998); Anthony Oliver-Smith, 'Theorizing Disasters', in Susanna M. Hoffman and Anthony Oliver-Smith (eds), *Catastrophe and Culture. The Anthropology of Disaster* (Santa Fe, NM: School of American Research Press, 2002), 23–47; Dieter Groh, Michael Kempe, and Franz Mauelshagen, 'Einleitung. Naturkatastrophen – wahrgenommen, gedeutet, dargestellt', in Groh, Kempe, and Mauelshagen (eds), *Naturkatastrophen. Beiträge zu ihrer Deutung, Wahrnehmung und Darstellung in Text und Bild von der Antike bis ins 20. Jahrhundert* (Tübingen: Gunter Narr, 2003), 11–33; Keith Smith and David N. Petley, *Environmental Hazards. Assessing Risk and Reducing Disaster* (London: Routledge, 2009).

7 Rohr, *Extreme Naturereignisse*, 50–57; Rohr, 'Floods of the Upper Danube River and Its Tributaries and Their Impact on Urban Economies (c. 1350–1600). The Examples of the Towns of Krems/Stein and Wels (Austria)', *Environment and History* 19: 2 (2013), 133–48, at 135; Christian Rohr, 'Disaster or Everyday Risk? Perceiving, Managing and Memorizing Floods in Medieval Central Europe', in Christopher M. Gerrard, Paolo Forlin, and Peter J. Brown (eds), *Waiting for the End of the World? New Perspectives on Natural Disasters in Medieval Europe* (London: Routledge, 2021), 201–17, at 203.

(a) the helplessness of humans when attempting to cope with damage through available means;
(b) an inability to explain and understand the event;
(c) material and personal suffering;
(d) the unexpectedness of the event, which depends on how prepared a society is for single or recurrent threats;
(e) whether a series of natural hazards occurs within a short period of time, thereby raising the vulnerability of those afflicted;
(f) symbolic connotations and patterns of interpretation, such as connections to natural disasters described in the Bible;
(g) the wider historical context in the form of economic, religious, or climatic crises.

In addition, for the last two centuries, the prominent presence of disaster reports in mass media will also be decisive, indicating how widely spread disaster perception is.

Unexpected and sudden natural hazards, such as earthquakes, storm surges, severe thunderstorms with hail, or ice floods, are typically perceived as disasters, because people do not have time to install an effective system of prevention, which, in turn, means that the number of victims will be higher. In some cases, vulnerability also plays a role – for example, if settlements are erected carelessly in dangerous places or unsuitable building materials are used.

If a society is prepared to cope with an environmental hazard, people will account for it throughout their daily life and their socio-economic system. Based on their communal experience, they adapt the design and layout of settlements and their behaviour in order to minimise risks. Where the level of resilience in such a society is sufficiently heightened, the result may be a 'culture of disaster'[8] or, more correctly, a 'culture of risk management' (in German: *Risikogesellschaften*)[9]. The widely accepted term *Risikogesellschaft* has been

8 Bankoff, *Cultures of Disaster.*

9 See Rohr, *Extreme Naturereignisse*, 201–398; Gerrit Jasper Schenk, 'Human Security in the Renaissance? *Securitas*, Infrastructure, Collective Goods and Natural Hazards in Tuscany and the Upper Rhine Valley', in Cornel Zwierlein, Rüdiger Graf, and Magnus Ressel (eds), *The Production of Human Security in Premodern and Contemporary History = Die Produktion von Human Security in Vormoderne und Zeitgeschichte* (Cologne: Quantum Information, 2010), 209–33 for riverine cultures; see also Manfred Jakubowski-Tiessen, *Sturmflut 1717. Die Bewältigung einer Naturkatastrophe in der Frühen Neuzeit* (Munich: Oldenbourg, 1992); Marie Luisa Allemeyer, *'Kein Land ohne Deich …!' Lebenswelten einer Küstengesellschaft in der frühen Neuzeit* (Göttingen: Vandenhoeck & Ruprecht, 2006); Adriaan M.J. de Kraker, 'Reconstruction of Storm Frequency in the North Sea Area of the Preindustrial Period, 1400–1625 and the Connection with Reconstructed

introduced by sociologist Ulrich Beck.[10] In addition, Niklaus Luhmann, another influential sociologist, made an important distinction between risk (in German: *Risiko*) undertaken consciously and danger/threat (in German: *Gefahr*), which is always useful for the study of historical societies exposed to natural hazards.[11]

For these communities with a 'culture of risk management', most natural hazards cease to be disasters at all, and their inhabitants understand the reasons for and indications of these extreme events. They undertake strategies of prevention. These can include building and regularly maintaining dykes, locating settlements on relatively secure ground, and adapting building techniques to the risk. For example, windows are not placed at ground floor level in order to prevent the ingress of floodwater, or roof constructions are designed to withstand heavy storms. As far as is possible, warning systems may be installed as a further preventative measure. As will be demonstrated in this study, precisely this kind of 'culture of risk management' existed among urban communities in premodern Central Europe.

Examples from Europe: Floods

I have shown in several previous studies that 'cultures of risk management' had been developed since the Middle Ages within communities living near the rivers and lakes in Central Europe.[12] These case studies focusing on the catchment area of the Danube River and its major tributaries in Austria and eastern Bavaria could carve out clear adaptation strategies and memory cultures of floods. Life close to waterways was always between benefit and risk. The risk of flooding thus had to be integrated into daily socioeconomic life, e.g. by the continuous acquisition of timber to maintain bridges and constructions for flood prevention. The communities living close to the rivers also undertook structural adaptation strategies: settlements were erected on the more secure inner side of river loops, as the example of the small but wealthy city of Laufen on the Salzach River north

Time Series of Temperatures', *History of Meteorology* 2 (2005), 51–69 for maritime coasts in Northern Germany and the Netherlands.

10 Ulrich Beck, *Risk Society. Towards a New Modernity* (New Delhi: Sage, 1992), originally published in German as *Risikogesellschaft. Auf dem Weg in eine andere Moderne* (Frankfurt am Main: Suhrkamp, 1986).

11 Niklaus Luhmann, *Risk. A Sociological Theory* (Berlin: De Gruyter, 1993), originally published in German as *Soziologie des Risikos* (Berlin: De Gruyter, 1991).

12 Rohr, *Extreme Naturereignisse*, 208–398; Rohr, 'Floods of the Upper Danube River'; Rohr, 'Disaster or Everyday Risk'.

of Salzburg shows. We will come back to this site later on. Historical city centres were mostly situated on at least slightly elevated places. Only crafts dependent on waterpower and water supply were necessarily exposed to flood risk. Important buildings close to the river were adapted to the worst case concerning the position of windows or the possible evacuation of storage rooms, etc. A good example is the new tollhouse in the village of Engelhartszell east of Passau, an important toll station along the Danube River since the Early Middle Ages. During the 'millennium flood' of 1501, the old tollhouse had been flooded nearly up to the roof. The larger new tollhouse beneath, erected over the sixteenth century, had windows only higher than the peak discharge of 1501.[13]

Flood marks served as signs of a vivid memory culture. They had been affixed to public buildings and/or other highly visible places and go back to disastrous worst-case scenarios, such as 1501 in the catchment area of the Danube River or even to 1342 in the catchment area of the Rhine and Main Rivers. However, they are useful only to some extent for hydrological reconstruction because the riverbed could have deepened, the inundation areas might have changed, and the entire environment such as fluvial forests served as retention areas. Human-made flood protection as well as bridges causing artificial obstacles may also have an influence on the water level during a flood. In some cases we have evidence that marble plates with flood marks had been removed and re-affixed on another place or even on the replacement building. We sometimes cannot reconstruct with certainty whether the level of the peak discharge expressed by a line, an arrow, a hand, etc. still fits with the historical reality. A combination of several source types is therefore essential for verification. The above-mentioned site of Engelhartszell situated in a canyon-like narrow passage of the Danube River breaking through hard granite can be taken as example where those concerns seem to have been negligible. Nevertheless, those flood marks are documents of a vivid memory culture reminding the people passing by of the most remarkable flood events in the past. They became something like a *memento naturae*, a signpost to consider the risk of recurring flooding when building a house close to the river or for prevention measures in general.[14]

13 For a historical image showing both tollhouses after the flood of 1954 (this view is today blocked by a high protective dam), see Rohr, *Extreme Naturereignisse*, 377; Rohr, 'Disaster or Everyday Risk', 207.

14 On the reliability of flood marks for peak discharge reconstructions, see in general Rohr, *Extreme Naturereignisse*, 89–91; Mathias Deutsch and Karl-Heinz Pörtge, *Hochwasser in Thüringen. Hochwassermarken und Hochwassergedenksteine* (Jena: Thüringer Landesanstalt für Umwelt und Geologie, 2018); Oliver Wetter, Christian Pfister, Rolf Weingartner, Jürg Luterbacher, Tom

In the nineteenth century, particularly in the second half, a significant paradigm shift took place. The straightening of many rivers on a far larger scale than in the centuries before became a public task designed to reduce the risk of flooding in the cities and in the countryside. In addition, the deepening of riverbeds enabled steam-powered ships to run on 'industrialised' river systems. The straightening of river courses meant that smaller and medium floods were less likely to burst their banks and cause damage. This apparently created a deceptive feeling of security, which led to a low estimation of the risk of flooding. Former flood plains became attractive new settlement places. Upper-middle-class villas were in many cases built close to the river because the unobstructed view and the green spaces near the riverbank conveyed a closeness to nature, while industrialisation was otherwise advancing. They were situated in the green fringes of the old town centres, but still close to them.[15]

However, they remained endangered and were not completely suitable as construction sites: the soft soil caused rifts in the houses, and groundwater problems were frequent. In general, carelessness was widely spread, because new flood protection measures were sometimes erected much later than the houses. This became evident when those newly erected quarters were severely affected by floods, such as in 1897 and 1899 in Bavaria and Austria and in 1910 in France and Switzerland.[16]

In July and August 1897 and in September 1899, two successive heavy floods in the Bavarian-Austrian catchment area of the Danube River (and beyond in Bohemia, Moravia, Silesia, and Central Germany) occurred, causing severe damage everywhere.[17] Many homeowners had obviously no

Reist, and Jürg Trösch, 'The Largest Floods in the High Rhine Basin Since 1268 Assessed from Documentary and Instrumental Evidence', *Hydrological Sciences Journal* 56: 5 (2011), 733–58.

15 See in detail Christian Rohr, 'Stadterweiterung versus Naturgefahr. Die Stadt Wels während der Hochwasserkatastrophen von 1897 und 1899', *Jahrbuch des Oberösterreichischen Musealvereines – Gesellschaft für Landeskunde* 157 (= *Jahrbuch des Musealvereines Wels* 36 (2012)), 551–74 (on the city of Wels, Austria, and the Traun River); Christian Rohr, 'Das Risiko im Bild. Frühe Naturkatastrophenfotografie als Quelle der Stadt- und Umweltgeschichte – die Beispiele Salzburg und Luzern', *Traverse. Zeitschrift für Geschichte* 21: 3 (2014), 73–82 (on the cities of Salzburg, Austria, and Lucerne, Switzerland); Christian Rohr, 'Urban Fringes. Conquering Riversides and Lakeshores in the Nineteenth Century – Examples from Austrian and Swiss Medium-Sized Cities', in Tim Soens, Dieter Schott, Michael Toyka-Seid, and Bert De Munck (eds), *Urbanizing Nature. Actors and Agency (Dis)Connecting Cities and Nature Since 1500* (New York: Routledge, 2019), 241–60 (on Wels, Austria, and Lucerne, Switzerland).

16 The Great Flood of 1910 will not be examined in more detail here. For Switzerland, see Rohr, 'Urban Fringes'. For Paris, this event has been analysed in detail by Jeffrey H. Jackson, *Paris Under Water. How the City of Light Survived the Great Flood of 1910* (New York: Palgrave Macmillan, 2010).

17 On these two severe flood events in the area of the lower Inn and Salzach Rivers, see in detail Eva-Maria Wiesner, 'Dann kam das Wasser ... Der Inn-Salzach-Bereich zwischen Hallein

Figure 14.1 The flood of 1899 in the city of Salzburg. Along the straightened Salzach River, the newly built villas as well as current construction sites are under water. Municipal Archives Salzburg, Sammlung Würthle, Sg. 320.112. Source: Fotoatelier Würthle

longer expected that such floods would still be possible after the river had been straightened. The flood of September 1899 in particular is excellently documented in photographs. Figure 14.1 shows the flood of 1899 in the city centre of Salzburg, where bourgeois villas were built along the Salzach River, which had been recently straightened as it flowed through the city centre.

Laufen (Bavaria) and Oberndorf (province of Salzburg), two small towns located about 15 kilometres north of Salzburg on the Salzach River, provide an interesting example with regard to the flood damage in 1899. While the damage inside the prominent Salzach loop, i.e. on the (Bavarian) Laufen side, was limited, the parts of Oberndorf that lay on the right bank of the Salzach River and thus on the outer edge of the loop were massively affected. The old centre of the settlement with around 200 houses – today the area around the so-called 'Silent Night district' – was relocated; new Oberndorf

und Schärding vor und nach den Hochwasserereignissen 1897 und 1899', unpublished MA thesis, Universität Salzburg, 2009. On the city of Schärding (Upper Austria), see Anna Gugerbauer and Ernst Dürr, *Vom Zorn des Inn. Hochwasserkatastrophen in Schärding und den bayerischen Nachbargemeinden* (Wernstein: Eduard Wiesner, 1999), in particular 47–69 (with an abundance of pictorial evidence).

was built around 600 metres upstream.[18] Likewise, the old bridge across the Salzach River was moved from its position behind the river bend to the new centre of Oberndorf upstream.[19] However, this relocation progressed only haltingly for both political and financial reasons. It was not until 1920, when the next severe flood again caused damage to (old) Oberndorf, that the project was swiftly completed. The new, upstream village of (new) Oberndorf was then not affected by the first severe flood of the post-war period in 1954. However, the absence of further severe floods in the second half of the twentieth century again led to carelessness. During the 1970s and 1980s, the former area of old Oberndorf again became densely built up with mainly single-family houses. However, the embankment could not provide sufficient protection during the Century Flood in 2002. Many residents were surprised by the intensity of the flooding and were correspondingly ill-prepared. The protective embankments, which were subsequently raised again, including a device for installing mobile flood protection walls, only partially fulfilled their purpose during the last severe flood at the beginning of June 2013.[20]

The example of Oberndorf fits within the diagnosis for which the climate historian Christian Pfister has coined the – controversial – term of a twentieth-century 'disaster gap'.[21] According to his assumption, technical flood protection and straightening of rivers has avoided small- and medium-scale floods. In addition, relatively few severe large-scale floods took place in the foothills of the Alps, e.g. in the Swiss Plateau between 1910 and 1987 or in the Bavarian and Austrian catchment area of the Danube River between 1954 and 2002. However, recent studies with a focus on small- and medium-size rivers in these areas have shown that on a more regional scale, severe and even disastrous

18 On the effects of the flood of September 1899 in Laufen/Oberndorf and the relocation of Oberndorf, see in detail Herbert Lämmermeyer, 'Das Hochwasser 1899 und die Verlegung Oberndorfs', in Heinz Dopsch and Hans Roth (eds), *Laufen und Oberndorf. 1250 Jahre Geschichte, Wirtschaft und Kultur an beiden Ufern der Salzach* (Laufen and Oberndorf: Eigenverlag der Stadt Laufen und der Marktgemeinde Oberndorf, 1998), 270–76; Horst Hieble, Herbert Lämmermeyer, and Heinz Schmidbauer, *Die Salzachbrücke zwischen Laufen und Oberndorf. Von der ersten Erwähnung eines Salzachüberganges im Jahre 1278 bis zur Gegenwart* (Laufen: Oberholzner Druck KG, 2003), in particular 41–55 and 61–69.

19 See in detail Manfred W. K. Fischer, 'Die "neue" Salzachbrücke. Projektdiskussion und Baugeschichte', in Dopsch and Roth, *Laufen und Oberndorf*, 475–79; Hieble, Lämmermeyer, and Schmidbauer, *Salzachbrücke*, 69–119.

20 Cf. Christian Rohr, 'Zum Umgang mit schweren Hochwassern an der unteren Salzach und am unteren Inn seit dem späten Mittelalter', in Wolfgang Wüst and Gisela Drossbach (eds), *Umwelt-, Klima- und Konsumgeschichte. Fallstudien zu Süddeutschland, Österreich und der Schweiz* (Berlin: Peter Lang, 2018), 433–67, at 461–66.

21 Christian Pfister, 'Die "Katastrophenlücke" des 20. Jahrhunderts und der Verlust traditionalen Risikobewusstseins', *GAIA – Ecological Perspectives for Science and Society* 18: 3 (2009), 239–46.

floods remained frequent; Pfister's diagnosis is therefore too superficial.[22] Nevertheless, carelessness concerning suitable settlement places became evident after World War II, in particular in the time of the 'economic miracle' (*Wirtschaftswunder*) of the 1950s and 1960s. The trend towards single-family houses erected in historically endangered areas ended only after the above-mentioned major floods of the late twentieth and early twenty-first centuries.

Another problem is caused by converting the use of historical buildings. In former centuries, the ground floor of buildings in endangered areas was mostly used for purposes that could be adapted quickly to an impending threat; furniture could be evacuated easily to the upper floors. Craftsmen like tanners lived and worked in the Matte Quarter in Bern (Switzerland), situated close to the Aare River. Now the ground floors are used by hairdressers, dentists, etc., whose equipment could not be removed during the disastrous flood of 2005. In sum, vulnerability has increased significantly to 'centennial floods' (HQ100 and higher). Integral flood protection measures have been discussed by experts since the 1970s in Switzerland, but they were politically enforceable only after the disastrous events of 1987 and 2005.[23]

Finally, the long tradition of flood memory by flood marks has been vanishing during the last decades. Numerous old flood marks have even been abolished on purpose, as the comment of an anonymous real estate agent from Regensburg (Bavaria) shows. After the disastrous flood of the Danube River in 2002, many affected house owners neither restored old flood marks nor affixed new ones, because this would have diminished the value of the property dramatically.[24]

Avalanches

Avalanches have always had a decisive influence on the life and death of people in alpine regions and have repeatedly led to massive demographic

22 For Switzerland, see the detailed case studies of Stephanie Summermatter, 'Die Prävention von Überschwemmungen durch das politische System der Schweiz von 1848 bis 1991', PhD dissertation, Universität Bern, 2017, https://boris.unibe.ch/id/eprint/97587 (accessed 30 March 2023); Melanie Salvisberg, *Der Hochwasserschutz an der Gürbe. Eine Herausforderung für Generationen (1855–2010)* (Basel: Schwabe, 2017).

23 For a detailed hydrological analysis of the 2005 flood in Switzerland, see Gian Reto Bezzola and Christoph Hegg (eds), *Ereignisanalyse Hochwasser 2005*, 2 vols (Bern and Birmensdorf: Bundesamt für Umwelt BAFU and Eidgenössische Forschungsanstalt für Wald, Schnee und Landschaft WSL, 2007–8).

24 Oral information provided by Austrian hydrologist Dr Heinz Wiesbauer (2013).

cuts for the mostly very small high alpine communities. Firstly, the resident population groups, such as the Walsers in the Swiss, Italian, and western Austrian Alps, who settled permanently in high valleys and sometimes even above the tree line since the late Middle Ages, were exposed to the risk of avalanches. Secondly, travellers such as merchants and pilgrims had to cross the alpine passes, even in snowy winters. Thirdly, since the fifteenth century, there were also miners who lived in high alpine settlements and were exposed to particular risk from avalanches on the way to their even higher gold and silver mines.

To assess which avalanches in history had a particularly devastating effect, it is first necessary to look at the different types and general framework of conditions, because in many cases these factors were decisive in determining whether the event was unexpected for the resident population. One main group is formed by snow avalanches (e.g. loose snow avalanches, slab avalanches, slush avalanches), which usually follow a fairly constant avalanche path and can be powerful because they remain on the ground but are usually predictable, provided that the corresponding empirical knowledge is available. Avalanche catastrophes with high numbers of victims resulting from such snow avalanches therefore tend to affect pass routes or are related to cases in which people without sufficient avalanche knowledge were in high alpine terrain.[25]

In contrast, powder snow avalanches are much more unpredictable. They consist of very loose snow that lifts off the ground increasingly quickly and eventually reaches speeds of up to 350 kilometres per hour. This means that even protective forests can be severely affected and rendered ineffective. In this way, powder snow avalanches can penetrate far into the valley and thus hit settlement areas, which are normally safe. The pressure wave mows down forests and destroys houses and other buildings; the highly compressed snow drastically reduces the chances of survival in the avalanche because the rescue work takes much more time. The majority of devastating avalanches in the Alps are probably caused by powder snow avalanches – unless explicitly mentioned otherwise in the sources.[26]

25 Christian Rohr, 'Sterben und Überleben. Lawinenkatastrophen in der Neuzeit', in Michael Kasper, Robert Rollinger, and Andreas Rudigier (eds), *Sterben in den Bergen. Realität – Inszenierung – Verarbeitung* (Vienna, Cologne, and Weimar: Böhlau, 2018), 135–59, at 135 and 144–47; Christian Rohr, 'Risikobewusstsein und Risikomanagement gegenüber der Lawinengefahr in hochalpinen Gesellschaften des Mittelalters und der Frühen Neuzeit', in Benjamin Scheller (ed.), *Kulturen des Risikos im Europa des Mittelalters und der Frühen Neuzeit* (Berlin: De Gruyter, 2019), 175–94.
26 On the types and the occurrence of avalanches in the Alps, see in detail Walter Ammann, Othmar Buser, and Usch Vollenwyder, *Lawinen* (Basel: Birkhäuser, 1997); Martin Engler and Jan

Several factors are decisive for avalanches: the slope of the terrain between 35 and 55 degrees, the amount and structure of snow, weather phenomena such as wind or solar radiation, and vegetation, such as forest cover, which has a decisive influence on whether avalanches can form or which path they take. The higher regions in the Western Alps tend to be more at risk than those in the Eastern Alps. Most records of damaging avalanches therefore originate from Savoy (France), the entire Swiss Alpine area, Vorarlberg (Austria), Tyrol (Austria, Italy), and the Hohe Tauern (Austria). The Eastern Alps east of the province of Salzburg (Austria) reach only a little above the tree line and are generally more densely forested than the higher-up areas of western Austria.

In many places in the Alps, risk cultures developed from the Middle Ages onwards, which had to counteract the 'White Death' with reactive measures based on local experience. Rescue systems were established along the pass routes and in the villages at risk of avalanches, so that many people could be rescued alive from avalanches. In addition, preventive measures such as protected forests and shelters were gradually introduced in many places. In order to protect the high alpine settlements, the authorities imposed restrictions on the felling of high alpine forests, probably at the explicit request of many inhabitants in the affected places. In Switzerland, as early as the fourteenth century, individual sections of forest were declared 'ban forests' by so-called ban letters, in which logging and cattle grazing were strictly forbidden, for example a triangular piece of forest above Andermatt (canton of Uri) in 1397.[27] Protective walls, the first avalanche barriers, were built in the Swiss spa town of Leukerbad after a devastating avalanche in 1518 and then on a larger scale at the beginning of the eighteenth century.[28] To protect high alpine mining, so-called snow collars (*Schneekrägen*) were built in the Salzburg mining districts from the mid-sixteenth century onwards. The entrance to the tunnel, which was often exposed, was protected by a porch covered with logs or stone slabs. This was especially important because the miners' lives were in acute danger should the entrance be buried by an avalanche.[29]

Protective structures for settlements and mines were continuously developed, e.g. in the form of splitting chocks, terracing in the start zone,

Mersch, *Die weiße Gefahr – Schnee und Lawinen* (Sulzberg: Martin Engler, 2001), 252–73.

27 Cf. Rohr, 'Risikobewusstsein', 191.

28 Cf. Isabel Furrer, 'Schadenslawinen im Oberwallis von 1500 bis 1900. Eine sozial- und kulturgeschichtliche Untersuchung', *Blätter aus der Walliser Geschichte* 51 (2019), 47–161, at 143–47.

29 Fritz Gruber, 'Lawinenschutzanlagen als Produkt des hochalpinen Bergbaus', *Technikgeschichte* 44 (1977), 203–12, at 204–5.

and ever more extensive avalanche protection walls. New types of houses, such as the Ebenhöch houses, which were widespread in large parts of Switzerland and probably appeared around 1500, increased the probability that people could survive the avalanche unharmed in their houses. In addition, the exact position of farmhouses and barns was increasingly optimised based on the knowledge of avalanche paths, as the example of the Vallée des Ormonts (canton of Vaud, Switzerland) shows.[30] Finally, in the writings of Placidus Spescha (1752–1833), an educated monk from the Grisons monastery of Disentis (Switzerland) with a high degree of local expert knowledge on avalanches, the first plans have been preserved, according to which endangered localities were to be relocated. They were based on the experiences from severe avalanches in the winter of 1816–17.[31]

Local cultures of remembrance can also be understood as forms of risk management, because they counteracted mental unpreparedness. This is evidenced by numerous *ex voto* commemorative tablets, for example from Lähn in the Außerfern (Tyrol, Austria) from 1726 or in the Rüti Chapel near St Gallenkirch, Montafon (Vorarlberg, Austria) from 1793 to 1830.[32] The Tschagguns Miracle Book from 1757 contains entries with stories of unexpected survival after avalanches, which was attributed to the grace of the Virgin Mary, who was particularly venerated at this Montafon pilgrimage site.[33]

30 On the technique of Ebenhöch houses, see Philippe Schoeneich, Denyse Raymond, and Mary-Claude Busset-Henchoz, 'Spaltkeil und Ebenhöch. Traditionelle Lawinen-Schutzbauten in den Waadtländer Voralpen', in Christian Pfister (ed.), *Am Tag danach. Zur Bewältigung von Naturkatastrophen in der Schweiz 1500–2000* (Bern: Paul Haupt, 2002), 147–52. For a focus on the architectural adaptation strategies in the canton of Valais, see Roland Flückiger-Seiler, 'Der Einfluss von Naturkatastrophen', in *Die Bauernhäuser des Kanton Wallis*, 3 vols, vol. 3.1: *Siedlungsformen und -anlagen im Wandel. Die traditionelle Walliser Landwirtschaft und ihre Bauten zwischen Rebberg, Maiensäss und Alp* (Visp: Rotten, 2011), 75–88. For the canton of Vaud, see Philippe Schoeneich and Mary-Claude Busset-Henchoz, *Les Ormonans et les Leysenouds face aux risques naturels. Représentation des risques naturels et stratégies d'occupation du territoire dans la Vallée des Ormonts (Préalps vaudoises)* (Zurich: Vdf, 1998), 59.

31 Disentis, Klosterarchiv, Ms. Pl Sp 5, 293–333. See Placidus Spescha, *Beschreibung der Val Tujetsch (1806)*, ed. Ursula Scholian Izeti, phot. Lucia Degonda (Zürich: Chronos, 2009), 317–31. On this passage, see also Christian Rohr, 'Placidus Spescha und seine Bedeutung für die historische Lawinenforschung', *Annalas da la societad retorumantscha* 127 (2014), 161–85, at 180.

32 For the example of Lähn, see Rohr, *Extreme Naturereignisse*, 413–14, colour plate 9; for the tablet from the Rüti chapel, see Andreas Rudigier, '"Das Bild zeigt Vertrautes und Fremdes. Ein unauffälliges Votivbild aus dem Bestand des vorarlberg museums mit bemerkenswerten Beziehungen', *Jahrbuch des Vorarlberger Landesmuseumsvereins* 2012 (2012), 32–49.

33 Cf. Klaus Beitl, 'Von "Schneelanen" und "reissendem Wasser". Berichte aus dem Tschaggunser "Mirakelbuch" von 1757', in Edith Hessenberger, Michael Kasper, Andreas Rudigier, and Bruno Winkler (eds), *Jahre der Heimsuchung. Historische Erzählbilder von Zerstörung und Not im Montafon* (Schruns: Heimatschutzverein Montafon, 2010), 207–13.

Furthermore, in the Montafon Valley, avalanche chronicles and avalanche letters were passed on from one generation to the next so that knowledge of the risk of avalanches was not lost.

In the late nineteenth century, the effects of avalanches were also recorded photographically for the first time and were partly reproduced via picture postcards. A series of five postcards documenting an avalanche disaster in Mittelberg (Kleinwalsertal Valley, Vorarlberg) on 31 January 1907 can serve as a good example.[34] These picture postcards were sold after catastrophic events to raise funds, as the inscription on the back of three of the five postcards shows:

> Friends of humanity are asked by the parish office mentioned below to purchase the pictures of the terrible avalanche disaster (31 January [19]07), damage 200,000 marks, in the hamlet of Ahorn, Kleinwalsertal, at 15 pence apiece. The proceeds are for the 5 rescued children who were deprived of their breadwinners and all their belongings. The parish office of Riezlern.[35]

The photos illustrate the destruction and recovery work in the hamlet of Ahorn (Fig. 14.2). Also noteworthy is the caption, which again illustrates the unexpectedness of the avalanche disaster: 'Partial view of the debris field of two houses and nine stables, 1000 metres long, 100 metres wide. These houses were considered a safe refuge for half a millennium in case of imminent avalanche risk'.[36] Another picture also shows nine recovered dead bodies, which are mentioned by name, another the salvage crew with the dead body of a twenty-year-old woman, who was only recovered after four days.[37] This is all the more remarkable because, in general, dead bodies were practically never depicted in early natural disaster photography before World War I. We can only speculate about the context in which these pictures were taken. Perhaps they are to be seen in the tradition of those

34 The photographs by Max Kessler, published by J. Heimhuber, have been preserved to this day in the Walser Museum Riezlern (inv. nos wamu-101 to wamu-105). Cf. Rohr, 'Sterben und Überleben', 154–56.

35 'Menschenfreunde bittet das gefertigte Pfarramt, die Bilder aus der schrecklichen Lawinenkatastrophe (31.1.07) Schaden 200.000 Mark, im Weiler Ahorn Kleinwalsertal, zu 15 Pfg. a Stück zu erwerben. Der Erlös für die 5 geretteten, ihrer Ernährer und ihrer ganzen Habe entblössten Kinder. Das Pfarramt Riezlern'.

36 Walsermuseum Riezlern, inv. no. wamu-104: 'Teilansicht des Trümmerfeldes v. 2 Häusern und 9 Ställen 1000m lang, 100 m breit. Diese Häuser galten ein halbes Jahrtausend als sichere Zufluchtsstätte bei drohenden Lawinengefahren'.

37 Walsermuseum Riezlern, inv. nos wamu-101 and wamu-102.

Figure 14.2 Picture postcard 'Partial view of the ruins of 2 houses and 9 stables ...' from Riezlern, Kleinwalsertal Valley, Vorarlberg (Austria). Walsermuseum Riezlern, wamu-104. Source: Verlag J. Heimhuber / Max Kessler

photographs of the late nineteenth century, showing the deceased laid out in the family circle. In any case, they illustrate the omnipresence of death under the avalanche in alpine communities.

In the last third of the nineteenth century, the major railway routes across and through the Alps were constructed. In Switzerland the Gotthard railway, with its 15-kilometre-long tunnel, even became a national symbol from its opening in 1882.[38] In Austria the Arlberg railway route, including a 10.6-kilometre-long tunnel, followed in 1884. Subsequent large railway projects in the Swiss Alps concentrated on a more westerly crossing of the Alps. Since 1906 trains were able to pass the Simplon route via a nearly 20-kilometre-long tunnel, and in 1913 the Lötschberg tunnel followed. In this way, the Swiss Plateau in the canton of Bern became connected with the upper Rhone valley (canton of Valais), the Val d'Ossola (Piedmont, Italy), and Milan, respectively. In addition, a regional transalpine railway network in the canton of Grisons was developed around 1900 to reach the main tourist hotspots such as Davos and St Moritz. All railway routes, however, were threatened by frequent avalanches. The need for the establishment of a

38 Cf. on the construction and the legacy of the Gotthard railway as a symbol of Swiss national identity Judith Schueler, *Materialising Identity. The Co-construction of the Gotthard Railway and Swiss National Identity* (Amsterdam: Amsterdam University Press, 2008).

large-scale protective system against avalanches became evident along the Gotthard and Arlberg lines after the snowy winter of 1887–88. Pioneers in technical avalanche protection such as Johann Wilhelm Fortunat Coaz in Switzerland and Vincenz Pollack in Austria were responsible for creating the first large-scale 'alpine landscapes of defence' in the start zone of the avalanches.[39]

These protective measures, like the large-scale straightening of rivers, had a paradoxical impact: avalanches on these major railway routes through the Alps became less threatening in 'normal' years, but carelessness increased. After the disastrous avalanche winter of 1950–51 in Switzerland and Austria, large-scale avalanche protection was erected quickly to prevent further casualties.[40] In Switzerland the federal and cantonal authorities decided to protect instead of giving up endangered villages such as St Antönien (canton of Grisons).[41] This process was also accompanied by extensive research by the already established Swiss Institute for Snow and Avalanche Research (SLF) in Davos.

This new 'security' in the high alpine areas was co-responsible for the establishment of large skiing resorts. The village of Galtür in the Tyrolean Paznaun Valley (Austria) had been a small and poor village founded by a Walser community. In the 1970s it became a popular ski destination for tourists, also due to its snow security at an altitude of 1584 metres a.s.l. Local memory culture had mostly vanished. Although the village had been hit by

39 Johann Wilhelm Fortunat Coaz, *Statistik und Verbau der Lawinen in den Schweizeralpen* (Bern: Stämpfli & Cie., 1910); Vincenz Pollack, *Ueber die Lawinen Oesterreichs und der Schweiz und deren Verbauungen. Vorträge gehalten im österreichischen Ingenieur- und Architekten-Verein in Wien am 29. November und 11. December 1890* (Vienna: Lehmann & Wentzel, 1891). On the beginning of 'Alpine landscapes of defence' in connection with the Lötschberg railway project, see Michael Falser, 'Alpine Landscapes of Defence. On Modern-Vernacular Avalanche Protection Systems in the Swiss Alps', in Schenk, *Historical Disaster Experiences*, 399–422. On the role of Johann Wilhelm Fortunat Coaz for the development of modern avalanche protection, see Michael Flütsch, *Johann Coaz als Begründer des Lawinenschutzes in der Schweiz* (Bern: Bern Open Publishing, 2023), https://www.hist.unibe.ch/forschung/publikationen/berner_studien_zur_geschichte/reihe_1_klima_und_naturgefahren/index_ger.html (accessed 30 March 2023).

40 Cf. M. de Quervain, T. Zingg, H.R. In der Gand, M. Schild, and H. Calörtscher, *Schnee und Lawinen in den Schweizeralpen. Winter 1950/51* (Davos: Eidg. Institut für Schnee- und Lawinenforschung, 1952); Martin Laternser and Walter J. Ammann, 'Der Lawinenwinter von 1951 und seine Auswirkungen auf den Lawinenschutz in der Schweiz', in Pfister, *Am Tag danach*, 153–68.

41 Cf. Ricky Umberg, '"Ohne besondere Schutzmassnahmen droht die Entvölkerung ganzer Talschaften". Die Lawinenschutzbautätigkeit in Anbruchgebieten im Lauf des 20. Jahrhunderts in der Schweiz – mit besonderer Betrachtung St. Antöniens und Andermatts', *Jahrbuch der Historischen Gesellschaft Graubünden* 150 (2020), 85–232.

numerous severe avalanches during the last centuries,[42] hotels and other buildings were erected close to the avalanche-prone slopes, even in the endangered 'red zone'. Local knowledge about endangered settlement places had been neglected. Protective forests were mostly missing, in particular from the north. In February 1999 the village became world-famous due to the Galtür Avalanche (1999). On 23 and 24 February 1999, two disastrous powder snow avalanches hit the village itself and a nearby hamlet called Valzur; the only road to the village was blocked, preventing access of emergency vehicles and food supply.

What made Galtür such a disaster and a world-wide media event? Firstly, unpreparedness and carelessness led to an underestimation of the risk of avalanches. Secondly, historical local knowledge was neglected in the pursuit of the economic profit deriving from tourism. Thirdly, this high vulnerability can be seen in a large number of casualties (38 killed, dozens injured) and destroyed houses. Finally, help from outside was impeded, because the road through the Paznaun Valley was blocked for five days and evacuation by helicopter was partly impossible due to bad weather.

A clear paradigm shift can be observed in the aftermath of the catastrophic event. More protective buildings made of steel were erected in the start zone of possible avalanches. A new protective wall (345 metres long, 19 metres high) at the bottom of the slope was built to protect the village itself. In addition, a new culture of disaster memory developed: the *Alpinarium*, a meeting place and new museum about daily life in high alpine areas and about the catastrophe of 1999, was integrated into the protective wall. The event itself is treated in scholarly literature as well as in a lengthy Wikipedia article.[43] In general, the event is present in both personal and collective memory, as Bernd Rieken has shown.[44]

However, a new disaster memory like the one in Galtür has not come back in general; there are still some examples of neglected and even consciously banished memory. This becomes evident in the example of Andermatt. The village in the canton of Uri (Switzerland) along the historical Gotthard pass route is one of the 'hotspots' for avalanches in the Alps and was severely destroyed in January 1951 despite its protective forest. After the large military barracks were abolished and the Gotthard motorway tunnel was opened in

42 According to local sources, nine people were killed by avalanches in 1613, another nine in 1622, and nineteen during the eighteenth century.

43 Wikipedia, '1999 Galtür avalanche', https://en.wikipedia.org/wiki/1999_Galtür_avalanche (accessed 30 March 2023).

44 Bernd Rieken, *Schatten über Galtür? Gespräche mit Einheimischen über die Lawine von 1999. Ein Beitrag zur Katastrophenforschung* (Münster and New York: Waxmann, 2010).

1980, the village suffered from economic decline. In the last fifteen years, Egyptian investor Samih Sawiris started to build the luxury hotel and private chalet resort *Andermatt Swiss Alps*[45], partly erected in places that were hit by avalanches in the past (e.g. the area of the former barracks with casualties in 1940). Obviously, the local authorities do not have any interest in memorialising the avalanche history of the village. In 2013, both in the German and English versions of the Andermatt Wikipedia article, none of the earlier avalanches was even mentioned. Now at least a very short notice refers to the more general article on the winter of 1950–51.[46]

Disaster Memory and Disaster Banishment in the USA

Finally, some glimpses at disaster memory and disaster banishment in the USA may serve as comparison. These insights start with the basic assumption that the European colonisation of the USA in the nineteenth century began parallel to the emergence of a general 'disaster banishment' in Western cultures. On the one hand, newly arriving settlers in the North American west often lacked long-term local knowledge of natural hazards. On the other hand, they established a 'rise up' mentality: after every disaster, a city should emerge larger and more spectacular than before. Furthermore, due to the high value of freedom enshrined in the American Constitution and in the general thinking, compulsory preventive measures seem to be mostly inadequate because legal requirements contradict freedom of action. Finally, hardly any forms of an active memory culture of extreme natural hazards, such as monuments and yearly commemorative celebrations, can be found.

The 1906 San Francisco Earthquake and Fire may serve as an illustrative example. The event developed into a worst-case scenario due to a lack of preparation and numerous adverse breakdowns. However, reconstruction works were seen as 'resurrection like a phoenix from the ashes', as Maynard Dixon's cover of the monthly journal *Sunset*, designed only three months after the disaster, shows (Fig. 14.3).[47] During the Panama-Pacific International Exhibition in 1915, San Francisco presented itself as a glamourous, newly

45 Andermatt Swiss Alps, https://www.andermatt-swissalps.ch/en (accessed 30 March 2023).

46 Wikipedia, 'Andermatt', https://de.wikipedia.org/wiki/Andermatt (accessed 30 March 2023).

47 On the pictorial evidence for the San Francisco Earthquake and Fire as well on its aftermath, see Jacob Birken, *Die kalifornische Institution. Fernwestliche Weltbilder um 1906* (Heidelberg: arthistoricum.net, 2018), https://books.ub.uni-heidelberg.de/arthistoricum/catalog/book/415 (accessed 30 March 2023), in particular 191–98 on the motif of female personifications of San Francisco rising from the ruins.

Figure 14.3 Maynard Dixon, *The New San Francisco*, cover of the monthly *Sunset* (June and July 1906). The Bancroft Library, University of California, Berkeley, F850.S95 v.17

erected metropolis to host this world's fair.[48] Up to now hardly any noteworthy memory culture has developed in San Francisco: no larger permanent exhibition, no historical walking trail recalls this crucial event in the history of the city. The famous panoramic photograph of the destroyed city by George R. Lawrence (achieved by using aerial kites) has become a popular poster

48 Cf. Laura A. Ackley, *San Francisco's Jewel City. The Panama-Pacific International Exposition of 1915* (Berkeley: Heyday, 2014).

Figure 14.4 Flooded House Museum, New Orleans. Source: Sandy Rosenthal / levees.org

for tourists, but nothing more. Hidden traces of a memory culture must be discovered, such as the 'Little Giant' (20th and Church Street), a historical hydrant in use during the fire following the earthquake and still operative today. Lotta's Fountain in San Francisco became a place of an annual remembrance of the event (18 April, 5:12 a.m.) for the survivors. It was erected in 1875 and remained one of the very few monuments withstanding the 1906 earthquake and fire. The last two survivors died after the 2015 anniversary. However, it is not well-known among most of the inhabitants of the city.

Hidden traces of a memory culture can also be found related to other natural disasters in the history of the USA. The Galveston Hurricane of 1900 is commemorated by a small memorial on the beach of Galveston, Texas.[49] However, it was erected only for the centennial anniversary in 2000. In Old North Dayton, Ohio, a flood memorial park remembering the flood of 1937 was created. It was designed in 1992 by Andrew Leicester but abolished already in 2006. This banishment of memorialising the flood fits with the diagnosis offered by Uwe Lübken, who described the flood of 1937 as a 'return of the banished' ('Rückkehr des Verdrängten').[50]

49 On the event, see Patricia B. Bixel and Elisabeth H. Turner, *Galveston and the 1900 Storm. Catastrophe and Catalyst* (Austin: Texas University Press, 2000).

50 Uwe Lübken, *Die Natur der Gefahr. Überschwemmungen am Ohio River im 19. und 20. Jahrhundert* (Göttingen: Vandenhoeck & Ruprecht, 2014), 206–38.

Hurricane Katrina hit the south coast of Louisiana, but also Florida, Mississippi, Alabama, and Georgia, in late August 2005. With gusts up to 344 kilometres per hour, it became one of the most devastating natural disasters in US history. After dyke breaches, up to 80 per cent of the urban area of New Orleans stood up to 7.60 metres under water. The more prosperous classes could mostly flee, whereas the poorer population stayed in their houses or fled to the Louisiana Superdome. In total, 1836 persons were killed, and numerous others went missing. The overall damage was estimated at around 100–125 billion USD. The catastrophic character of the event became even more dramatic due to inadequate individual prevention measures and looting in the city.[51]

In the aftermath of Katrina 2005 (and of other hurricanes after 2000), the discussion about compulsory building codes was raised. Guidebooks for hurricane-proof building exist for most of the affected states in the southern USA, mostly published by insurance companies. However, they are not executable for private houses, but only for public buildings such as schools, hospitals, or hotels. It is obvious that detailed regulations contradict the American mentality, in particular in traditionally conservative states in the south.

Memory cultures after natural disasters remain very fragmentary in the USA. Few museums or commemorative tablets dedicated to natural disasters exist – unlike political events such as 09/11 that have been prominently commemorated. Short-term 'remembrance' was created via YouTube, a platform established only some months before the event. A mentality of 'Rise up and never look back!' can also be seen related to hurricanes. Nevertheless, some first signs of a more conscious memorial culture after 2005 have become visible: In New Orleans, a large Hurricane Katrina Memorial was opened in 2006. A commemorative tablet stands on the 17th Street Canal Floodwall, and a building, which had been completely flooded and severely damaged, now serves as a Flooded House Museum due to an initiative of levees.org, a private commemorative platform founded by civic activist Sandy Rosenthal (Fig. 14.4).[52]

51 On hurricanes in New Orleans and southern Louisiana from the early eighteenth century until 2012, see Eleonora Rohland, *Changes in the Air. Hurricanes in New Orleans from 1718 to the Present* (New York and Oxford: Berghahn, 2019), although without dealing with aspects of memory cultures. On the management and aftermath of the flood, see Arjen Boin, Christer Brown, and James A. Richardson, *Managing Hurricane Katrina. Lessons from a Megacrisis* (Baton Rouge: Louisiana State University Press, 2019).

52 The private museum was opened in 2019: Flooded House Museum on levees.org, https://levees.org/flooded-house-museum/ (accessed 30 March 2023).

Conclusions and Outlook: On the Way to a Banished/Eroded Disaster Memory and Back?

In many societies that are exposed to recurring natural risk, 'cultures of disaster' or rather 'cultures of risk management' can be identified, be it 'flood cultures' on the rivers of Central Europe and on the North Sea coast in the premodern era or 'earthquake cultures' in the Mediterranean countries. Numerous forms of memory were omnipresent in these cultures as a form of 'mental prevention', from clearly visible flood marks to house chronicles about avalanches. With the 'taming' of natural hazards through river straightening and protective structures of all kinds, especially since the nineteenth century, small and medium-sized events have generally been avoided, but serious events then became even more devastating by hitting an unprepared population. A prolonged absence of extreme events, a 'disaster gap' (Christian Pfister), could thus significantly increase the catastrophic nature of a new event. The belief that technical precautions could be used to get a grip on natural hazards was shaken in many places towards the end of the twentieth century.

Disaster memory in premodern societies eroded if no one could remember an extreme event, i.e. after seventy to eighty years. This time of erosion has become increasingly short, probably reduced to one generation (twenty to thirty years) or even less in our present-day media society (*Mediengesellschaft*). Besides this erosion of disaster memory, some examples of a conscious banishment of disaster memory have been highlighted. This phenomenon is evident beginning in the late nineteenth century. It may demand further research to examine whether banishing or suppressing disasters from the mind is typical for rich Western societies relying on insurance systems.

Thus, in conclusion, the crucial question is whether the local population has learned nothing or too little from its centuries-long disaster history. Firstly, a central thesis here is that flood protection, river straightening, and protective measures against other natural hazards in the past were able to prevent smaller and medium-sized damaging events but also made people careless in the face of rarer major events. Secondly, the development of a new culture of disaster memory should be promoted. This cannot necessarily be done only through the (re)installation of flood marks, memorials, or similar forms of remembrance; it also requires platforms that utilize new media. In Switzerland the Mobiliar Lab at the Oeschger Centre for Climate Change Research, together with Mobiliar Insurance Company, launched the *Collective Flood Memory* platform in May 2018, where both experts and laypeople can upload their images of current or historical flood events in Switzerland.[53] The precise geo-referencing of the images by specialists makes it possible to reconstruct the flood risk for many locations in Switzerland with pinpoint accuracy. Thirdly, new forms of knowledge transfer are needed between newcomers and people whose families have lived in the area for a long time (perhaps for centuries) and who know a lot about historical flood events through family tradition.[54] A higher sensitivity due to awareness of the climate crisis may help to draw attention to more disaster memory again.

About the Author

Christian Rohr is Full Professor of Environmental and Climate History at the University of Bern, working both at the Institute of History and the interdisciplinary Oeschger Centre for Climate Change Research. He studied history and ancient and medieval Latin at the University of Vienna, where he obtained his PhD in 1995. As Assistant Professor and later Associate Professor of Medieval History at the Institute of History at the University of Salzburg (1996–2010), his research started to concentrate on nature-induced disasters, climatic fluctuations, and conflicts over natural resources in the medieval and early modern period. Since 2010 he has widened his scope to include modern and contemporary environmental history.

53 Platform 'Überschwemmungsgedächtnis', Oeschger Centre for Climate Change Research, Universität Bern, https://ueberschwemmungsgedaechtnis.hochwasserrisiko.ch/de/home (accessed 30 March 2023).

54 Rohr, 'Umgang', 466–67.